I0868603

The Morality of Software-Defined Warfare:

Just War Theory, Army Medicine, and AI

Piotr Waldemar Pietrzak
https://orcid.org/0000-0003-0464-1991
Vistula University, Poland

Evangelos Ioannis Koumparoudis
https://orcid.org/0000-0002-1068-4376
Bulgarian Academy of Sciences, Bulgaria

IGI Global
Scientific Publishing
Publishing Tomorrow's Research Today

Published in the United States of America by
IGI Global Scientific Publishing
701 East Chocolate Avenue
Hershey, PA, 17033, USA
Tel: 717-533-8845 | Fax: 717-533-7115
Website: https://www.igi-global.com E-mail: cust@igi-global.com

Library of Congress Cataloging-in-Publication Data

Names: Pietrzak, Piotr editor | Koumparoudis, Evangelos editor
Title: The morality of software-defined warfare : just war theory, army medicine, and AI / edited by: Piotr Waldemar Pietrzak, Evangelos Ioannis Koumparoudis.
Description: Hershey, PA : IGI Global Scientific Publishing, [2026] | Includes bibliographical references and index. | Summary: "This book contributes to the debate on the use of new technologies in software-defined warfare. It thoroughly scrutinizes and compartmentalizes the ethical dimensions of decision-making processes related to the emerging AI-driven software-defined developments in warfare"-- Provided by publisher.
Identifiers: LCCN 2026003463 (print) | LCCN 2026003464 (ebook) | ISBN 9798337362649 hardcover | ISBN 9798337362656 paperback | ISBN 9798337362663 ebook
Subjects: LCSH: Artificial intelligence--Military applications | Artificial intelligence--Medical applications | Medicine, Military--Technological innovations | Artificial intelligence--Moral and ethical aspects
Classification: LCC UG479 .M67 2026 (print) | LCC UG479 (ebook)
LC record available at https://lccn.loc.gov/2026003463
LC ebook record available at https://lccn.loc.gov/2026003464
British Cataloguing in Publication Data
A Cataloguing in Publication record for this book is available from the British Library.

Editorial Advisory Board

Table of Contents

Detailed Table of Contents

This chapter offers a brief overview of selected efforts by the United States, Russia, and China in harnessing new technology. The United States is unquestionably a leader and pioneer in industry, leveraging information superiority to achieve strategic advantage, enhanced situational awareness, and self-synchronization. American military scientists proposed the concept of software-defined warfare (SDW) in 2017 and presented it as a paradigm shift from traditional hardware-centric approaches to software-centric warfare within the broader framework of Project Maven's operational network-centric capabilities. SDW's adoption can be seen as the most sophisticated, holistic change in the US's modus operandi, visible across multiple domains, and an investment that the Department of Defense has, each year on every budget since 2017, proportionally increased despite circumstances and budgetary constraints. For a long time, Russia and China lagged in harnessing this potential until leveraging it became a state doctrine in those countries as well.

The 21st century has brought about a fundamental transformation in the character of armed conflict due to rapid technological development. In particular, the introduction of drones and autonomous weaponry introduces new challenges to the conventional constructs and internal definitions of the Just War Theory. The critical analysis of scholarly discourse on Artificial Intelligence (AI) reveals that this technology challenges the core principles of jus in bello and jus ad bellum. The chapter will utilise a descriptive conceptual analysis and a critical analysis based on the available empirical evidence in the existing academic literature. Although the strategic benefit of drones and Lethal Autonomous Weapons Systems (LAWS) is obvious, the consequences should be given the same grave consideration.

The chapter explores the development of Lethal Autonomous Weapons Systems (LAWS), also known as killer robots. The first task is to establish what is motivating the evolution and operationalisation of weapon systems functioning in the absence of meaningful human control, that is, the decision-making of machines. The study is related to the motives behind this growth and the subsequent effects of these motives on military behaviour and the character of the conflict. Synthesising the theories of Antinomic Democratic Peace and Adoption Capacity Theory, the incentives for their utilisation and the mechanisms of their adoption have been investigated. Two key conclusions were obtained from the qualitative content analysis. The chapter's conclusions indicate that LAWS will likely accelerate their growth at an intermediate pace, potentially disrupting the world's balance of power.

The academic literature on the development of artificial intelligence applications in the military industry primarily focuses on technical requirements and legal and ethical paradigms. Nonetheless, there is a gap in the scholarly literature regarding an essential element of any military action: the role of command officers. The aim is to examine Indian military officers' perceptions and reasoning regarding the integration of AI into the military decision-making process. It examines the logic of their approval or rejection of the usage of AI on the battlefield. The analysis introduces a methodological innovation to the war science study vignette survey, enabling the determination of the dynamics between humans and AI across increasing levels of

existence, distinct from other beings, is marked by self-awareness and intentionality, shaping its ontological and ethical framework. Heidegger, emphasizing fundamental ontology, asserts that existence precedes essence. For Dasein is inseparable from meaning, constituting the foundation of reality through immanence. Il y a, in this view, signifies a persistent presence, a 'will to power' grounded in the categories of Dasein, leaving no space for an existence beyond being. Levinas, by contrast, reconfigures il y a through ethical transcendence, shifting from an ontological to an ethical framework. He exteriorizes the subject by situating being within an eschatological and messianic horizon that extends towards infinity.

The moral consideration involved with autonomous weapon systems is a highly debated topic in both academic and social circles. As system development and research continue, various non-governmental bodies propose a total ban on technology. The chapter critically assesses deontological claims that prevent the use of autonomous weapon systems through argument analysis. This chapter is done to find the validity and strength of the deontological case against these systems. The results show that, despite a greater catalogic abundance of influential deontological arguments against the use of autonomous weapon systems than in favour of it, the arguments against the technology are qualitatively poor in terms of tenability and relevance.

Artificial Intelligence (AI) is quickly reshaping the healthcare on battlefield by making the use of AI better in decision-making, in bettering the management of casualties, and making healthcare staff in the high-stress, resource-constrained settings. The rapidity and precision with which medical response is provided is important in saving lives in war in modern-time. AI technologies, including predictive analytics, autonomous systems, and decision-support tools, make triage, diagnosis and treatment to be faster even in extreme conditions. As an example, AI algorithms embedded into wearable sensors may continually test the vital signs of soldiers and identify the initial signs

Preface

INTRODUCTION

The 1991 edition of the *Penguin Encyclopedia of Modern Warfare,* by Kenneth Mackey and William Woodhouse, brilliantly explains the nature of the Cold War. However, it is also a product of its time. When the Soviet Union collapsed, the nature of conflict changed beyond recognition. Many irredentist movements began using a new arsenal of unconventional strategies and tactics. Neither the authors of this publication nor anyone else was ready for what was coming, as new conflicts unpredictably mushroomed. Moreover, this publication fails to mention artificial intelligence (AI), electronic warfare, network-centric warfare, new generation non-linear warfare, or software-defined warfare (SDW), which did not yet exist except as futuristic ideas for science fiction novels and blockbuster movies such as *The Terminator* (1984).

Today, in contrast, defense trade fairs sell technological innovations such as software-defined solutions, robotics, AI, drones, cyborgs, and holograms that define the ever-changing nature of the seventh generation of warfare and improve internal communications. AI has radically changed warfare. Nonetheless, commentators often focus on surface phenomena such as the use of unmanned aerial vehicles, treating them as a sign of a new era in which technological advancements will bring ever more rapid innovation to a world in which human soldiers will be replaced by sophisticated multifunctional cyborgs, drones, and other AI-driven killing machines capable of destroying or subjugating entire regions.

The integration of software-defined systems is well underway and is likely to revolutionize every aspect of the modern theater of war and expand its scope. Although this technology is new and demands new thinking, it is consistent with a perennial drive for innovation. New technologies for more effective warfare might also be implemented without proper ethical oversight.

Naturally, logic dictates that even under such circumstances, jus in bello, the principles governing the justice of wartime, should apply. In combat, one's priority is to survive and avoid harm. Nonetheless, one's response should be proportionate and humane, which means that surrendering soldiers, POWs, and civilians are protected, in theory, as dictated by the Geneva Conventions. However, practice in this sense has been far from perfect in recent wars in Syria, Ukraine, Nagorno-Karabakh, Gaza, Sudan, Lebanon, and Iran, where basic laws of engagement have been repeatedly violated when military necessity dictated it.

Meanwhile, overwhelming new weapons systems innovations of the last three decades have not reduced risks but rather amplified them in modern warzones. Their built-in safeguards are switched off by irresponsible, greedy, and clueless humans who treat war as romantic or useful to their short-term interests.

Those conflicts provide many examples of new AI-driven systems directing unmanned drones and other cyborgs into battle, remotely and with limited human input. The problem becomes more acute when drones become so independent that the humans who produce, procure, and program them no longer control them once they are deployed in combat.

Rather than improving the situation for non-combatants, there have been many blatant examples of severe violations of humanitarian law. Various dilemmas can be examined with three carefully selected moral positions: 1. Kantian deontology is a moral theory paying special attention to intentionality in such settings and is also useful for scrutinizing arbitrariness and the instrumentality of any given decision to distinguish whether it is morally right or wrong. Accordingly, deontological logic could stipulate that drones and robots operated by AI should be controlled by a device capable of independently assessing whether an attack is morally right or wrong in accordance with a set of rules and regulations. The collateral damage that it could produce should be assessed and limited before action. 2. Machiavellian consequentialism focuses on the consequences of any given action; intentions are of lesser importance. This applies to the actions of independent AIs, robots, drone operators, or human soldiers. 3. Finally, for the adventurous, Benthamite utilitarianism would assert that actions, whether taken by humans or AI, are right only if they promote the greatest happiness of the greatest number of people involved.

Although soldiers are familiar with their ethical code and are expected to act humanely, they cannot be expected to engage in sophisticated ethical debates when their lives are on the line. However, more is expected of their superiors and the political leaders who decide to wage war: that they take full responsibility for their actions, even if those actions are taken in an emergency. Just War Theory suggests that the same commitment should be expected from responsible AI and software developers and that indiscriminate killing should never be accepted.

Existing literature covers some aspects of this dynamic process, but vast areas of science, technology, and warfare are still uncharted. There is a notable gap in the literature on the moral dimension of SDW, which is yet to be addressed holistically. *The Morality of Software-Defined Warfare: Just War Theory, Army Medicine, and Artificial Intelligence* (2026) discusses the new ways in which militaries around the world seek a strategic edge by integrating AI and supplements those deliberations with serious debates and a more holistic civilian perspective. This publication scrutinizes seemingly unrelated debates across various fields and brings them together to integrate various aspects of the modern military in one system that is both effective on the battlefield and accountable to just war theorists and ethicists who scrutinize conflict both during wartime (jus in bello) and its aftermath (jus post bellum).

There are moral issues with the military use of AI regarding accountability and responsibility that this publication addresses at the conceptual foundations with elements of Just War Theory. One timely approach includes arguments from Michel Walzer's "Just & Unjust Targeted Killing & Drone Warfare", which argues that soldiers are intelligent and can refuse an order, despite the consequences. Many recent conflicts, such as those in Ukraine, Sudan and Iran, have shown that while AI can make systems more lethal, such as with kamikaze drones, it also can protect civilians and soldiers, countering threats ranging from physical assaults to cybersecurity breaches. Deliberating whether the use of unmanned aerial vehicles is moral is not the essence of this project. As we enter the seventh generation of warfare with its automated SDW innovations, we are also starting down the slippery slope of outsourcing moral decisions to algorithm-driven machines that can kill on a scale that was unimaginable a few years ago. Therefore, we need a more vigorous civilian oversight than that which barely exists at present.

The publication addresses ethical responsibility and the practical applications of AI in conflict management and resolution. It analyses the dynamics and challenges of human-machine collaboration in modern warfare, as well as the ethical considerations of integrating AI and human decision-making processes. Finally, it also deals with the use of AI in the military, raising ethical and legal concerns regarding the level of AI autonomy. These will contribute to pressing debates on the use of AI and software-defined warfare, scrutinizing the latest advancements in AI, robotics, and cognitive and operational domains within NATO's multilevel governance framework.

This is a product of deep, contemplative engagement with research that remains highly speculative and secretive. *The Morality of Software-Defined Warfare: Just War Theory, Army Medicine, and AI* (2026) analyzes human-machine collaboration in modern warfare and the ethical considerations of integrating AI and human decision-making processes. This publication also assesses moral responsibility for combat actions and algorithm-driven drones, checks their robotics, equips them with the correct algorithmic and operational systems, and sends them to the battlefield. It also assesses the moral dimension of using rescue drones to collect the wounded from the battlefield. This project also discusses the PTSD symptoms of drone operators and other belligerents operating in the new SDW environment. As AI is the future of warfare, it and other new military-related inventions are discussed critically from the perspectives of just war theory and political philosophy. This publication discusses the evolution from the historical practice of leaving injured soldiers on the battlefield until the cessation of hostilities to the contemporary deployment of instant assistance using AI and robotics in portable operating theaters meters away from combat zones where medical practitioners operate from remote locations.

Finally, this publication also emphasizes the oversight of AI in warfare and medicine, advocating a more responsible, ethical deployment of AI systems, even in contemporary war zones. It unlocks the moral complexities of AI in warfare and medicine, essential for conflict management specialists, medical practitioners, just war theorists, and philosophers. It provides us with tools to examine the historical evolution and contemporary applications of AI and robotics in warfare and medicine, from traditional battlefield practices to modern AI-driven interventions. Those themes have attracted a formidable group of IR theorists, political philosophers, legal scholars, and just war theorists to conceptualize software-defined warfare and brainstorm to expand their horizons by challenging preconceived notions in international relations, just war theory, military studies, political philosophies, and military medicine.

CHAPTER OVERVIEWS

This publication consists of thirteen chapters:

- **Chapter 1,** "Project Maven, Software-Defined Warfare, and their Impact on China and Russia's Strategic Culture," authored by myself, offers a brief overview of selected efforts by the United States, Russia, and China to harness new technology. This chapter suggests that the United States is unquestionably a leader and pioneer in industry, leveraging information superiority to achieve strategic advantage, enhanced situational awareness, and self-synchronization. For a long time, Russia and China lagged in harnessing this potential until it became state doctrine in those countries as well.
- **Chapter 2,** "The Algorithm of Warfare: Challenges to Jus ad Bellum and Jus in Bello in the Age of AI," by Keshav Madhav, Aditya Agrawal, Anushree Shrikant Mukte, and Raunaq Chawla, discusses the introduction of drones and autonomous weaponry, which introduces new challenges to the conven-

tional constructs and internal definitions of the Just War Theory. The critical analysis of scholarly discourse on AI reveals that it challenges the core principles of jus in bello and jus ad bellum.

- **Chapter 3,** titled "The Emergence and Implications of Lethal Autonomous Weapon Systems: Motivations, Adoption, and Impact on Global Military Dynamics," by Tulip Suman, Anuj Kumar Sinha, Akash Bag, and Ananya Bharadwaj, explores the development of Lethal Autonomous Weapons Systems (LAWS), also known as killer robots. The first task is to establish what is motivating the evolution and operationalization of weapon systems functioning in the absence of meaningful human control, that is, the decision-making of machines. The study examines the motives behind this growth and their subsequent effects on military behavior and the character of conflict. Synthesizing the theories of Antinomic Democratic Peace and Adoption Capacity Theory, the incentives for their utilization and the mechanisms of their adoption are investigated. Two key conclusions obtained from qualitative content analysis indicate that LAWS will likely accelerate their growth at an intermediate pace, potentially disrupting the world's balance of power.
- **Chapter 4,** "The Role of Artificial Intelligence in Military Decisions: Making Trust, Ethics, and the Future of Warfare," by Riday Mukherjee, Krishna Panda, Sarada Saha, and Ashutosh Kumar, suggests that there is a gap in the scholarly literature regarding an essential element of any military action: the role of command officers. The aim is to examine Indian military officers' perceptions and reasoning regarding the integration of AI into military decision-making. It examines the logic of their approval or rejection of the use of AI on the battlefield. The analysis introduces a methodological innovation to the war science study vignette survey, enabling the determination of the dynamics between humans and AI across three scenarios, each with increasing levels of threat to human life. The authors also suggest that empirical studies indicate that officers retain confidence in the use of AI in military missions, regardless of the mission's risk profile or complexity, provided that principles such as human command remain in place.
- **Chapter 5,** titled "Political Strength and Cognitive Warfare: How Democracies Radicalize" by Iskren Ivanov, argues that the driving forces behind post-Cold War democratization assumed that building accountable and transparent institutions would secure peace. The failure of transitional regimes and polarization within democratic societies have resulted in a survivorship bias for the North Atlantic Treaty Organization, as such polarization within Western societies presents a considerable risk to their national security. The purpose of this chapter is to introduce a structural model of how cognitive warfare undermines political strength, thereby radicalizing liberal democracies. The

study explains how cognitive warfare affects the strategic perceptions of the major NATO allies, bringing the alliance one step closer to domestic fragmentation and institutional instability.

- **Chapter 6,** titled "Meaningful Human Control and Morality: Implementing Advanced Control Directives for Autonomous Systems," by Prateek Mishra, Bhanu Pratap Singh, Pranjal Khare, and Sapna Singh, asserts that modern warfare is accelerating as conventional military forces, autonomous systems, and human-machine combinations are integrated. This development presents an essential quandary: establishing moral and legal causation within systems under non-standard conditions of human time perception. To counter this, the question is posed as to whether humans can reciprocate independent actions in real time through a prior contract, thereby arranging future conduct. It is especially applicable in situations that are too fast or too slow for humans to comprehend, where constantly concentrating on every action is not feasible. Leveraging the medical-legal precedent of advance care directives, the framework suggests a so-called advance control directive (ACD) to help pivot the rigorous deliberation needed to become accountable in the first place. From this perspective, this development creates an autonomy command, a framework of power designed and justified by creating these orders before implementation.
- **Chapter 7**, titled "Moral Code: Ethical Software Engineering for AI-Driven Battlefield Healthcare," by Hanh Le, maintains that as warfare becomes increasingly defined by algorithms and autonomous systems, the moral weight of software development extends far beyond lines of code. Nowhere is this more apparent than in AI-driven battlefield healthcare, where diagnostic models, triage algorithms, and telemedicine platforms can determine who lives or dies. This chapter explores the ethical, technical, and human dimensions of software engineering for military medical AI. It argues that secure, explainable, and accountable software systems are indispensable for preserving trust, upholding international humanitarian law, and ensuring the moral integrity of code deployed in combat zones. By examining software verification, human-in-the-loop design, and ethical auditing frameworks, this chapter offers a blueprint for moral resilience in the age of software-defined warfare.
- **Chapter 8,** titled "Humanitarian Dilemmas of AI's Role in Warfare and Medicine: Insights from India and the United States," authored by Naresh Prajapati, Anwesha Ghosh, and Nathan Darren Manuk, examines the humanitarian dilemmas of AI in warfare and medicine, with a focus on India and the United States. It explores how AI technologies enhance diagnostics, treatment, and defense operations while raising profound ethical concerns around accountability, equity, and human oversight. By comparing AI adop-

tion across two contrasting national contexts, the chapter highlights systemic risks, including data bias, misdiagnosis, and the use of autonomous weapons. It argues for responsible innovation rooted in ethical foresight, inclusive policy, and sustained human agency, ensuring that AI serves, rather than displaces, human values.

- **Chapter 9,** titled "From Nothingness to *Il y a*: Recontextualizing 'Just War' in Heidegger and Levinas," by Stavros Panagiotou, maintains that Martin Heidegger and Emmanuel Levinas engage in a profound exploration of the tension between *il y a* (there is) and nothingness in their philosophy of war. This chapter examines the philosophical roots of war, tracing its derivation from nothingness to *il y a* through the perspectives of Heidegger and Levinas. Human existence, distinct from that of other beings, is marked by self-awareness and intentionality, shaping its ontological and ethical framework. Heidegger, emphasizing fundamental ontology, asserts that existence precedes essence. Dasein is inseparable from meaning, constituting the foundation of reality through immanence. *Il y a*, in this view, signifies a persistent presence, a will to power grounded in the categories of Dasein, leaving no space for an existence beyond being. Levinas, by contrast, reconfigures *il y a* through ethical transcendence, shifting from an ontological to an ethical framework. He exteriorizes the subject by situating being within an eschatological messianic horizon that extends towards infinity.
- **Chapter 10**, titled "Ethical and Deontological Challenges of Autonomous Weapon Systems: A Critical Analysis of the Responsibility Gap and Dignity Arguments," authored by Prashant Rahangdale, Shashank Nande, Ankita Kumari, and Prachi Priyamvada, discusses the moral considerations involved with autonomous weapon systems, which is a highly debated topic in both academic and social circles. As system development and research continue, various non-governmental bodies propose a total ban on technology. The chapter critically assesses deontological claims prohibiting the use of autonomous weapon systems by analyzing the arguments. This chapter assesses the validity and strength of the deontological case against these systems. The results show that, despite the greater abundance of influential deontological arguments against the use of autonomous weapon systems than in favor, the arguments against technology are qualitatively poor in terms of tenability and relevance.
- **Chapter 11,** titled "How Can AI Help in Battlefield Healthcare" by Apoorav Sharma, Lovleen Marwaha, and Richa Singh, maintains that AI is quickly reshaping healthcare on the battlefield by improving decision-making and the management of casualties in high-stress, resource-constrained settings. The rapidity and precision of medical responses are important for saving lives

in modern warfare. AI technologies, including predictive analytics, autonomous systems, and decision-support tools, enable faster triage, diagnosis, and treatment even in extreme conditions. For example, AI algorithms embedded in wearable sensors may continually monitor soldiers' vital signs, identify early signs of trauma or internal bleeding, and automatically notify medics or command centers. Machine learning algorithms can also analyze large volumes of battlefield data to forecast injury patterns and evacuation routes and distribute medical resources effectively.

- **Chapter 12,** titled “AI Driven Advances in Nuclear Medicine from Image Analysis to Personalized Treatment,” authored by Regan R, Nithya V, Thabeswar A, Yokeshvaran S, suggests that AI is transforming nuclear medicine by boosting diagnostic precision, image analysis efficiency, and patient outcomes. AI-based techniques, especially deep learning and machine learning algorithms, have demonstrated maximum efficiency in oncology, neurology, and cardiology. Applications range from preclinical disease detection to tumor segmentation and customized treatment planning. Constraints remain data availability, regulatory clearance, and model interpretability. This paper focuses on AI applications in nuclear medicine imaging, provides an overview of the models, discusses datasets and evaluation metrics, and summarizes recent developments and challenges. Emerging research directions include explainability, multimodal imaging integration, and ethics to enable AI's large-scale adoption in clinics.
- **Chapter 13,** titled “Responsibility and Hospitality in the Doctor-Patient Relationship: Jus in Bello and Jus post Bellum Cases,” authored by Evangelos Ioannis Koumparoudis, moves away from deontological and utilitarian ethics usually proposed in the doctor- patient relationship, as applied to in bello and post-bellum cases. By contrast, the emphasis is on responsibility and hospitality, as proposed by thinkers such as Levinas and Derrida. The relationship becomes Other-directed and asymmetrical: the Doctor is directed by the Other-suffering patient, refugee, or even hostile enemy, and he is responsible for all of them. This fact is further elaborated through Derrida's notion of conditional and unconditional hospitality: the doctor offers hospitality even when it may turn into hostipality, providing a host to a hostile person. The doctor-patient relationship is also viewed through the lens of new technologies in army medicine, as applied to the philosophies of the two great thinkers.

Piotr Waldemar Pietrzak
Vistula University, Warsaw, Poland

Chapter 1
Project Maven, Software-Defined Warfare, and Their Impact on China and Russia's Strategic Culture

Piotr W. Pietrzak
http://orcid.org/0000-0003-0464-1991
Vistula University, Warsaw, Poland

ABSTRACT

This chapter offers a brief overview of selected efforts by the United States, Russia, and China in harnessing new technology. The United States is unquestionably a leader and pioneer in industry, leveraging information superiority to achieve strategic advantage, enhanced situational awareness, and self-synchronization. American military scientists proposed the concept of software-defined warfare (SDW) in 2017 and presented it as a paradigm shift from traditional hardware-centric approaches to software-centric warfare within the broader framework of Project Maven's operational network-centric capabilities. SDW's adoption can be seen as the most sophisticated, holistic change in the US's modus operandi, visible across multiple domains, and an investment that the Department of Defense has, each year on every budget since 2017, proportionally increased despite circumstances and budgetary constraints. For a long time, Russia and China lagged in harnessing this potential until leveraging it became a state doctrine in those countries as well.

DOI: 10.4018/979-8-3373-6264-9.ch001

INTRODUCTION

During a US presidential debate in 2012, Mitt Romney, the Republican Party's candidate, accused his opponent and the sitting president, Barack Obama, of mismanaging and weakening the American military, arguing that the US Navy had fewer ships then than in 1916. At that time, most conflicts were still platform-centric, so it seemed like a credible accusation of negligence or mismanagement that would prove that Obama was not prepared to be commander in chief. However, this political stunt backfired on Romney because the Republican presidential candidate and his political advisors had not done their homework on the nature of the modern military; Obama's response destroyed his opponent's credibility and helped him secure his second term:

> *You mentioned the Navy, that we have fewer ships than we did in 1916. Well, Governor, we also have fewer horses and bayonets, because the nature of our military has changed ... We have these things called aircraft carriers, where planes land on them ... for the question is not a game of Battleship, where we're counting ships (Obama, 2012, 32:15).*

Many political commentators assert that this response sank Romney's presidential chances and turned the American public against him; even those who had questioned Obama's track record realized that Romney might not have been ready to make challenging strategic decisions. Obama's rebuttal showed that Romney's traditional way of approaching strategic challenges was not up to the challenges of the job.

This chapter's main focus is not the brutal nature of US political campaigns but rather to illustrate that Obama's "fewer horses and bayonets" was not only a brilliant rhetorical stance but also an important insight that many insiders ignored at that time. In 2012, the US was struggling to find an exit strategy for the conflicts in Afghanistan and Iraq, and it turned to a very unconventional modus operandi. The US military brass realized that having a technologically advanced army with unquestionable superiority in every sphere did not guarantee success on the battlefield. Modern leaders cannot prioritize outdated indicators such as a simple count of tanks, fighter jets, submarines, vessels, or aircraft carriers, which may be used as an effective deterrent against a traditional adversary but may not be useful against an unconventional actor that effectively uses the ever-changing circumstances of modern warfare to its advantage, thinking outside the box and catching its adversary off-guard.

Even during peacetime, leaders should envision future needs, while the military must be a driving force of innovation, ingenuity, forward thinking, and innovative approaches to war. Simply accumulating more modern, technologically-advanced tanks, fighter jets, personal armor carriers, and howitzers has no place in modern strategic thinking. Since 2012, there have also been systemic changes in strategic thinking that are clearly visible in the military strategy of most NATO states, as well as in that of their adversaries, which is consistent with a strategic integration of comprehensive software solutions. These fully activate the hardware potential and more effectively utilize the stockpiled arsenal to achieve the best results in the first phase of the conflict, deterring an aggressor faster and more effectively (Hartzell, 2023).

According to popular perception, the adoption of modern drones, software-defined technology, and artificial intelligence in contemporary warfare marks twenty-first-century warfare as unprecedentedly sophisticated. New reconnaissance drones are killing machines capable of carrying a heavy payload to strike targets hundreds of miles away. Supervised autonomous weapons, or human-on-the-loop systems, are autonomous weapon systems that enable human operators to intervene and terminate engagements before unacceptable levels of damage occur. They independently select and attack targets according to their programs while humans retain full supervision of all operations and can override the systems, if necessary, within a limited time period. On the other hand, the semi-autonomous weapon or human-in-the-loop system, once activated, engages only individual targets or specific target groups selected by a human operator. Examples include homing munitions that, once launched to a particular target location, search for and attack pre-programmed targets within the area. Finally, the fully autonomous weapon or human-out-of-the-loop system, once activated, can select and engage targets without further intervention by a human operator. Examples include loitering weapons that, once launched, search for and attack their intended targets over a specified area without any further human intervention and weapon systems that autonomously use electronic jamming to disrupt communications.

The adoption of software-defined systems in a reformed and upgraded chain of command still occurs within a traditional structure that is adapting to those changes, purchasing systems that offer the best utility, the highest combat readiness, and the greatest possible technological advantage over enemy forces by keeping pace with their advances (Mulchandani & Shanahan, 2022; CSIS 2023).

However, in the vast history of warfare, the emergence of drones and the networked environment in which they operate are but the blink of an eye in the broader context of the emergence of geostrategic ontologies and strategic ecosystems that redefine modern conflict. What matters in this context are strategic decisions and the systems that operate behind the scenes, how they are integrated, the novelty that those more holistic ontological structures bring to the decision-making table, how useful they are for surprising one's enemy, and how effectively they bring the conflict to a desired conclusion.

The evolution of the strategic environment depends on many factors, including traditional and more modern pressures that impact the structure, culture, and modus operandi of the modern army. Armies may show a different agility in embracing the evolutionary or revolutionary changes that are desirable or forced upon them. The adoption of innovations may be faster than in previous decades, but it depends on pressures from without, as the propensity to adopt advanced technologies may be low during peacetime and much higher when war looms. The pace of the abandonment of traditional hard power systems, units, and solutions may vary depending on factors, circumstances, and financial constraints.

Although the nature of the military has changed over the last 30 years, inventors are modernizing traditional hardware and making existing systems more agile. Armies did not just stop purchasing traditional equipment; the push for broader integration of the new software defined ecosystem has become more important on the procurement side, as spending on traditional units still consists of 80–85% of military budgets.

The transition to software-centric warfare is evident, as the constraints of the modern theater of war demand that armies seek a strategic edge by harnessing new developments in AI and software-based applications across all strategic environments. Leading scholars emphasize that new developments are linked to advances in information and communication, supported by secure, encrypted VPNs, enhanced antivirus software, and other innovations that are part of broader evolutionary changes (Mulchandani & Shanahan, 2022; CSIS 2023; Hartzell, 2023).

Although this chapter focuses on software-defined systems within the doctrine of network-centered warfare, those changes are occuring in an environment that is also transitioning from the sixth to the seventh and eighth generations of warfare rather than in the speculative realm of future warfare. To fully understand the changes that accompany the adoption of autonomous systems, they must be seen as taking place within strategic environments that may be fluid but are still largely confined to the twelve traditional sides/dimensions of an unconventional strategic environment (Hartley, 2021, p. 65).

It is also certain that a linear software-based revolution toward a fully software-based army and a total abandonment of the traditional army will never happen. The progress towards software-defined integration is a slow, incremental, and balanced evolution towards a hardware-based army that is already more capable of taking advantage of the most comprehensive and efficient modern communication techniques that amplify its functionality and enhance its strategic potential.

Finally, the reader will be introduced to a different, more humanitarian and human-centric perspective that will rearticulate new developments in the military and new approaches in conflict management in relation to just war theory and debates related to hybrid warfare, active measures, and information warfare across four leading disciplines dedicated to thoroughly scrutinizing the ethical dimensions of decision-making processes related to the emerging AI-driven software-defined developments in warfare and medicine (Walzer, 2016).

RESEARCH METHODOLOGY

The chapter should be treated as an imperfect theoretical departure that explores some of the most interesting innovations in modern warfare that could lead to more meaningful discoveries in the future. Its most obvious limitation is a lack of access to classified information from the US, Russian, and Chinese governments, as well to data related to SDW-related operational procedures, plans, strategies, and contingency plans. Despite this limitation, accessible data can be synthesized and integrated into debates in international theory and security studies to build strong civilian oversight of new developments in the military for the benefit of future inquiry.

The research addresses some of the most advanced military strategies and doctrines currently in the public arena, while much information on SDW remains classified or restricted and therefore not obtainable without clearance. This may be challenging, both conceptually and methodologically, especially for those who are not government officials. For this reason, particular caution has been exercised when dealing with highly speculative material, which has been independently verified across three to four layers. Naturally, the publication of findings that could not be independently verified was postponed.

Much methodological and conceptual energy in this project was devoted to translating documents from Russian and Chinese, addressing the speculative nature of many of the Center for Strategic and International Studies' (CSIS) assumptions and clearly distinguishing speculation from verifiable information. It was necessary to treat various claims with skepticism throughout this project. Ample independent information, such as commentaries, budgetary information, and congressional hearings, make it possible to verify substantial information related to SDW. Other sources are published academic evaluations and speculative reports that are in the public domain either intentionally or because of leaks obtained from multiple sources and independently verified.

According to Peter Robertson, when it comes to budget matters, comparisons related to three different countries predominantly measure "inputs, not outputs; flows and not stocks; and we need to take into account that many non-budget factors such as experience, moral and organizational efficiency will affect capabilities"; moreover, there are severe data limitations, including comparisons of quality. Also, it is possible that defense budgets of some countries may be intentionally under-reported in order to obfuscate (Robertson, 2024).

To capture the essence of this project, an ontology *in statu nascendi* is proposed as an inclusive methodological tool that makes it easier to address information that is otherwise uncritically reproduced by scholars and less-specialized media outlets. Conceptually, the main modus operandi was to confront unverified knowledge from the public domain with the specialized literature engaging in broader inquiries into seventh- and eighth-generation warfare. Some of this work is by its nature more speculative and futuristic, combining seemingly unrelated debates from various fields into one specialized interdisciplinary inquiry supplemented by methods from IR theory, which was designed to embrace more systematic, interdisciplinary, and contemplative IR-related deliberations, breaching the main conceptual, methodological, analytical and transformative differences between rival approaches, theories, and traditions that should be better integrated in theory and practice (Pawłuszko, 2025; Pietrzak, 2024, pp. 1-27).

This chapter is imperfect as a result of limited access to certain information, which will be corrected over time. The most crucial data relate to the period between 2001 and 2017, whereas the data from 2017 to 2026 is still sealed and is unlikely to be declassified anytime soon, so by default the information in this short study represents only a fraction of what could provide us with more insight into the adoption of SDW in the US. Even less reliable is information gathered about Russia and China because these countries are even more secretive and less inclined to reveal this important intelligence due to their more authoritarian socio-political systems. When assertions are based on speculation, I make sure to acknowledge this.

LITERATURE REVIEW

To prepare this chapter, several documents related to Software Defined Warfare were studied; most were produced by CSIS or by researchers affiliated with this think tank. The main document was the 2022 CSIS report, "Software-Defined Warfare," by Mulchandani and Shanahan (2022), which argues that the US Department of Defense (DOD) must transition from a hardware-centric to a software-centric force to maintain tactical advantages. It emphasizes rapid, continuous updates to software and AI models to outpace adversaries. The content of this report guided me to approach SDW as a paradigmatic shift in modern warfare and question my understanding of the seventh generation of warfare that coincides with an adoption of the doctrine of network-centric warfare (NCW), which, according to Raymond McConoly "seeks to translate an information advantage, enabled in part by information technology, into a competitive advantage through the robust computer networking of well-geographically informed dispersed forces" (Alderman, 2015; McConoly, 2021). To gain a strategic advantage, the military must be more agile to better utilize new tactics, techniques, and procedures that a fully or even partially networked force can employ to create a decisive warfighting advantage. A careful study of the congressional hearings of Gregory C. Allen, as well as other key public servants and political figures who have developed firsthand experience in dealing with American, Chinese, and Russian strategic culture within the emerging NCW and SDF, informed this research (Allen, 2024). I also used various reports produced by the Jamestown Foundation, especially one dedicated to Russia's SDW. Still, most of the data used to research SDW comes from the public domain, collected from publicly available documents on the US's SDW and from commentaries on similar systems in Russia and China. I treat those governmental reports, insights, independent analyses, and documents as imperfect but reliable sources of information, as they were produced by retired military officers and policymakers; comprehensive CSIS reports are for public consumption, so they must withstand scrutiny. Nonetheless, I compare information from these sources with other information obtained from independent researchers affiliated with governments, think tanks, and universities, independent military strategists, political commentators, and civilians dealing with military oversight in NATO countries as well as researchers from countries such as India, Pakistan, Afghanistan and Australia that tend to be very critical about the actions of the US, Russia, and China (Mulchandani & Shanahan, 2022).

THE TRANSITION TO THE SEVENTH GENERATION OF WARFARE

Evolution Towards Software-Defined-Warfare

Suggestions that modern warfare should embrace innovation, agile adoption of advanced technologies, and coordination are not new. A similar push for the holistic modernization of contemporary warfare across multiple domains, such as land, sea, cyberspace, and the information domain, was characteristic of the United States, the Soviet Union, and their allies after 1945. The first noticeable difference is a departure from the Cold War rhetoric advocating upgrading nuclear stockpiles and less scaremongering about nuclear war and mutually-assured destruction. The second noticeable difference is the focus on precision, defense, and technological integration, which is clearly visible in the recent developments of systems and technological advancements, specifically in long-range air defense systems. In an age dominated by the fear of nuclear deterrence, systems like the newest versions of Patriots that transition from blast-fragmentation to hit-to-kill technology, such as PAC-3 Missile Segment Enhancement (MSE), LTAMDS (Lower Tier Air and Missile Defense Sensor), IBCS (Integrated Battle Command System), and the

Russians' S-400s and S-500s, are developed to protect against attacks. Finally, the third noticeable difference between the Cold War period and now is that China is replacing Russia as the US's main strategic competitor in all areas. Nonetheless, Russia's neoimperialist ambitions in Georgia, Ukraine, Syria, and Africa clearly show that it should still be considered an active threat to the Western world, not only because of its conventional and nuclear weapons but also its SDF skills; the side that adapts faster and demonstrates the greatest agility, including rapidly updating and promulgating field software and AI models, is likely to have a significant competitive advantage (Amilakhvari & Baghaturia, 2024; Mulchandani & Shanahan, 2022).

Project Maven

Most of the literature presents software-defined warfare as a paradigm shift from traditional hardware-centric approaches to a software-centric model, developed to control, manage, and adapt sophisticated IT systems to better coordinate and integrate multiple military systems that previously lacked adequate coordination. The origin of this term largely developed within the US Department of Defense's specialized unit, Project Maven (officially the Algorithmic Warfare Cross Functional Team), launched in April 2017 to accelerate the adoption of machine learning and data integration across US military intelligence workflows. The program operates under the National Geospatial-Intelligence Agency and encompasses multiple applications across the Department of Defense, spanning military operation targeting support, data integration and visualization for analysts, and training machine learning models on labeled datasets of military assets and infrastructure. Today's wars are still not fully automated; they are fought by human soldiers, not cyborgs, and they are not dominated by semi-autonomous weapons systems. Even if weapons systems achieve far greater autonomy by 2050, they will remain supervised and deployed by humans who will design and equip them with AI and algorithm-driven software and deploy them according to their strategic goals and needs. In this respect, the strategic interactions between the major players in the international system as a specific hypothesis related to their policies regarding SDW are not testable or fully comparable due to the lack of transparency and accountability in Russia and China's software-defined systems. Existing literature also shows that China's and Russia's military officials compartmentalize and approach military studies in a different manner, and the project was also designed to address these pre-existing gaps in literature and interlink them with similar debates in these three countries to strengthen security and military studies in general (Mulchandani & Shanahan, 2022).

The unit that oversees the implementation of software-defined warfare in the US military is the Department of Defense (or, according to Donald Trump's renaming, the Department of War)'s Project Maven (officially the Algorithmic Warfare

Cross-Functional Team, often abbreviated AWCFT). This unit was formed on April 26, 2017, by the Department of Defense's Center for Strategic and International Studies (CSIS).

According to a report by the inaugural chief technology officer of the CIA, Nand Mulchandani, and former director of the Department of Defense's (DOD) Joint AI Center, General Jack Shanahan, what led them to form this specialized unit and shift towards treating software as the primary driver of military capability was Deep Blue's success against world chess champion Garry Kasparov in their 1997 match, which demonstrated the effectiveness of combining powerful hardware, advanced search algorithms, and expert-designed evaluation functions to compete at the highest level. Deep Blue, developed by IBM, was powered by an algorithm based on artificial neural networks with representation learning, which can be supervised, semi-supervised, or unsupervised. This revealed that the future belongs to algorithms that combine several techniques and heuristics, including advanced search algorithms and evaluation functions. Deep Blue incorporated extensive book databases, endgames, and sophisticated evaluation functions that assessed the strength and value of various features of the board position. The algorithm that led to this success was the alpha-beta algorithm, which explores possible moves and evaluates their outcomes by considering the resulting board positions. It employed a highly optimized parallel search algorithm to explore millions of positions per second. Compared to today's algorithms, it was rather simplistic, but it made Mulchandani and Shanahan realize that the DOD needed a new design and architecture that would allow it to be more flexible, scale on demand, and dynamically adapt to changing conditions. The only way to stay competitive in this new war-fighting environment is to use the most sophisticated technology, and more specifically, software. This shift led to revolutionary and at times ruthless moves in a DOD that at the time was burdened by Industrial Age, hardware-centric practices that were not responsive enough to meet the challenges of next-generation war fighting systems These changes made them more agile in harvesting machine learning techniques and automating the tedious and time-consuming tasks involved in analyzing visual data, which enabled human decision makers to focus on higher-level assessments and strategic assessments. Utilizing recent discoveries to attain a competitive edge in future conflicts depends on quickly updating field systems, not just quality hardware (Mulchandani & Shanahan, 2022).

Since the inception of this unit, its chief manager Robert O. Work has pushed for accelerated adoption of machine learning and data integration across US military intelligence workflows. The other key people reporting to him were retired Gen. Jack Shanahan, who became the inaugural director of the Joint Artificial Intelligence Center, and Nand Mulchandani, chief technology officer, civil servant, and entrepreneur, who coordinated targeting support, data integration, and visualization for analysts, and training machine learning models on labeled datasets of military assets and infrastructure. The key objective is to harness AI and machine learning in the defense and security domain to augment human capabilities and improve the efficiency of data analysis and interpretation in military operations (Mulchandani & Shanahan, 2022).

Subsequently, key stakeholders provided the DOD with insights and recommendations on how to prepare the army for seventh- or eighth-generation war-fighting systems. Meanwhile, the main responsibility of Project Maven is to allow the United States to retain its dominant position in the future by using the most potent weapon available: technology, and more specifically, software (Mulchandani & Shanahan, 2022). Advanced SDW technology is designed to establish and maintain secure trust relationships at the device level within military systems. Ensuring high-speed processing of large datasets can significantly reduce decision-making time during the planning and execution of operations. Project Maven's main responsibilities are to adopt machine learning and data integration for military intelligence workflows and increase the efficiency of troop and weapons management by compartmentalizing and activating a networked battlefield in which data moves at the speed of light to connect not only sensors to shooters, but also the totality of deployed forces and platforms. Because of rapid technological developments and geopolitical competition, DOD must ensure it maintains the ability to rapidly adapt to changing threats and environments, securing seamless integration and communication between different systems and platforms. Its creators suggest that SDW's main purpose is to enable military leaders to make informed, swift, and accurate decisions based on the existing network of sophisticated military systems integrated under one command to enhance threat response or a potential pre-emptive attack. SDW represents a paradigm shift from traditional hardware-centric approaches to warfare towards a software-centric model (Mulchandani, 2022), which entails the more complete interconnection of decisionmakers' awareness of the ontological structure and compartmentalization of the twelve traditional domains of modern warfare than ever before. The partial shift from a traditional hardware-centric to a software-centric model necessitates AI and machine learning integration to optimize operations across the land, sea, air, space, and cyber domains so as to enable an unparalleled flexibility, adaptability, and intelligence in military systems, integrating cutting-edge technologies across multi-domain and joint forces environments through secure device-level trust relationships with real-time controls to capitalize on existing leverage and benefit from strategic asymmetry of the new information age (Alderman, 2015; Mulchandani & Shanahan, 2022).

Of course, some of those declarations cannot be fully trusted, as we don't have access to classified US government information. However, those technologies are built for the benefit of the society at large, and the access to some information related to these projects is in the public domain. A strong civil society, international organizations, and leading think tanks such as CSIS have access to information to conduct their research and verify it.

Even though we do not have firsthand access to DOD data, we can follow the money to make our own independent assessments as to whether those technologies are legitimate. DOD's 2026 budget shows Maven-style efforts do not necessarily appear as large procurement lines. They are distributed, relatively small, and research, development, test, and evaluation (RDT&E) -heavy; nonetheless, the budget clearly indicates that the US high command is investing heavily in multi-domain interoperability of secure communication channels, real-time data processing, and robust, reliable information and devices. Five to seven percent of 2026 DOD spending is allocated to software-defense warfare programs. We anticipate that expenditures related to R&D, development, and the procurement of new technologies will grow in the coming years. The 2026 budget also shows that the DOD is taking steps to enhance their ability to withstand and recover from cyberattacks and other disruptions, training and supporting teams leveraging cutting-edge crypto-agnostic techniques and real-time controls to ensure the authenticity, integrity, and confidentiality of data and communications. Those expenditures together with the R&D, long-cycle engineering, and continuous iteration of those systems might make the actual figure 15–20% of the military expenditures, which is still marginal relative to hardware-centric efforts, but they have been rapidly growing since 2020, as the proportion of software spending to hardware spending is clearly growing. This is visible not only in US DOD spending but also among other NATO states, which are strongly influenced by multi-domain coordination in their operations and spending. Adoption of this new framework by all NATO states to improve the coordination of organizational and intergovernmental security necessitates the adoption of similar spending patterns. Access to and acting on important information is of crucial importance for the stability of the military ecosystem (U.S. Department of Defense, 2026).

China's Project Maven

China's development of AI warfare is often depicted by American scholars as a threat to the US's strategic advantage over its main geostrategic competitor. A report produced for CSIS by Matthew P. Funaiole and Brian Hurt suggests that China's defense spending has increased thirteen-fold over the 30 years from 1995 to 2024. Jiao (2026) estimates that China's military invested one to two percent of its $250 billion 2025 defense budget in AI-related research and development, which would mean $2.5 –5 billion. She claims that China is not only a leader in AI but also exports its AI solutions around the world in accordance with the Global AI Governance Initiative (2023) and the Shanghai Declaration on Global AI Governance

(2024), which shapes the way it is perceived around the world. Jiao (2026) maintains that this coordinated China AI strategy combines substantial state investment, tight regulation, and deep collaboration with major tech firms and local governments both in China and abroad. She also writes that AI is central to China's civil-military fusion strategy, which channels university research and private-sector innovation into defense. According to Koichiro Takagi, the People's Liberation Army (PLA) refers to this as "intelligentized warfare   a new paradigm of conflict that moves beyond informatized warfare by leveraging artificial intelligence (AI), autonomous systems, big data, and cognitive operations to dominate the future battlefield (Takagi, 2025)." From Jiao's perspective, "it also leverages AI to improve decision-making, battlefield awareness, logistics, and cyber operations". This is especially visible in the Chinese Army's strategy to further develop:

1.) Drone swarms — palm-sized reconnaissance drones and larger jet-powered swarms for surveillance or payload delivery. 2.) AI-driven logistics — predictive maintenance and automated supply chains to enhance operational efficiency. 3.) Autonomous guidance systems — AI integration into vehicles and missiles for real-time navigation and targeting. 4.) Cyber and cognitive operations — AI-enabled tools for information dominance and psychological operations (Jiao, 2026).

Even if it cannot be fully verified at what stage of implementation of its SDW strategy China is currently, as we do not have reliable access to classified information, US scholars tend to assume that China launched its own Project Maven at the same time as the DOD. Gregory C. Allen (2024) asserts that "the Chinese military embarked on its AI modernization campaign at roughly the same time as the United States in 2017, and they encountered the same problem that Project Maven uncovered: a desperate need for AI chips." Allen also writes that in the early 2010s China began working toward adopting the National AI Strategy, which would make it an AI-enabled authoritarian superpower. Allen with other CSIS scholars produces regular reports on software-defined warfare and AI developments in China; recent ones assert that China has invested heavily in developing its own SDW for a considerable period, and there is ample evidence to support this. He regularly testifies before Congress, most recently in late 2025. Those reports depict China's efforts in following the US's lead in this critically important area. However imperfect and one-sided those reports tend to be, they consistently reveal important patterns in China's modus operandi. Allen, in 2025 Senate hearings, confirmed that China is engaged in designing and deploying next-generation, Internet-scale warfighting software and hardware systems, which are portrayed as a threat to the US and the Western way of life and are consistent with the perception of China as the US's strongest geostrategic competitor. In most CSIS reports, China is presented as a hawkish, uncontrollable revisionist power eager to escalate global tensions to conquer an island that is more skilled at producing cutting-edge semiconductors and microchips.

Taiwan produces over 60% of the world's semiconductors and over 90% of the most advanced ones, so its tech and artificial intelligence position is critical to Western industries, and its autonomy is crucial for the well-being of the Western world. Taiwan is a geographic choke point of great importance to global trade, which is one of the main reasons that the US military stays nimble in the Indo-Pacific and US administrations are successful; even Donald Trump declared that the US is determined to protect Taiwan's sovereignty (Fann & Bursi, 2023).

As China's ambition is to gain control of Taiwan, Taiwan has several contingency strategies to avert this drastic change in the Indo-Pacific's geopolitical dynamics.

Most researchers associated with DOD, Project Maven, and CSIS, as well as independent strategic scholars, adopt the position, in line with Graham Allison's Thucydides trap, that further Chinese growth will lead to war when it threatens to displace an existing great power as a regional or international hegemon (Allison, 2017). As a result, most of those reports read as if they were informed solely by the realist school of IR thought. The only difference is between emphases more on Nicollo Machiavelli or Hans Morgenthau's notion of national interest, or on John Mearsheimer's offensive realism or Kenneth Waltz's defensive realism, the latter two of which argue that after the collapse of the Soviet Union, China emerged as the main competitor of the United States, whereas Russia became too weak to compete with the United States in either conventional or software-defined warfare. Such a standpoint is consistent with a broader post-1991 geopolitical imperative that is clearly dominant in the US establishment (Mearsheimer, 2001; Morgenthau, 1978; Neacsu, 2010).

Those strategists perceive China's military rise as a strategic challenge, prompting decoupling, the formation of new defense alliances, and greater defense spending. However, according to Peter Robertson, "the claims over the size of China's real military spending relative to the US are contested and range widely from around one quarter of US spending to near parity" (2024). Robertson also argues that purchasing power parity (PPP) comparisons previously used to assess Soviet military spending show they can be very misleading. He asserts that "the use of appropriate relative prices and unbiased economic measurement techniques greatly reduces the range of reasonable estimates." When it comes to budgetary estimates, in comparisons of the three countries one can measure predominantly inputs, not outputs; flows and not stocks. Many non-budget factors such as experience and moral and organizational efficiency will also affect capabilities, and there are severe data limitations, including comparisons of quality. The defense budget may also be substantially underreported in official data (Robertson, 2024).

In this respect, Gregory C. Allen's 2025 Senate hearing can provide more context on how the Chinese government is building and utilizing AI systems for military purposes. However, China's approach to SDW is unique, as it is using a different conceptual ontological framework to project its SDW power. Its SDW belongs to a broader information warfare domain that is not directly under the Ministry of National Defense (MND, , Guófáng Bù) but rather the Central Military Commission (CMC, , Zhōngyāng Jūnshì Wěiyuánhuì), chaired by China's president, the supreme military authority. The CMC controls the People's Liberation Army (PLA), including the army, navy, air force, rocket force, and strategic support force (Funaiole & Hart, 2025).

China is investing heavily, particularly in drone technology, developing a fleet of underwater drones that gather information in the Indian Ocean and the South China Sea and around Taiwan. In China, all equivalent institutions report directly to the president, Xi Jinping, who exercises indirect control over all units on behalf of the more than 100 million members of the Chinese Communist Party. The regime's approach to information warfare is holistic; enhanced public security and law enforcement purposes are integrated at the discretion of the party, which has exercised direct absolute command over the People's Liberation Army, the Central Military Commission, and the Military-Civil Fusion of Strategy (MCF) since Mao's Cultural Revolution. For SDW to be effective, it needs a fundamentally scalable architecture to build on in which it can be tested, which China is doing in the South China Sea to project its influence. This technology is deployed to coordinate the irregular maritime militias, "little blue men," to assert and expand its control over an increasingly significant area of disputed and reclaimed islands and reefs in the strategically important area.

China's military commanders must use sophisticated information systems to coordinate the actions of its vast paramilitary force of 4 million in three main forces: the People's Armed Police, the militia, and the reserve force, who act as auxiliaries to the PLA.

China has deployed its software-defined capability to ernhance its military and economic superiority by creating hundreds of artificial islands designed for military and economic purposes in the South China Sea to turn this strategically contested area into green (i.e. coastal) waters by building artificial islands, asserting control over the surrounding area and islands.

Xi and other CCP officials are adopting a more precise intelligent tracking, recognition, and analysis project, often referred to as Sharp Eyes. Like the Maven Project, whose procedures and operational data are classified, outside observers are not privy to relevant Chinese information; this autocratic regime does not overshare strategically pivotal information in the public domain. Nonetheless, it is evident that China is one of the most advanced users of AI for social and political control in the world and the leader of industry in its AI surveillance system, which uses video analytics and facial recognition technologies and scans and flags even subtle criticism of the state in real time across digital platforms to identify individuals and track their movements. These tools go far beyond old keyword censorship and can detect nuanced content flagged as sensitive.

This is often described by experts as surveillance authoritarianism because it augments state power with automated detection and control, which is why it has drawn attention and raised concerns regarding privacy rights the potential misuse of personal data both within China and internationally. As recent events have shown, harnessing AI technology to gain an advantage over adversaries, both domestic and international, and sharing this technology with one's allies may have real international consequences. Chinese surveillance and facial recognition technology was used in Iran to kill more than 40,000 anti-government protesters in January 2026, blind another 10,000, and ultimately crush those protests. China is not preoccupied with their responsibility in this matter and continues to supply Iran with surface-to-air missile batteries as Tehran rapidly rebuilds defenses destroyed by Israel during their Twelve-Day War (June 13–24, 2025).

In the aftermath of that war, China, despite an international arms embargo on Iran, supplied Iran with jets, ballistic missiles, air defense, anti-stealth radars, anti-ship capabilities, C-802 anti-ship missiles, and YJ-12 or CM-302 systems to help it detect stealth aircraft and to enhance maritime capabilities, denouncing the use of force in the Middle East. However, recent events in Iran have proven that those systems failed to intercept aircraft and missiles. Chinese systems designed to detect stealth aircraft were reportedly overwhelmed by more advanced Israeli and American electronic technology.

Russia's Project Maven

There are three leading opinions on Russia's SDW these days. The first is that it either doesn't exist or is backward, underdeveloped, or neglected. To prove that Russia is still lagging behind the West and Chinese SDW and AI development, scholars claimed that Russia heavily relies on post-Socialist conventional technology and remains oblivious to the shifting paradigm towards SDW because its political leadership knows no better. Some CSIS experts can be categorized as such. In his recent 2025 Senate hearing, Gregory C. Allen was so preoccupied with China-related issues that he almost neglected to mention Russia. However, since 2022, as Russia has been changing its attitudes to SDW, its expertise in this area is no longer underestimated by many leading analysts. Naturally, it is in Russia's interest to present itself as a novice. Vladimir Prikhvatilov of the Russian Academy of Military Science asserted in 2024 that for now Russia has no chance to catch up with the

Chinese or the Americans in the use of AI (Powers, 2025). Those statements might be a premeditated decoy, as the facts on the ground do not support this humility.

The second leading opinion about Russia's SDW's readiness is that it is shifting its military strategy and arsenal toward the information age, but slowly, as its strategic culture is still not ready for the full integration of software-defined systems that are still considered an insignificant part of its more holistic information strategy. Scholars who present this position often maintain that Russia traditionally emphasizes the development and procurement of conventional and nuclear weapons, cognitive warfare, and active measures and hybrid techniques. Russia is using SDW, but military hardliners focus more on achieving more compound effects. This position is not completely wrong.

Finally, a last group of scholars argues that Russia's SDW and AI expertise are underestimated, especially since 2022, and Russia has not revealed its full SDW potential. It is in Russia's interest to prove that its strategy is ineffective in Ukraine so that it will be underestimated potential future conflicts. However, underestimating Russia is dangerous, as it can wreak havoc not only on Ukraine but also on Central and Eastern Europe. It is not unusual to underestimate Russia's strength. Despite Zbigniew Brzezinski's pre-2014 warnings about Russia's growing imperialist ambitions, leading American political analysts kept dismissing Russia as backward when it comes to AI and SDW. However, the war in Ukraine has been an unprecedented testing ground for Russia's AI weapon systems, which has persuaded its political and military leadership to completely change its outdated strategic culture. Napoleon and Hitler made the mistake of underestimating or even looking down on the Russians, which brought them to their doom. When the war in Ukraine ends, Russia will reemerge as a formidable power, rebuilding its military capabilities and learning from its experiences.

The Russians are nihilistically existentialist and extremely impatient by nature, so they are resourceful in quickly coming up with solutions to their own shortages because they have no other choice in order to survive, even if and especially when they are pushed to their existential limits. Their leaders are not doing them any favors. Traditionally, Russia shows a propensity to enter war completely or partly unprepared, and over time it adjusts and enhances its capabilities at great human cost and resource expenditure. Once the Russian society is pushed to its limits, it shows unbreakable ingenuity in overwhelming its opponents. For Russian leaders, it is all about contingency planning and pursuing national interests, driven by an imperial desire for expansion. They push their society to make enormous sacrifices because they do not particularly value human life.

Although details of Russia's AI military initiatives may not be publicly available, the Russian Ministry of Defense does impressive damage control for the Ministry of Propaganda. The units responsible for information and cognitive warfare are divided between Military Intelligence (GRU-GU), Security Services (FSB & SVR), the armed forces and specialised units, Propaganda, the media apparatus, the Russian Orthodox Church, and Strategic Command, all of which report directly to the president. In recent years, Russia has been investing heavily in AI research and development through academic institutions and government-backed initiatives to bolster its military capabilities and enhance its technological edge (Hagen, 2024; Miller 2023).

Like China, Russia uses AI tools and employs its information warfare for facial recognition and crowd monitoring to track protesters and opposition activists while online information campaigns are used to influence foreign elections and public discourse. Authoritarian regimes are quickly advancing in surveillance, censorship, and AI-enabled control, which gives them strong domestic and strategic leverage. As in China, there is no transparency or accountability in Russia's Ministry of Defense; the information on Russia's SDW is from American and Ukrainian intelligence and independent researchers. The war in Ukraine has shown that while AI can be used to optimize the lethal capabilities of systems, such as kamikaze drones, it also can save civilian and military lives and counter various threats, from physical assaults to cybersecurity breaches. AI applications there extend to forensic data-collection operations behind enemy lines and can compel enemy combatants to surrender. Moreover, AI-driven medical advancements improve access to innovative treatments for soldiers and prisoners of war.

At the time of the formation of Project Maven in Washington, the Russians also declared their own project, and similarly to China, this was not a coincidence but a reaction to their main geostrategic rival. Vladimir Putin officially called for a national AI task force and strategy to develop Russian generative AI and other capabilities in early 2017, and by early 2019, Russia had formed its National Strategy on the Development of Artificial Intelligence with a strategic plan to meet its objectives by 2030. Although there was seemingly no direct equivalent to Project Maven, the Russian military and defense industry actively worked on AI and autonomous systems. In late 2022, the focus changed, as software-defined aspects of information warfare could no longer be ignored. Since then, Russia has been investing in AI military technologies, including for intelligence gathering, surveillance, and autonomous weapon systems. Although details are not publicly available, those changes can be seen in its conduct of the war in Ukraine.

Those findings are consistent with Jamestown Foundation findings that "the use of AI in Russia had a late start, with Russian companies falling short of North American or Chinese corporations working in the sector. This image, however, may be deceptive." Their specialists maintain that

Following Russia's full-scale invasion of Ukraine, Moscow's emphasis on the development of AI, both for military and civilian purposes, has become more

prominent. (…) The most recent data suggest that Russia is now among the top 10 countries in the world in terms of AI integration throughout various spheres of public life and the economy (Powers, March 3, 2025, citing: Lenta.ru, March 13, 2024).

There is no end in sight for the war of attrition in Ukraine, in which each side gradually wears down the other through small- and large-scale actions rather than a fully integrated, seventh-generation software-defined war. When Putin started this conflict in 2022, he might not have expected such determination from Ukrainian society to defend itself. Thanks to its involvement in Ukraine, Russia is rapidly gaining essential experience in SDW. Its army is not fully automated, and it has suffered 1.5 to 1.8 million human casualties (killed and wounded). The years of war since 2014 are longer than the First and Second World Wars combined. Paradoxically, the longer this conflict lasts, the more likely it is that Russia will be able to redesign its systems to ensure success in future conflicts. The war in Ukraine might be the biggest laboratory for warfare since the Spanish Civil War or the Syrian Arab Spring; its nature changes bewilderingly rapidly. Russia both tests new technology and dispenses with outdated technology inherited from the Soviet Union by using it on the battlefield. When Russia runs out of its own technology and weapons, it uses stocks from its allies, North Korea, Iran, and China.

In early 2023, Russia purchased tens of thousands of Iranian Shahed-136 kamikaze drones along with other drones. Russia's increased use of Iranian drones to attack Ukrainian civilian infrastructure is not a sign of Russia's weakness in this area, but a sign of growing awareness that it must develop its own holistically interconnected systems and integrate them with existing operational systems. These purchases caused some to conclude that its defense industry can't keep up with demand. The Soviets used this strategy in the 1940s when Hitler tried to conquer Russia, and they were very successful, supported by American and British Land-Lease programs. By 1943, the Russian army could strike back and push the Wehrmacht thousands of kilometers back to Berlin. In an old imperial strategy, Russia does not prepare before the war starts but adjusts to the needs of the confrontation while waging it. The longer the conflict lasts, the more prepared Russia becomes, even if it suffers countless casualties.

Since 2022, there has been a clear, consistent shift toward investment in SDW systems and AI military technologies for intelligence gathering, surveillance, and autonomous weapon systems. Today, Russia produces its own drones on Iranian licenses and, in return, is supplying Iran with intelligence to protect it from ongoing Israeli and US attacks. Russia is not dependent on Iranian drones; it is planning to restore its mass production capabilities thousands of kilometres away from Ukraine to protect itself from increasingly frequent Ukrainian reprisals. Russia is capitalizing on its unique software-based solutions that integrate its military arsenal more closely with central command. Several independent political analysts observe that the Russian military has persistently fine-tuned, adapted, and improved AI-enabled and -enhanced systems for immediate deployment while its scientists are actively developing and using its AI technologies, both in civilian and military applications, albeit with different capabilities and limitations than those of Western and Chinese AI programs (Powers, 2025). However, Russia does not typically use the term software-defined warfare in its official public doctrine, nor does it assert that it engages in hybrid or cognitive warfare. Instead, it calls these approaches a blending of non-kinetic and kinetic tools (Wither, 2023). In similar discussions on the seventh or eighth generation of warfare, Russians have been using more poetic-sounding terms, such as tip of the cyber spear operations, which means SDW. The latest report by the Jamestown Foundation asserts that those debates were launched by Russian academics and military specialists long before 2014 and were entangled in broader, more holistic discussions on the future of warfare in general, the future of misinformation, the new face of electronic warfare, new generation non-linear warfare (novoye pokoleniye voyny, nelineynaya voyna) mixing political, economic, informational, and technological measures aimed at degrading, dividing, and disrupting, often preceding or accompanying conventional military actions (Alderman, 2015; Powers, 2025).

The first AI development can be traced back to early Soviet experiments in the 1960 in the Zelenograd-based "science city", whose proto-AI technologies were tested and incorporated into the Soviet defense industrial complex and helped to develop proto-AI systems such as the P-700 Granit naval anti-ship cruise missiles, which were upscaled in surface-to-surface and submarine-launched variants and capable of targeting both sea- and ground-based targets. After the collapse of the Soviet Union, those efforts were derailed because of a lack of funding. The Jamestown Foundation report also argues that "Russia has been quite slow to prioritize its UAV development, but ultimately, thanks to those efforts, Russia is in possession of modern offensive drones to precisely strike targets deep inside the enemy territory" (Powers, 2025). Russia relies on simpler, easier-to-produce Iranian drones to save money; it has not managed to create products that would be cheap enough to procure en masse. Russia changed its modus operandi after February 2022, when it encountered difficulties with certain types of Western weaponry and equipment (Powers, 2025). Still, this technological development must have been foreseen since the Russians must have known that their SDW systems were backward, and they pushed for the war to gain momentum and the necessary funding to bridge this gap, which could only happen during wartime.

Russian leaders have learned a great deal from the war in Ukraine due to their adversary's agility in this new environment. Since 2022, Russia has managed to adopt, replicate, or develop its own AI-based systems for target recognition, analysis of satellite imagery, and autonomous unmanned aerial vehicles (UAVs). The problems with deployment of those sophisticated systems and lack of funding for launching new dedicated satellites that could amplify their effectiveness are causing Russia to resort to improvisation, geostrategic brinkmanship, and controlled chaos without scruples about sacrificing the lives of military personnel.

Many scholars argue that even in the early days of the 2022 invasion, the Ukrainians supported by the US were better equipped to surprise the invaders who should have had the advantage of surprise; they were supported by American satellite intelligence provided by Maven Smart Systems and the CIA, so they were more apprised of the approaching Russian battalions than the Russians themselves, who had to capture their first Starlinks to have the same level of technology. Only after Starlink switched off the borrowed Starlinks in Russia in February 2022 did retaliatory strikes against Ukraine decrease. Politicians in Russia would still defend their decision; without this negative experience, Russia would not have understood the scale of the technological gap between itself and the Western powers (Entous, 2025; Fontes & Kamminga, 2023).

Starting wars without being properly equipped to do so is the typical Russian modus operandi. They expose their army to their rival, observe them, and adopt a similarly ruthless strategy for the nation to survive. During the Second World War, especially during the Siege of Leningrad and the Battle of Stalingrad, the Red Army, facing widespread shortages, sent men into battle without weapons or ammunition and left soldiers to cope under extreme circumstances. Since 2023, integrating AI technologies into military systems has become state doctrine, focusing on AI-based systems for target recognition, satellite imagery analysis, and autonomous UAVs. The term information warfare is often used to describe software-defined warfare strategically aligned to shape the governance and geostrategic orientation of a target state. Incorporating AI into various aspects of its defense capabilities became a new norm for Moscow, and despite sanctions limiting access to advanced microelectronics, the Russian military has already managed or is in the process of finding a way to upgrade its much slower secure data networking, develop air and missile defence systems to locate and destroy flying objects, and develop software ecosystems that are more flexible than Western SDW without succumbing to dependence on centralized command structures. The Russians are also enhancing their ability to collect, process, and systematize large amounts of information and data to create a strategic ecosystem that counters enemy drone systems, supported by autonomous recognition and response, unmanned UAVs, and drone swarms. The new Russian approach to asymmetric warfare involves the use of military and non-military means to achieve its military goals with new and old electronic warfare techniques, using Russian ingenuity as a primary combat arm of its cyber and kinetic operations, closely observing its enemies, listening to podcasts dedicated to new technologies, developing new ways of jamming radars and enemy drones and satellites, improving long-range communications disruption, and using drones to spoof cellular networks. Russia invested heavily in military departments and research units that develop a systematic approach to rapidly reprogramming jamming profiles, adaptive signal targeting, and integration with battlefield intelligence. Military scientists build systems that disrupt GPS, which in turn affects drones. They develop automated battle management systems and rapid iterations of targeting software, incorporating sensor data in real time with the use of automation and AI in command systems and unified tactical management systems. They use AI features such as target recognition, swarm capabilities, and autonomous targeting and autonomous behavior in Russian-made loitering kamikaze drones such as the Lancet-3 and the ZALA Lancet. Subsequently, more coordinated efforts have been made to apply AI in command and control, decision support, and automated systems to prioritize the digital and information space. In the tip of the spear, cyber operations, misinformation, and electronic warfare are used to degrade, divide, and disrupt, often preceding or accompanying conventional military action to prioritize the digital and information space and modification of firmware (Powers, 2025).

Some assert that that the Gerasimov Doctrine is currently in play in Russia, but others argue no such doctrine was invented by Valery Gerasimov, the chief of the general staff of the Russian armed forces. They argue that his geostrategic ideas are just those of the Primakov Doctrine, a holistic, multi-domain approach to modern conflict. Primakov was a brilliant geostrategic thinker who was often called the Russian Kissinger; his long-term vision was comprehensive as to how to counter American unipolarity. He suggested collaborating with traditional rivals of the US, such as China, Iran, and India, and promoting alternative global institutions. He argued that Russia should lead a coalition opposing the US-led international system and use any means to bring another multilateral world into existence, including non-linear warfare (Nelineynaya voyna), which uses political, economic, informational, and technological measures. Primakov was among the first to propose such sharp-power policy tools as cognitive warfare, hybrid warfare, active measures, cyberattacks, propaganda, and fake campaigns to achieve Russia's goals and objectives long before they were introduced into everyday practice. His was not the hybrid warfare known in the West, first proposed by Frank Hoffman (Bankov, 2024, p. 23), which employs political warfare and blends conventional, irregular, and cyber warfare with other influences. The emphasis is often placed on cognitive warfare, which weaponizes public opinion and alters the thinking and cognitive processes of its targets to influence public and governmental policy and destabilize public institutions (Backes & Swab, 2019). Primakov's concept of hybrid warfare encompasses the concepts of the war of controlled chaos and reflexive control (Refleksivnoye upravleniye): feeding an opponent selected information to make them voluntarily make decisions beneficial to Russian interests. This could involve information operations, political manipulation, economic coercion, and support for proxy forces to exert influence and achieve strategic objectives. Primakov's definition is often associated with Russia's tactics in the annexation of Crimea and the ongoing conflict in Ukraine, where a combination of military force, covert operations, and information warfare has been employed to achieve Russia's political objectives (Jauhiainen & Erbsen, 2023; Rădulescu, 2026).

APPLES AND ORANGES

That the three nations with the most technologically advanced militaries, the US, China, and Russia, started the shift towards SDW and treat it as an important element of a broader information war is to be expected because of their competitive and innovative strategic cultures. The motivations of these unquestionable leaders in adopting SDW-driven solutions are defined by different political systems, values, circumstances, economic factors, structural constraints, internal governance, attitudes towards innovation and agility, moral considerations, geostrategic planning, and societal doxa. One might understand recent developments in SDW from a broader perspective beyond the struggle of superpowers in a world where many new technologies have changed the way nations wage war. Those technologies have become more accessible and will inevitably change the dynamics and the nature of modern conflicts. For the last two decades, AI has transformed modern warfare beyond recognition. The combination of new technologies, software, algorithms, and computing power is one of the most critical factors in preparing a nation for an offensive or defensive campaign (Kahl, 2026; Tsereteli, 2023).

Russia's invasion of Ukraine has shown that Ukraine needs machine learning, AI algorithmic warfare, and new high-tech weapons to counter an otherwise overwhelming enemy force. The latest weapons, modern intelligence, and data transmission, control, and destruction systems, fuelled by efficient AI components obtained from the US, the UK, the Baltic states, Sweden, Poland, and other Western countries have helped Ukraine defend itself against Russian aggression. Still, we need to consider several important dilemmas about modern conflict elsewhere (Chayka, 2022).

This research embraces ontological roots to reinvigorate IR's exploration of uncharted scientific territory. That the US is still an unquestionable leader in technological innovation is affirmed by both its allies and adversaries, and it is forecasted to preserve its leading role for the foreseeable future. Strategic concepts have less to do with the latest drone technology and more to do with the desire of the US's competitors to develop the most advanced integrated professional communication systems under a single command. The Israeli Iron Dome and Ukraine's Network-

Centric Defense Systems are examples of similar innovations outside the US (Belkin et al. 2024; U.S. Congress, 2026). They are often inspired by US innovations, but their level of integration is different. Countries that care about their security have no choice but to harness advances and incorporate new elements into their military strategies and systems so that they can be more integrated and agile. The US's main competitors, China and Russia, hope to achieve similar or even better results despite the more authoritarian nature of their strategic culture and political systems, which does not prevent them from skillfully copying the US's solutions and adjusting them to their own operations, as Russia has in Ukraine (Khurshudyan et al., 2022).

THE PROLIFERATION OF CERTAIN ELEMENTS OF SOFTWARE-DEFINED WARFARE

The Americans clearly started the era of software-defined warfare; the Chinese and Russians are quickly catching up by both copying US approaches and devising their own novel solutions. They are slightly more ruthless in using SDW against their own citizens. The West, though technologically advanced, is politically divided and constrained by norms, laws, and ethics that slow the adoption of centralized AI for security and military purposes. Strategically, authoritarian states have short-term advantages in coercion and speed (Koumparoudis, 2024).

The West hopes to retain its innovation, economic dynamism, and coalition power to offset these advantages over time. Not only China and Russia but other state and non-state actors are catching up fast; the 2020s has seen major worldwide proliferation of SDW elements. New software, algorithmic, and computing technologies are flooding the military market. Russia's invasion of Ukraine caused a global rise in military spending. The application of AI and other advanced technology to the war has greatly changed its course, transforming the Ukrainian army from one of the weakest in Europe into its most technologically advanced (Koumparoudis, 2025). Other examples of software-defined warfare by non-state actors such as Hamas, on October 7th, 2023 (Ben-Israel, 2025; Muduli et al., 2025; Segell, 2025), and Boko Haram should also be considered in future studies (Campbell, 2014; Council on Foreign Relations, 2025). Similarly, the Houthis use these tactics against the Yemeni, Saudi, and Israeli governments, as did Ethiopians during the latest civil war between the government and the Tigray minority. Opposition leaders in Iran targeted the regime's CCTV cameras in December 2025 and January 2026. A Pakistan-supported militia used AI tools against India in 2025 attacks. Turkish President Erdogan acts with impunity using sophisticated SDW and cognitive tactics to target Kurds in post-Assad Syria (Ivanov & Dronzina, 2025; Pietrzak, 2019 & 2021; Suman et al., 2025). Regimes in Uganda and Iran use SDW in a manner that resembles sharp power policies, completely blocking internet access to achieve political goals. Conservative governments in Central and Eastern Europe use the Israeli counterinsurgency network Pegasus to spy on their political opponents, and Israel uses its Iron Dome technology against civilians in the Gaza Strip (Vasundhara, 2026). Each of these examples deserves further attention in future projects, but overall, they demonstrate that the US, Chinese, and Russian governments do not enjoy a SDW monopoly. Selected case studies should soon be carefully reviewed to improve this imperfect, speculative research, which asks more questions than it provides answers. Naturally, this aligns with the main ontological agenda that fuels the brainstorming capacity of the in statu nascendi ontology and our broader network of scholars, and we will return to this theme soon to explore it in greater depth (Pietrzak, 2024).

CONCLUSION

This chapter offered a brief overview of selected efforts by the US, Russia, and China to harness new systems of software-defined warfare. Department of Defense officials are determined to adopt a software-defined framework to establish a new approach to modern warfare, which is a paradigm shift and a revolution in conflict management, moving the traditional hardware-centric military toward a software-centric model. The adoption of AI-driven semi-autonomous systems clearly makes the US military the leader in various traditional domains of operations: the systems that support high command in integrating and deploy missile and air defines; linking military communication systems; and optimizing intelligent ammunition, which can independently identify targets and determine trajectories. SDW helps military commanders better display optical electronic information to identify persons, choose locations, and diagnose the condition of service members; counter cyber threats; and take into account weather conditions when planning military operations, even in complete darkness.

Although the US remains the unquestionable leader in SDW, China is following the US's lead and pursuing its own efforts to design and deploy next-generation, Internet-scale warfighting software and to equip all of its existing hardware systems with it. Their focus is visibly more holistic, as Chinese officials show a higher propensity to fully embrace a new information war, already dedicating more than 2% of their military budget to integrating SDW. This worries those Americans influenced by John Mearsheimer's *The Tragedy of Great Power Politics*, which argues that China, not Russia, became the US's main geostrategic competitor following the Soviet Union's collapse. Nonetheless, Russia is not far behind, which is especially evident since 2022 in Ukraine (Mearsheimer, 2001).

Russia is learning quickly from its recent strategic blunder and overreliance on military hardware and is investing heavily in automated command-and-control systems to operate effectively in the integrated information environment. Decision-makers in Russia use different terminology for SDW, treating it as an integral part of their electronic warfare, which encompasses both the cyber and kinetic layered architectures.

Most literature suggests that Russia failed to take SDW seriously before 2022; it was less interested in developing software-integrated solutions and more interested in strengthening and prioritizing the development of conventional warfare, along with its cognitive warfare, active measures, and its own interpretation of hybrid conflict, as those measures were easier to apply against adversaries.

Russia's military adventure in Ukraine changed everything about Russia's strategic culture, necessitating the accelerated development of a more holistic approach by aggressively adjusting its systems and strategically adopting those that have proven most successful in Ukraine, the US, and China. The heavier the losses Ukraine inflicts on Russia, the more motivated Russia becomes to improve its SDW and harvest innovation to better adapt to 21st-century military operations. Paradoxically, despite a series of strategic misfortunes in Ukraine, Russia is likely to replace the Soviet weapons that were lost or captured in Ukraine with new, modern equipment. Thus, Russia should not be underestimated or treated as a failed state that is ruled by unimaginative people.

On the contrary, the three analyzed countries operate within a broader context of efforts for innovation, integration, and enhanced communication. Even though both Russia's and China's military strategists are determined to leverage the newest information technology to compete with the US in turning information superiority into a decisive combat advantage and attaining their respective shared situational awareness and self-synchronization, their respective strategies differ considerably from the ones that are followed by the US Department of Defense.

The strategic competition between the United States, China, and Russia for technological superiority in the military realm is the logical continuation of their desire to control their spheres of influence and is already becoming a new Cold War. From a theoretical perspective, the nature of modern warfare has clearly changed, yet the logic of structural offensive and defensive realism best describes this competition, for the systems in question are adopted to gain strategic advantage over competitors.

Those strategies can be treated as "liquid" and "competing horizons" driven by consequentialist and inherently efficient and potentially deadly effective sets of adversarial interactions created and maintained for the purpose of attaining military superiority at all costs and at all times without consideration for the ethical dimension of just war theory that theoretically should govern the rules of engagement even in a software-defined theatre of war (Freiberger, 2013).

REFERENCES

Alderman, R. (2015). Looking toward 7th and 8th generation warfare. Military Embedded Systems, 7 July. Available at: https://militaryembedded.com/unmanned/isr/looking-toward-7th-and-8th-generation-warfare (Accessed: 14.04.2026).

Allen, G. C. (2023). China's Pursuit of Defense Technologies: Implications for U.S. and Multilateral Export Control and Investment Screening Regimes. https://www.uscc.gov/sites/default/files/2023-04/Gregory_Allen_Testimony.pdf

Allen, G. C. (2024, October 15). Blocking China's Access to AI Chips Matters to U.S. National Security. Center for Strategic and International Studies. https://www.csis.org/analysis/blocking-chinas-access-ai-chips-matters-us-national-security

Amilakhvari, L., & Baghaturia, O. (2024). Russia's Militaristic Rhetoric, Imperialism, and Expansion: Wars in Georgia and Ukraine. In Pietrzak, P. (Ed.), *Dealing With Regional Conflicts of Global Importance* (pp. 161–181). IGI Global Scientific Publishing., DOI: 10.4018/978-1-6684-9467-7.ch008

Backes, O., & Swab, A. (2019). *Cognitive Warfare. The Russian Threat to Election Integrity in the Baltic States.* Belfer Center for Science and International Affairs.

Bankov, B. (2024). Hybrid Warfare: How to Escape the Conceptual Gray-Zone. *Journal of Strategic Security*, *17*(1), 1–23. https://digitalcommons.usf.edu/jss/vol17/iss1/1. DOI: 10.5038/1944-0472.17.1.2118

Belkin, P., Bowen, A. S., Nelson, R. M., & Welt, C. (2024, December 23). Russia's war against Ukraine: U.S. policy and the role of Congress. Congressional Research Service.

Ben-Israel, G. M. (2025). Deconstructing Warfare: Redefining Superpowers, Battlefields, Wars, and Armies. In Pietrzak, P. (Ed.), *International Relations Theory and Philosophical Political Insights Into Conflict Management* (pp. 175–208). IGI Global Scientific Publishing., DOI: 10.4018/979-8-3693-9626-1.ch008

Campbell, J. (2014, November 18). U.S. policy to counter Nigeria's Boko Haram. *Council on Foreign Relations*. https://www.cfr.org/reports/us-policy-counter-nigerias-boko-haram

Center for Strategic & International Studies. (2023, April 18). *Testify with Gregory C. Allen: China's Pursuit of Defense Technologies* [Video]. YouTube. https://www.youtube.com/watch?v=eSudeO55On0

Center for Strategic & International Studies. (2023, September 25). Testify with Greg Allen: Advanced Technology - Examining Threats to National Security [Video]. YouTube. https://www.youtube.com/watch?v=fL_8WmGoo_o

Chayka, K. (2022, March 3). Ukraine becomes the world's "First TikTok war." The New Yorker. https://www.newyorker.com/culture/infinite-scroll/watching-the-worlds-first-tiktok-war

Council on Foreign Relations. (2025, November 3). Boko Haram. https://www.cfr.org/keywords/boko-haram

Entous, A. (2025, March 30). The secret history of America's involvement in the Ukraine War. The New York Times. https://web.archive.org/web/20250330021832/https://www.nytimes.com/interactive/2025/03/29/world/europe/us-ukraine-military-war-wiesbaden.html

Fann, K., & Bursi, C. (2023). *Protecting the Porcupine: Why Taiwan Matters.* Blog. Center for International Affairs at Havard University.

Fontes, R., & Kamminga, J. (2023, March 24). Ukraine A Living Lab for AI Warfare. https://www.nationaldefensemagazine.org/articles/2023/3/24/ukraine-a-living-lab-for-ai-warfare

Freiberger, E. (2013, July 18). Just War Theory and the Ethics of Drone Warfare. *E-International Relations.* https://www.e-ir.info/2013/07/18/just-war-theory-and-the-ethics-of-drone-warfare/

Funaiole, M. P., & Hart, B. (2025, September 2). China's Military in 10 Charts. Center for Strategic and International Studies. https://www.csis.org/analysis/chinas-military-10-charts

Gray, C. S. (1999). Strategic culture as context: The first generation of theory strikes back. *Review of International Studies*, *25*(1), 49–69. DOI: 10.1017/S0260210599000492

Hagen, R. A. (2024, February 16). Cognitive Warfare, Cybersecurity, and the AI Challenge. https://www.linkedin.com/pulse/cognitive-warfare-cybersecurity-ai-challenge-raymond-andr%C3%A8-hagen-fvsef

Hartley, D. S. (2021). *An ontology of modern conflict: including conventional combat and unconventional conflict* (1st ed. 2021). Springer., DOI: 10.1007/978-3-030-53214-7

Hartzell, C. L. (2023, July 10). Future weapons technology of 2040. Army University Press. https://www.armyupress.army.mil/Journals/NCO-Journal/Archives/2023/July/Future-Weapons-Technology-of-2040/

Ivanov, I., & Dronzina, T. (2025). In ISIS We Trust: Why Does Radicalization Still Matter? In Pietrzak, P. (Ed.), *Exploring the Implications of Local and Regional Conflicts* (pp. 127–150). IGI Global Scientific Publishing., DOI: 10.4018/979-8-3693-2877-4.ch005

Jauhiainen, J., & Erbsen, H. (2023, April 3). Multilevel governance in the temporal protection and integration of Ukrainians within the European Union: The case of Estonia. *Journal of European Integration*, *45*(3), 413–430. DOI: 10.1080/07036337.2023.2190109

Jiao, A. (2026, January 15). China AI Strategy: Policy, Regulation & Global. *Impact in, 2025*, •••. https://ashleydudarenok.com/china-ai-strategy/

Kahl, C. H. (2026, February 6). The myth of the AI race: Neither America nor China can achieve true tech dominance. *Foreign Affairs*.

Khurshudyan, I., Ilyushina, M., & Khudov, K. (2022, December 2). Ukraine and Russia are fighting the "first full-scale drone war." *The Washington Post*. https://www.washingtonpost.com/world/2022/12/02/drones-russia-ukraine-air-war/

Kirilova, N. V. (2024). Regional Conflict Prevention and Perceived Power Competition: Six Elements of Power. In Pietrzak, P. (Ed.), *Analyzing Global Responses to Contemporary Regional Conflicts* (pp. 252–274). IGI Global., DOI: 10.4018/979-8-3693-2837-8.ch012

Koumparoudis, E. I. (2024). Thou Shall Not Kill: The Ethics of AI in Contemporary Warfare. In Pietrzak, P. (Ed.), *Dealing with Regional Conflicts of Global Importance* (pp. 299–313). IGI Global Scientific Publishing., DOI: 10.4018/978-1-6684-9467-7.ch015

Koumparoudis, E. I. (2025). Russian Invasion in Ukraine: A Software-Defined Warfare. In Pietrzak, P. (Ed.), *International Relations Theory and Philosophical Political Insights Into Conflict Management* (pp. 53–68). IGI Global Scientific Publishing., DOI: 10.4018/979-8-3693-9626-1.ch003

McConoly. R. (2021, June 21). "What is Network-Centric Warfare? Naval Post-Naval News and Information" ttps://navalpost.com/what-is-network-centric-warfare.

Mearsheimer, J. J. (2001). *The Tragedy of Great Power Politics*. W. W. Norton.

Miller, S. (2023, September 1). Cognitive warfare: An ethical analysis. *Ethics and Information Technology*, *25*(3), 46. Advance online publication. DOI: 10.1007/s10676-023-09717-7

Morgenthau, H. J. (1978). *Politics Among Nations: The Struggle of Power and Peace* (5th ed.). Alfred A. Knopf.

Muduli, H. S., Kishor, A., Bag, A., & Bharadwaj, A. (2025). Assessing the EU's Quest for a Two-State Solution in the Israel-Palestine Conflict: Realism, Liberalism, and the Challenges of Peacemaking. In Pietrzak, P. (Ed.), *Exploring the Implications of Local and Regional Conflicts* (pp. 27–56). IGI Global Scientific Publishing., DOI: 10.4018/979-8-3693-2877-4.ch002

Mulchandani, N., & Shanahan, L. G. J. N. (2022). *Software-Defined Warfare: Architecting the DOD's transition to the digital age*. https://www.csis.org/analysis/software-defined-warfare-architecting-dods-transition-digital-age

Neacsu, M. (2010). *Hans J. Morgenthau's Theory of International Relation, Disenchantment and Re-enchantment*. Palgrave Macmillan. DOI: 10.1057/9780230245761

Obama, B. (2012, October 22). Second presidential debate [Video]. YouTube. https://www.youtube.com/watch?v=ZDDJLuIREAQ

Pawłuszko, T. (2025). The War in Ukraine and the Limits of International Relations Theory. In Pietrzak, P. (Ed.), *Exploring the Implications of Local and Regional Conflicts* (pp. 401–418). IGI Global Scientific Publishing., DOI: 10.4018/979-8-3693-2877-4.ch015

Pietrzak, P. (2019). "The Syrian Conflict (2011–ongoing): How a Perfectly Winnable Uprising has been transformed into a Civil War, Only to End up as a Ferocious Proxy War of Global Importance," *In Statu Nascendi – Journal of Political Philosophy and International Relations* Vol. 2, No. 1 2019, p. 34–95, ISBN: 9783838213095.

Pietrzak, P. (2021). *On the Idea of Humanitarian Intervention: A New Compartmentalization of IR Theories* (1st ed.). Ibidem Verlag.

Pietrzak, P. (2024). The Sixth Great Debate in International Relations Theory Revolves Around Clarity: Ontology in Statu Nascendi as a Cutting-Edge Tool for Conflict Management. In Pietrzak, P. (Ed.), *Dealing with Regional Conflicts of Global Importance* (pp. 1–27). IGI Global Scientific Publishing., DOI: 10.4018/978-1-6684-9467-7.ch001

Powers, M. (2025, March 3). Russia capitalizes on development of artificial intelligence in its military strategy. *Jamestown Foundation*. https://jamestown.org

Rădulescu, B. (2026, March 26). Cognitive hegemony: the true objective of Russia's new active measures. Intelligence Info. https://www.intelligenceinfo.org/en/cognitive-hegemony-the-true-objective-of-russias-new-active-measures/

Robertson, P. (2024, December 17). China's military rise: Comparative military spending in China and the US. CEPR. https://cepr.org/voxeu/columns/chinas-military-rise-comparative-military-spending-china-and-us

Segell, G. (2025). Israel vs. Hamas Expands to Eight Fronts. In Pietrzak, P. (Ed.), *Exploring the Implications of Local and Regional Conflicts* (pp. 1–26). IGI Global Scientific Publishing., DOI: 10.4018/979-8-3693-2877-4.ch001

Suman, T., Pandey, S., Bag, A., & Sharma, P. (2025). Russian and Turkish Intervention in the Syrian Civil War: A Comparative Analysis (2011-2022). In Pietrzak, P. (Ed.), *Exploring the Implications of Local and Regional Conflicts* (pp. 97–126). IGI Global Scientific Publishing., DOI: 10.4018/979-8-3693-2877-4.ch004

Takagi, K. (2025, January 27). Is the PLA overestimating the potential of artificial intelligence? National Defense University. https://www.ndu.edu/News/Article-View/Article/4116123/is-the-pla-overestimating-the-potential-of-artificial-intelligence

Tsereteli, A. (2023, January 13). War in Europe: Use of Technologies in the Russia-Ukraine War. https://www.freiheit.org/south-caucasus/use-technologies-russia-ukraine-war

U.S. Congress. (2026, April 10). U.S. Security Assistance to Ukraine. https://www.congress.gov/crs-product/IF12040

U.S. Department of Defense, Office of the Under Secretary of Defense (Comptroller/Chief Financial Officer). (2026, July). *Program acquisition cost by weapon system: Fiscal year 2026 budget request.*

Vasundhara. (2026, March 11). *Iron Dome Air Defence Missile System.* Army Technology. https://www.army-technology.com/projects/iron-dome/

Walzer, M. (2016). Just & Unjust Targeted Killing & Drone Warfare. *Daedalus*, *145*(4), 12–24. https://www.jstor.org/stable/24916780. DOI: 10.1162/DAED_a_00408

Wither, J. K., Partnership for Peace Consortium of Defense Academies and Security Studies Institutes, & Creative Commons. (2023). Hybrid Warfare revisited: A Battle of 'Buzzwords.' Connections: *The Quarterly Journal*, 7–27. https://doi.org/.DOI: 10.11610/Connections.22.1.02

KEY TERMS AND DEFINITIONS

Collateral Damage: "injury inflicted on something other than an intended target. Specifically: civilian casualties of a military operation" (Source: Merriam Webster Online Dictionary),

Electronic Warfare: One of the interpretations of SDW (preferred also by Russian scholars), that is often referred to as the Tip of the Spear, cyber operations, misinformation, and electronic warfare are used to degrade, divide, and disrupt, often preceding or accompanying conventional military action to prioritize the digital and information space, modification of firmware (Powers, 2025).

A Great Power: is a state that enjoys a significant economic, political, and military advantage over other countries and can use its power over world diplomacy. Other nations must consider its opinions before taking diplomatic or military action.

Intelligentized Warfare: One of the interpretations of SDW (preferred also by Chinese scholars), "a new paradigm of conflict that moves beyond 'informatized' warfare by leveraging artificial intelligence (AI), autonomous systems, big data, and cognitive operations to dominate the future battlefield (Takagi, 2025).

Modus Operandi: A particular way or method of doing something. How perpetrators of any given act operate.

Paradigm: "a philosophical and theoretical framework of a scientific school or discipline within which theories, laws, and generalizations and the experiments performed in support of them are formulated" (Source: Merriam Webster Online Dictionary),

Regional Power: is a country that is very powerful and influential in its region.

Software-Defined Warfare (SDW): is a military approach prioritizing software as the core driver of capability, shifting away from rigid, hardware-centric systems. It involves developing modular, adaptable systems that use AI and real-time data to rapidly upgrade functionality and improve cross-domain interoperability, allowing forces to innovate faster than adversaries (Mulchandani & Shanahan, 2022).

Sharp Power: "The ability of an IR actor to influence other IR actors through the possession, access, and exchange of information" (Kirilova, 2024).

Strategic Culture: "Strategic culture refers to the socially transmitted, persistent, and shared set of beliefs, values, attitudes, and habits of mind regarding the use of force, developed from a nation's history, geography, and experience." (Gray, 1999, p. 49-69).

Chapter 2

The Algorithm of Warfare:

Challenges to Jus ad Bellum and Jus in Bello in the Age of AI

Keshav Madhav
http://orcid.org/0009-0002-7960-0166
School of Law, University of Petroleum and Energy Studies, Dehradun, India

Aditya Agrawal
Jindal Global Law School, O.P. Jindal Global University, India

Anushree Shrikant Mukte
http://orcid.org/0009-0008-6912-151X
Symbiosis Law School, Nagpur, Symbiosis International University, Pune, India

Raunaq Chawla
O.P. Jindal Global University, India

ABSTRACT

The 21st century has brought about a fundamental transformation in the character of armed conflict due to rapid technological development. In particular, the introduction of drones and autonomous weaponry introduces new challenges to the conventional constructs and internal definitions of the Just War Theory. The critical analysis of scholarly discourse on Artificial Intelligence (AI) reveals that this technology challenges the core principles of jus in bello and jus ad bellum. The chapter will utilise a descriptive conceptual analysis and a critical analysis based

DOI: 10.4018/979-8-3373-6264-9.ch002

on the available empirical evidence in the existing academic literature. Although the strategic benefit of drones and Lethal Autonomous Weapons Systems (LAWS) is obvious, the consequences should be given the same grave consideration.

INTRODUCTION

Historically, war has undergone profound change, but the fundamental strategic goal remains to employ asymmetrical warfare, reducing casualties among friends and eliminating the human element at risk, while demonstrating great power. This goal became a reality after breakthroughs in modern technology, and drone attacks and self-learning (AI) algorithms have ushered in a new stage of military confrontation. An important event that happened was on January 3, 2020, when U.S. President Donald Trump authorised an airstrike with an uncrewed combat aerial vehicle, or a drone strike, a Predator, targeting Qassam Suleimani, the head of the Iranian Revolutionary Guard Corps. The outcome of this operation was the elimination of zero civilian casualties (Lushenko, 2021) and the realisation of the effectiveness of drones in carrying out precision attacks and surveillance without putting military personnel in danger. Nevertheless, such technological functions require essential consideration of their outcomes. The classical Just War model, which operates within the binary aggressor/defender paradigm in which offensive war is unjust and defensive war is just, is becoming less ad hoc (Gentry, 2014). An offensive pre-emption is becoming normal in modern warfare, and the necessity of using it to defend effectively (Gentry, 2014). Although the conditions of Just War are changing, the use of autonomous weapons has been subject to a strident backlash because it puts strains on the existing principles of jus ad bellum and jus in bello.

In 2014, the United Nations met in Geneva to discuss a possible ban on such systems, citing their notorious complexity and the controversial nature of the laws of war (Sloan, 2015). This has resulted in a polarised debate. Although robotics has proved its usefulness in counter-terrorism through its alleged potential to reduce damage, the application of the technology raises serious moral issues regarding the principles of distinction and proportionality (Brunstetter & Braun, 2011). Since the implementation of robotics and AI in warfare is a comparatively novel phenomenon, the scholarly literature is still underdeveloped and can be discussed at the international level. It is up to future theorists to bring Just War Theory back into modern-day reality by reinstating its moral dimensions in their proper place. In turn, this paper aims to examine how the concept of AI as a military doctrine questions fundamental principles, including the distinction principle, the principle of proportionality, the necessity principle, the principle of cause, the principle of last resort, the principle of probability of success, and the principle of proper authority. Moreover, other types of ethical externalities beyond the traditional theory should also be explored, as the nature of war evolves rapidly. It has also changed the character of combatants (Brunstetter & Braun, 2011). According to the U.S. Department of Defence, autonomous AI lethal weapon systems can be defined as the combat systems that are able to perform military duties without any human intervention (Department of Defence, 2012).

BACKGROUND

The drones were first used in the 1990s during the Balkan wars: in Bosnia and Kosovo, drones were mainly used as surveillance tools. At that period, the United States considered the targeted killings in the form of a direct breach of the International Humanitarian Law. Nonetheless, events of September 11 enabled a change of this perception and the policy was changed to authorise certain extrajudicial killings (Brunstetter & Braun, 2011). Autonomous deadly robots are completely changing the conventional understanding of war. Technological advancements are shifting battlefield operations to distant facilities, many miles apart. This enables military members to operate missions through a computer interface that is disconnected from reality. Unlike the tradition of the past, when soldiers spent months in trenches, a combat soldier in modern-day warfare can go to work in the morning, fly missions abroad, and return to their families in the evening (Sparrow, 2009). Such a metamorphosis of contemporary warfare poses a danger to the ethical norms laid down in the Just War tradition (Lushenko, 2021). Walzer states that the morality-based justification of killing in war is based on conceptual equality and mutual risk perception (Johansson, 2015). These moral equalities are further complicated by the introduction of robotics into combat, since a robot, unlike a human, does not bear the same risk of death. Units that have been destroyed are merely replaced.

Counterarguments suggest that robots can be more effective in bringing war (Royakkers & van Est, 2015). Can robots make rational ethical decisions and act in accordance with the rules of engagement without committing war crimes, unlike human beings, who are subject to cognitive stress, self-interest, and emotional changes (Royakkers & van Est, 2015)? Also, lethal autonomous robots offer greater surveillance capabilities, with the ability to remain in place for days and refuel when required (Brunstetter & Braun, 2011). Lethal autonomous weapons systems cause a record of asymmetrical violence (Kaempf, 2014). Drones have the advantage of handling various functions such as targeting international terrorists and offering security to the troops on the ground, as well as reducing civilian fatalities (Schulzke, 2011). The utility has questioned the need for a human factor in warfare, presenting humans as the system's weak link (Sloan, 2015). By 2009, the U.S. Air Force was training more drone operators than fighter pilots (Helmore, 2009). The Pentagon developed a research program to build autonomous weapon systems in 2013 (Sloan, 2015), and the U.S. Air Force forecasts the introduction of totally autonomous fleets by 2050 (Royakkers & van Est, 2015).

These systems have brought the United States' general interpretation of International Humanitarian Law (IHL) into question, as it has defined its own interpretation of the traditional Just War theory (Kaempf, 2014). Following increasing dissatisfaction with IHL breaches, an unofficial conference was convened in Geneva in 2014 to deliberate on the prohibition of lethal unmanned systems (Royakkers & van Est, 2015). Although intelligent robotic systems are believed to be high-potential inventions, a weapon without some ethical attribute may have an opposite intention (Dodig-Crnkovic & Çürüklü, 2012). To provide an example, in Yemen, the purchase of lethal unmanned weapons by violent non-state actors has been challenging to the Saudi authority (Lushenko, 2021). Nonetheless, little has been achieved towards a ban due to low public awareness and opposition from the big superpowers. The concept of AI technology being introduced as a military doctrine is a modern phenomenon that has posed a tremendous challenge to the conventional understanding of the Just War theory. The use of robotics and AI weapons has transformed the conventional concept of war and is the next phase of its development. These technologies emerged in recent decades and, as such, represent a modern concept that introduces significant challenges actively debated on an international scale. This paper aims to explain how the development of AI technology has exposed the basic tenets of the conventional theory of Just War. This is accomplished by critically analysing the existing academic debates. Moreover, the research examines ethics beyond the Just War principles, focusing on the special problems faced by drone operators arising from this phenomenon. So, there are two research questions introduced:

a) What is the primary question that AI technology has brought to the principles of traditional Just War theory?
b) What are the ethical issues that drone operators are currently experiencing due to the availability of AI technology?

METHODOLOGY

The present research is based on a Qualitative Text Analysis. Its main goal is to provide an analytical overview of the academic discourse on how the advent of Artificial Intelligence (AI) as a military doctrine undermines the premises of the Just War Theory, in particular the notions of jus in bello and jus ad bellum. During the research, the critical aspects of each principle are highlighted and debated. This leads to the conclusion that two methodological strategies, namely descriptive idea analysis and critical analysis of ideas, complement each other to constitute the methodological basis. Simple concepts, presumptions, and arguments leave traces in empirical readings and studies. The role of an idea analysis is to compare, describe and debate on the implicit knowledge of such empirical content. Initially, in this process, there will be a contradiction of hypothetical assumptions to the underlying structure of ideas in the content to create a persuasive interpretation. It can be argued that an idea analysis is the most suitable methodological approach because it serves to depict, explain, and argue for the advent of AI technology and its threats to the core perception of Just War Theory (Pietrzak, 2024).

In addition to these general characteristics, an idea analysis can have a wide range of uses, including descriptive, explanatory, and critical. A descriptive idea analysis tries to outline the basic ideas of an issue. According to it, such descriptions are a mandatory measure during the analysis of the idea, as they serve as the foundation for all other aspirations the researcher has for their study. Because the intention is to explain and characterise the current scholarly controversy about AI technology and its threat to the Just War Theory, the descriptive methodology will be the primary one (Pietrzak, 2023). Nonetheless, summarising an ongoing discussion with already available literature is not enough to make a thesis. Thus, there are cases when a critical analysis of concepts complements the descriptive method. In specific research on ideas, the above-mentioned purpose is to critically analyse the content under study and take a position on the material under consideration (Pietrzak, 2022). This critical analysis aims to assess the validity of the materials. Validity needs to be checked, since any academic text can be interpreted and carries the author's assumptions. This way, the internal consistency of the empirical material is tested against Just War Theory, and the criticism is provided in depth. Lastly, the personal thoughts on the future problems of AI technology are provided and critiqued in detail in the conclusion.

The analysis will be separated into four sections. The initial part of the paper will discuss individual acts under the Just War framework: namely, the jus in bello concepts, and will discuss how the use of AI technology has posed challenges and possible benefits to the concepts of distinction, proportionality, and necessity (Lazar, 2016). The second part concentrates on the moral judgment of war in general, or jus ad bellum, taking into consideration cause, last resort, proportionality, the probability of success, and the proper authority (Lazar, 2016). The third segment discusses viewpoints outside the Just War Theory, including how AI technology poses a distinct challenge for drone operators and raises numerous ethical concerns. The paper concludes with a discussion that presents the findings and offers personal suggestions for future scepticism regarding the subject. To perform a properly organised analysis of ideas, it is reasonable to evaluate and discuss the relevant empirical material in the subject. Therefore, the selection of material on AI implementation in modern warfare relies on its relevance to the concepts of Just War Theory, jus in bello, and jus ad bellum. In the process, individual meanings are constructed by analysing the hidden meanings of texts. The primary sources are the discussions in articles, journals, books, official American government documents, IHL databases, United Nations documents, presidential transcripts, and theories explaining Just War Theory, as the aim is to systematise the academic debate.

THEORETICAL FRAMEWORK

Just War Theory

Michael Walzer is arguably one of the most renowned theorists of Just War Theory, with a tremendous impact on philosophy, political science, international law, and military practice (Lazar, 2016). In his 1977 masterpiece, Just and Unjust Wars, the author explains why and how wars are fought (Moseley, 2004). War is viewed as a flow of events that requires moral reflection before, during, and after. Just War Theory offers a prescription for war doctrine, consisting of conditions intended to make state conduct in war just and equitable (Moseley, 2004). Although the theory primarily governs relations between states rather than individuals, it has application to individuals, especially in the context of jus in bello (Lazar, 2016). This development of terrorism and Violent Non-State Actors (VNSA) poses serious problems for the conventional use of these terms in modern times. The principle of justified war concerns the morality of individual acts performed in the course of war, evaluating the fairness of specific actions (Moseley, 2004). The second principle, jus ad bellum, concerns the overall moral analysis of resorting to war and sets out the conditions states must consult before deciding whether to use military force (Moseley, 2004). Researchers believe that the two principles must be viewed separately and, therefore, an armed conflict may be justified in its entry, but unfair in its conduct (Lazar, 2016). An example of this is that resorting to war might be acceptable, yet the acts committed in the course of war might violate international law. On the other hand, a contradictory situation is also feasible (Lazar, 2016). The following parts will sum up the particular principles that are enshrined in both jus in bello and in jus ad bellum.

Jus in Bello

The rights to life and liberty that civilians enjoy give them immunity to the direct impact of the war (Lazar, 2016). The law of war under international law only allows the use of force against combatants and not against civilians. The concept of liability is a key point in the principle of distinction. Combatants lose their immunity and become legitimate, liable targets by engaging in hostilities (Lazar, 2016). This liability terminates in case a combatant either surrenders, is captured or is incapacitated in a manner that he or she does not present a lethal force. The Prisoners of War (POWs) should be handled in accordance with the Geneva Convention in case of surrender and eventually revert to civilian status (Moseley, 2004). According to the principle of proportionality in jus in bello, the military force used against civilians (or civilian infrastructure) may not be disproportionate to the direct military benefit expected to be achieved as a result of an attack on a legitimate target (Lazar, 2016). Civilians can also be killed due to collateral damage in war that is regarded as a legitimate consequence of warfare, unlike the principle of distinction that firmly prohibits attacking any civilian being (Lazar, 2016). According to the principle of proportionality, the unintentional killing can be deemed justified provided that the military action produces a good in its totality (Lazar, 2016). Nevertheless, civilians cannot be attacked at random to achieve strategic goals. Proportionality basically demands finding a balance between the harm expected and the possibility of preventing more evil (Lazar, 2016). Similar to the principle of distinction, this doctrine has been very much criticised on the issue of negligence of indiscriminate and disproportionate strikes against civilians and property, especially in the conflict in the Middle East.

The necessity doctrine holds that only possible precautions should be taken to prevent unnecessary suffering, and that reasonable force should be used to get the opponent to yield and end the war (Lazar, 2016). Like proportionality, the principle seeks to reduce civilian casualties and property damage while achieving military goals. The primary dilemma of necessity is how much risk the own troops should take in order to do the least damage to innocent (Lazar, 2016, Necessity, para. 3). It is, however, difficult to estimate the exact rate of force needed to bring the threat to a standstill without endangering the lives of one's own soldiers to severe fatalities. As a result, the doctrine is ambiguous about whether there is sufficient force or about avoiding personal casualties.

Jus Ad Bellum

Wars are also reasonable, based on the idea that they prevent excessive death and destruction (Lazar, 2016). The classic conception of the Just War Theory has identified two grounds for supporting the principle of Cause: humanitarian interventions and war as national defence. The self-defence doctrine is legal when a nation's sovereignty or national borders are violated by aggression (Moseley, 2004). This is the principle that can be applied against destructive motives and can tolerate pre-emptive measures concerning the expected threats (Moseley, 2004).

Another fundamental justification for the use of military power is humanitarian intervention. Modern understanding of Just cause, which was offered in the year 1993 in US Catholic Conference says: The force should only be used to correct a grave, public evil that is, aggression or gross violation of fundamental human rights of entire people (USCCB 1993, Just War: New Questions, para. 4). In as much as such standpoints demonstrate the justification to engage in war in accordance to a Just Cause (Lazar, 2016), it still presents significant concerns on how much humanitarian interventions and self-defense can be critiqued and justified (Lazar, 2016). Contemporary foreign policy interventions are often criticised, however, for failing to create peace or inoculate against the spread of violence in the international arena, as well as for the question of whether preventive wars are proportional to the threat they are expected to avert.

The Last Resort doctrine holds that all possible alternatives and negotiation options should be carefully considered and analysed before resorting to war (Moseley, 2004). War has come with enormous human and structural losses and economic and political impacts. Therefore, the decision to start a struggle cannot be misjudged, as it is usually hard to stop once the war has started (Moseley, 2004). If a conflict can be de-escalated through diplomatic efforts, these alternatives should be given priority (Lazar, 2016). Pre-emptive strikes are a pressing modern concern in terms of this principle (Lazar, 2016). We shall discuss that the requirements of the Last Resort can be met prior to an attack (Lazar, 2016). To a great extent, current warfare is dynamic, and it is not always possible to predict strikes with accuracy. As a result, the idea of taking precautionary measures to avoid domestic suffering by means of strict political regulation to prevent the abuse of preventive war with intolerant purposes through the idea of the Last Resort may be seen as permitting states to exercise necessary precautionary measures (Lazar, 2016).

The proportionality of jus ad bellum requires the use of force to be commensurate with the expected extent of evils or harms required without excessive destruction or acts of revenge (Moseley, 2004). This doctrine denounces invasion in itself. For example, when Country A invades Country B, the first invasion should be commensurate with the threat and not inflict excessive suffering. The retaliatory attacks by Country B must be adequate to take back lost ground without the use of excessive force or the invasion of the lands of the intruder, unless it is the sole means to discourage further retaliations (Moseley, 2004). Jus ad bellum proportionality puts more emphasis on the military operation itself rather than individual strikes, unlike the jus in bello proportionality. The Probability of Success doctrine requires a fair probability of success to justify waging war (Moseley, 2004). A country should always consider the costs and benefits it may face before resorting to conflict (Moseley, 2004). War cannot be justified on the one hand when its cost is too high, and on the other hand, when it is impossible to win. Nevertheless, there are problems: defending the nation on its own or defending the population of a besieged country against an oppressing army can imply a purpose, even though the likelihood of success is minimal (Moseley, 2004). Wars cannot be fought to the best of one's knowledge; hence, their outcomes can never be predicted comprehensively. International law holds that only sovereign states have the right to wage war, which non-state actors do not (Lazar, 2016). Statist identity is the only situation in which national defence and the representation of soldiers can be accepted (Lazar, 2016). A soldier who has been fighting based on state legitimacy has the proper authority to fight under international law. Proper Authority means that a governing authority that involves war assumes responsibility for the resources of the population and the state's reputation (Lazar, 2016). Illegal vice, therefore, creates a lot of moral and legal problems both domestically and internationally. It is hard, even to imagine, any incident with legal implications or specific justice without transparency within a political system (Dorbolo, 2001).

JUS IN BELLO IN THE AGE OF AI

The Principle of Distinction

Probably the most important principle of the Just War Theory is the distinction between civilians and combatants. Therefore, the application of the principle of distinction to AI technologies and lethal autonomous weapons is one of the most debated areas in the field, with multiple arguments on both sides of the spectrum. Since war is always life-threatening, having a programmed weapon system with the power to decide who lives and dies according to a coded algorithm reveals itself as an uphill challenge to several ethical, legal, and moral interpretations of jus in bello principles. To ensure that autonomous systems aim at people lawfully, it entails a myriad of requirements to ensure that such individuals are targeted in locations that comply with International Humanitarian Law (IHL) (Lushenko, 2021). Nonetheless, the technical difficulty of programming these requirements is great since autonomous machines cannot morally reason, which is an essential element of dynamic warfare (Schulzke, 2011). Naturally, a machine might be programmed to abide by the rules of ethics, but this does not imply that it is a moral creature, as explained by Singer. Software code does not contain a moral code; zeros and ones have no moral significance (Singer quoted in Schulzke, 2011, p. 300). Since the context-dependence and dynamism of distinction in warfare are extreme, extraordinary moral judgment is also a condition for a robot's legitimate strike (Sparrow, 2009). At present, no system exists that can think of such extensive moral reasoning.

There is a legal complication relating to surrendering combatants that is specifically difficult to recognise. In both versions of Hague Conventions, under article 23, section II (1899) and IV (1907), respectively: To kill or wound an enemy who has laid down his arms, or has no means of defence, and by surrendering himself at his discretion has made himself some such a target is expressly forbidden (ICRC, n.d., (c), para. 1). Robots do not possess moral arguments and therefore, differentiating an enemy combatant who has surrendered is a significant cognitive problem to the machine. Lethal Autonomous Weapons Systems (LAWS), in contrast to human soldiers, can think morally, have empathy, and make judgments in their use of force, including surrendering enemies (Horowitz, 2016). This aspect is further complicated in asymmetric warfare, where the terrorists often live among civilians, thus making it an uphill task to fight terrorists without harming civilians using a different arm (Brunstetter & Braun, 2011). On top of that, the locals bearing arms for local defence, as in tribes in Afghanistan and Pakistan, confuse the distinctions (Sparrow, 2009). Besides, with the physical distance of the combatants growing due to the implemented lethal AI usage, there exists the risk of combatants changing their behaviour in terms of the use of force. foes can be seen as nothing more than pixels on a monitor rather than actual human beings; thus, the threat of drone operators shooting haphazardly does not conform to the rules of jus in bello (Sparrow, 2009).

Nevertheless, autonomous and semi-autonomous systems have clear advantages in deployment. In the 1990s, a NATO airstrike on a civilian bus convoy was a human error, thinking it was a tank convoy during the time of the Balkan War (Lushenko, 2021). It is human nature for pilots to fear death, which can cause them to strike impulsively in the middle of battle. On the other hand, since drones lack cognitive emotions, including stress, emotion, and fear, they can be developed to show hesitation in the application of force, thus preventing the loss of civilians (Brunstetter & Braun, 2011). This technical quality enables robots to meet the requirements of jus in bello without the involvement of human-like background knowledge (Schulzke, 2011). Moreover, drone-delivered video content can help monitor human operators, who will then be accountable for any attack on the legitimacy of the target (Schulzke, 2011). These systems can acquire key information to support decision-making, and their surveillance capabilities can enhance the ability to differentiate between combatants and civilians, especially in conventional state-on-state wars where armies use insignia (Brunstetter & Braun, 2011). The problem of the drones and autonomous capabilities in order to recognise the civilians and the soldiers is a somewhat contentious issue, where there is no moral statement, as there is some evidence on each side. Nevertheless, with technology now changing the face of war and robots becoming an alternative to human soldiers, there is a possibility that, at some point in the future, nations will be pushed to attack civilians to force them to surrender and guarantee victory (Johansson, 2015).

The Principle of Proportionality

Similar to the principle of distinction, proportionality is another area of concern that raises many controversies in the application of AI weapon systems during war. The definition of the level of force that is proportionate to an attack is a highly contested issue among scholars and military organisations when applied to drones and autonomous robots. Consideration of the justification for any strike, whether in terms of magnitude and risk, entails many ethical, moral, and situational aspects (Sparrow, 2009). The military officers have a responsibility to estimate the appropriate proportion of military force against the target; too much force may result in collateral damage to civilians and infrastructure, and too little force may jeopardise the military staff and negatively affect their missions. Advocates of drone use maintain that technological readiness enables them to hit their target more accurately than conventional planes and thus reduces the chances of excessive collateral damage (Brunstetter & Braun, 2011). Moreover, the accuracy of drone attacks may completely replace ground invasions or extensive bombing campaigns, thereby diminishing the overall relative balance of the forces utilised (Brunstetter & Braun, 2011). As mentioned in the distinction, soldiers must make life-or-death decisions in a split second. On the other hand, drone operators operate under safe conditions and, to that end, have support teams, which, in retrospect, supports their ability to make better proportionality decisions since they are not exposed to the dangers of combat (Brunstetter & Braun, 2011).

Moreover, drone operators are less prone to committing war crimes than ground soldiers during fighting because they do not get such emotional experiences of anger, stress, or anxiety (Sparrow, 2009). According to a report on soldiers involved in Operation Iraqi Freedom, about half of those interviewed do not think that all non-combatants must be treated with dignity and respect and about half also feel that the use of torture can be used as an option for retrieving intelligence about insurgents (Office of the Surgeon General, 2006). It is also possible to envision future situations in which drones can serve as ethical governors, offering guidance on how to use weapons to prevent indiscriminate attacks or to warn the operator of potential violations (Royakkers & van Est, 2015). Nonetheless, a change of aerial bombing to drone strikes can be paradoxical, as the latter can result in a higher number of civilian victims because such systems are used more often (Royakkers & van Est, 2015). The other issue of concern is the fact that lethal autonomous drones can be used to make an easy kill. This lack of consideration for life and property in the pursuit of specific goals or the need to secure soldiers in the name of national security may fail to account for the moral and ethical aspects in justifying a strike (Sloan, 2015). The Geneva Conventions state that civilian loss of life during an operation is acceptable as long as it is not unwarranted or higher than necessary to achieve the military goals (Vogel, 2011). Nonetheless, the legal formalities involved in interpreting this in the context of national security are intricate. According to statistics released by New America, between 2004 and 2018, 245-303 civilians died due to drone attack strikes in Pakistan, and 211-328 more had not been confirmed (New America, 2018). Although semi-autonomous robots commanded by humans can take into account ethical, moral, and legal concerns before evaluating proportionality, the potential of totally autonomous robots having complete control over proportionality remains a debatable issue to be witnessed in the foreseeable future.

The Principle of Necessity

The utilitarian approach, which can be defined as the pursuit of the greatest good for the greatest number and the reduction of harm to the lowest possible level, is, unfortunately, hardly achievable in a conflict (Johansson, 2015). Smith has indicated that over the last ten years, there has been a tremendous shift from restraint of violence to the justification of force (Smith, 2002). This is one of the shifts in the ethical focus towards the legal. The United States has exhibited readiness to sacrifice the life of a civilian for national security and strategic considerations (Vogel, 2011). As a result, the development of drone technology suggests that in the future, this technology can be used mainly to reinforce military purposes and not to attain humanitarian goals. The war on terror initiated by the U.S. has created enormous security effects on the globe. Whereby, countries that are victims of radical terrorism cannot effectively control their internal security, countries like the U.S. contend that it is their role to intervene to guarantee stability in the region and to ensure that they do not continue compromising their sovereignty (Brunstetter & Braun, 2011). The question of drones' necessity confronts the basic theory of the legality of war, becoming a paradigm of maximised military utility and strategic benefit. This makes it hard to know whether they are necessary, since they cannot determine future attacks on a nation's sovereignty. To answer Brooks, how can one decide whether or not lethal force is essential in averting a potential future attack, of which the person does not know? (Brooks, 2014, p.95). Even though the modern concept of necessity in the classical theory of Just War is contentious in the contemporary global arena, drones offer the opportunity to use just enough force to threaten an enemy into surrendering without causing many casualties among local fighting troops.

JUS AD BELLUM IN THE AGE OF AI

The Principle of Just Cause

The two that make up the doctrine of Just Cause are the national defence and humanitarian interventions aimed at enhancing sovereignty. Nonetheless, the contribution of drones and AI technology to upholding these principles is an ambiguous topic with enormous political consequences. Sovereignty promotes harmony in the community and fosters collaboration among states to pursue mutual interests of security, stability, and the establishment of standard norms (Lushenko, 2021). It is important to note that the US has also used drones to aid politically unstable countries, thus upholding their sovereignty. As an example, the U.S. uses drones in Somalia to aid the Federal Government in having control over the situation in Mogadishu (Lushenko, 2021). Therefore, the international community arguably has a duty to intervene by force to place states in positions of sovereignty and legitimacy, thereby sharing in the benefits of the world order (Gentry, 2014). The international arena is a complex and dynamic world, with threats that cross borders. The self-defence doctrine and the preventive force doctrine have been developed in response to increasing threats to national security, challenging the principle of Just Cause. In one of the speeches regarding the U.S. drone policy, Barack Obama said, We are at war with an organisation that currently would kill as many Americans as possible did we not stop them first. This is then a just war: a war conducted on equal terms, as a last resort, and in self-defence. (Mills, 2013, para. 32) The U.S. policy justifies the application of drones in war defence by labelling it as a matter of self-protection. It denies the terrorist factions the power to organise and carry out intercontinental attacks (Brunstetter & Braun, 2011).

Nevertheless, when there is a risk that states will act contrary to the principle of general morality to pursue strategic interests and national goals, abusive practices are bound to occur, thereby undermining the general perception of Just Cause (Buchanan & Keohane, 2004). Moreover, other states may imitate the U.S. policy and justify their drone deployments as self-defence while pursuing their national interests (Hajjar, 2019). In the absence of proper rules and regulations on drones and lethal autonomous weapon systems, nations will probably misuse these advantages to benefit their own countries. War, in a broad sense, is equally very expensive, irrespective of whether the combat is superior. The loss of national service members is an unpleasant aspect that will increasingly be a reason for a crowd to rise and protest. By disconnecting the human element from the loop, there could be a reduction of the reasons not to go to war (Sharkey, 2012). An illusion of loss of life and human marginalisation is brought about using drones. These factors can significantly affect how the public perceives war (Johansson, 2015). Indicatively, in

January 2003, thousands of American troops and civilians were lost in the intervention by the Iraqi soldiers. The use of drones has since received widespread support as a preferred option (Lushenko, 2021). In the case of the military commanders, the use of drones is a solution, as they can reduce the number of dead people and eliminate the use of massive invasions that usually lead to mass mutilations (Sharkey, 2012). The distance at which lethal force can be made without putting soldiers at risk has been hugely tantalising to the U.S. to push the design of lethal unmanned weapons to a greater extent (Hajjar, 2019). Horowitz believes that numerous issues that accompany drones and remote weapons are highly problematic to address, as the incentives to wage war are made easy and inexpensive, the principles of jus ad bellum are being undermined, since Artificial Intelligence has no moral sense and fails to meet the demands of Just Cause (Horowitz, 2016).

The Principle of Last Resort

The doctrine of the Last Resort ensures that, even non-military approaches, such as diplomatic and/or arbitration, and counteractions, such as political influence and/or economic sanctions, must be thoroughly considered and evaluated before an attack is considered and justified (Brunstetter & Braun, 2011). The international realm, however, is marked by a dynamic environment and the lack of a supreme judicial body, making the notion of mutual dependence between states idealistic. Thus, there exists a realist approach that is characterized by the prioritisation of national security and national self-interest in interpreting foreign affairs, hence making it challenging to apply the principle of the Last Resort on Lethal Autonomous Weapons Systems (LAWS). The unique strength of drones and LAWS is that they will make it harder to have an actual threat without having to launch violent retaliation in the form of a full-scale military attack, which may increase the cost of a Last resort (Brunstetter & Braun, 2011). However, according to what was mentioned in the article on Proportionality, the possibility of killing and getting away with killing insurgents through the use of drones might actually be seen as an incentive to countries using violence as a faster means of attaining their negotiating objectives. This restraint was therefore minimised, which might lead to a failure to follow the jus ad bellum principle that force should be a last resort. Still, in cases where the precaution is necessary for state security or in response to a breach of International Humanitarian Law (IHL), the use of drones is usually considered a means of defence. It may be accepted as an action of Last Resort (Gentry, 2014). Nevertheless, serious concerns remain about justice. Though unintentional civilian deaths and wrongful targeting are considered the usual state of affairs during war, the use of drones to prevent threats directly leads to a procedural injustice: a threat target is not provided with a chance to defend themselves in court in a just manner (Royakkers & van Est, 2015).

The Principle of Proportionality

The exact degree of proportionality in the scale of a conflict is hard to determine and arguably unimplementable. According to Brooks, both necessity and proportionality are to be considered within the contaminated categories of simply hypothetical worst-case scenarios (theoretically, any suspect of terror can be on the brink of carrying another gigantic attack of the magnitude of 9/11) (Brooks, 2014, p.95). As discussed in the jus in bello, the Geneva Convention does not prohibit loss of life in war; hence, it is tolerated when it is not irresponsible or unreasonable to the military purpose (Vogel, 2011). Nonetheless, applying these interpretations to prevent the perceived threat is complicated, involving legal formalities that lawyers frequently discuss. According to some of the scholars, the U.S. invasions of Iraq and Afghanistan were hugely disproportional in relation to the original suffering caused by 9/11 (Brunstetter & Braun, 2011). The subsequent civilian death toll, destruction of the property and aftermath of the invasion imposed very massive setbacks in the recovery processes of both governments. Brown says that only extraordinary reasons can justify the gateway to war (Brown, 2011). Because drones lack the capacity for moral judgment, it is doubtful whether they are capable of making exceptionally justified judgments. According to proportionality, any drone authorised to declare war or to take an independent position would raise serious legal, ethical, and moral issues. In addition, the measures that can either be used to compute or quantify disproportionate suffering do not exist at the moment (Schulzke, 2011). There are several independent Lethal Autonomous Weapons Systems (LAWS), but they are mostly restricted to self-defence. An example is the autonomous sentry gun SGR-A1, deployed on the South Korean side of the Demilitarised Zone, designed to detect trespassing and respond to assaults in the event of a border violation (Velez-Green, 2015). This weapon could also recognise surrendering intruders by physically identifying combatants who were disposing of weapons or placing their hands on them (Velez-Green, 2015). The implementation of these systems can be argued to increase the barrier to violence between states; the fear of immediate retaliation may deter states from going to war, thereby lowering the probability of war (Velez-Green, 2015). Lastly, as addressed in jus in bello, drones and LAWS can be used in precision-based strikes, which could be more proportional to the particular menace and may not require a broader war.

The Principle of Probability of Success

The process of war involves a series of incidents that require much reflection before, during, and after the course of action. The likelihood of success and victory is a prerequisite for any country that chooses to engage in hostilities. In the absence of precautions, the chances of failure are likely to depend on the sequence of activities undertaken during the war, which ultimately determines its outcome. Next, technological progress has enabled more proportional responses during the execution of the strike, reducing collateral damage to people and property. In addition, drones and autonomous weapons have the intelligence capacity to effectively gather information on hostile insurgents, thereby denying them a haven and reducing their retaliatory capacity (Brunstetter & Braun, 2011). The killing of high-profile officials and terrorist commanders through drones can be said to increase the prospect of determining a just war since it eliminates the elucidation of massive attacks. Additionally, humans have the traits of stress, fear, and exhaustion; during the heat of battle, these cognitive tendencies result in disproportionate attacks. Having eliminated the human element from the loop, drones are arguably better suited to high-stress environments, as they are insensible to emotions that could affect their ability to act proportionately (Royakkers & van Est, 2015). Therefore, it can be assumed that drones increase the chances of effectively waging a war with the destruction being commensurate through the utilisation of drones.

Nevertheless, there are serious problems with drones as well. English philosopher Henry Sidgwick opposed war, arguing that it should be carried out without reflecting resentment or vindictive urges (Johansson, 2015). Avoiding cruelty and unfair treatment is a critical factor that a nation should consider when engaging in war to lay the foundation of a peaceful and calm future (Johansson, 2015). On the other hand, as Norris says, the role of Drones in war is not to limit it but to proceed with a vicious cycle of terror and violence. (Norris 2020, Not a way to peace, para. 2) Norris states that the unethical character of drones and their incapability to judge people and generate feelings of hate and fear leads to the development of terrorism and vengeance desires in the local population (Norris, 2020). Thus, it can be assumed that the effects of drone attacks and the loss of family members lead to a large number of recruits. According to the words of Baitullah, a Pashtun commander of the Pakistan Taliban, each drone attack offers three or four suicide bombers who mostly find their origin in families that have lost people as a consequence of drone airstrikes (Ghosh & Thompson, 2009, p. 2). Effective relationships between the coalition and the local public sector are an important element in identifying a successful operation (Gentry, 2014). It is necessary to develop a sustainable relationship with the local population, grounded in cross-cultural competence and empathy, along with knowledge of local cultural values, religious beliefs, and expectations (Gentry, 2014). Drone and autonomous systems lack moral reasoning, so they are arguably incapable of conveying these important elements. As a result, it is arguably impossible to establish any reciprocal arrangement or cooperation with the local population. The thought of the drone bargaining with the house occupant who has lost her property evokes a sense of alienation; Sharkey says you cannot find justice in a machine that has no emotion (Sharkey, 2012). According to the proponents, drone attacks are a necessity in fighting terrorism. This argument holds in the current international context, which continues to focus on the war on terror. Nevertheless, it is also true to state that the exploitation of drones provokes and elevates the risk of terrorism to the global level (Hajjar, 2019). As Hajjar also describes, in case drone attacks are the cure, it is also an extension of the illness (Hajjar, 2019, p. 947).

The Principle of Proper Authority

Proper authority is also a doctrine obligatory in war and defines the legitimate use of force and accountability, as opposed to lawless acts beyond the authority of international law. The use of drones threatens these core tenets of legitimacy and authority as it may lead to the destabilisation of what is defined as legitimate warfare. The problem of accountability is immensely complex because drones lack moral and ethical reasoning in their judgments (Horowitz, 2016). Adding to this complexity, it is not in the law of the day that an organisation like the CIA is obligated to disclose any classified information, nor does it require formal presidential authorisation before executing an attack (Brunstetter & Braun, 2011). This allows operations to be carried out in secrecy, without being observed by outsiders. In the absence of transparency, it is impossible to design a situation in which legal repercussions or the existence of a working authority is needed. Moreover, the U.S. State Department has ignored efforts by the outside elements to interrogate this clandestine (Brunstetter & Braun, 2011). Thus, critics believe that drones are breaching international law, relating to extrajudicial murder and terrorism, because every nation is willing to launch strikes without revealing vital information that may have legal ramifications.

Another complication of a necessary authority is based on the U.S. military intervention in states such as Pakistan and Yemen without the official declaration of war. The endless scope of the U.S. War on Terror, which, though not declared, is not officially proclaimed and was not agreed to by the states subjected to the terroristic attacks, creates a blur in defining the concept of war and the principle of appropriate authority in the jus ad bellum concept (Brunstetter & Braun, 2011). In the Charter of the United Nations, the right of nations to self-defence (whether individual or joint) under the presence of an armed attack against a Member of the United Nations was considered inherent and nothing on the current Charter was to be damaging the natural right to self-defense until measures were undertaken by the Security Council such that to keep international peace and security (United Nations, n.d., art. 51). The rule of law, as interpreted in the context of international terrorism, is arguably appropriate to provide sufficient legal rationale in the context of states employing drones and lethal autonomous weapon system in self-defense (Brooks, 2014). On the other hand, the non-state actors produce and buy commercial drones and turn them into deadly weapons. As stated by Chávez, according to numerous sources such as policy reports, military literature, and open-source media, there are an estimated 57 Violent Non-State Actors (VNSAs) that have acquired UAVs (Chávez & Swed, 2021). Furthermore, parties such as the Houthis have been able to build deadly drones through 3D printing enabled by technological advancements (Lushenko, 2021). Having such potent weapons systems confronts states at the level of revolution and poses serious questions about national security. In this way, Lethal Autonomous Weapons Systems (LAWS) and the secrecy surrounding them will mark another stage in war, threatening the idea of appropriate authority and potentially altering the traditional understanding of legitimate authority in war.

ACCOUNTABILITY AND THE MORAL BUFFER OF AI

This chapter considers the distance the soldier is no longer on the battlefield and how the development of autonomous AI algorithms and drones could have potentially unethical impacts from the drone operator's perspective. War has fundamentally changed the way soldiers view warfare. With the growth of technology, the physical distance between soldiers and the battlefield has grown significantly. As a result, many changes in the rationale of the fighters have changed the sense of accountability, responsibility, and duty of the soldier. Also, other psychological and moral considerations have opened new questions with the advent of drones and the Lethal Autonomous Weapons Systems (LAWS). With the existence of drones and autonomous robots, it is expected that a moral buffer will be created between soldiers and their sense of accountability and responsibility (Sparrow, 2009). Detert et al. state that people who have little personal power over specific actions tend to accept the result, which may also include negative ethical or moral charges (Detert et al., 2008). Therefore, on a functional level, when an AI algorithm decides and orders a strike, human operators may not identify with that action and may not want to take responsibility for the results of drone actions (Horowitz, 2016).

The future of the possibility of such acts is that they will chain down the key tenets of the reasons behind killing, namely, the concept of sharing the risk of killing between combatants (Johansson, 2015). Besides, the question of accountability regarding drones is more complex, particularly when a drone is an autonomous agent, which makes tracing liability nearly impossible (Sloan, 2015). Whom to hold responsible in case of the violation against International Humanitarian Law- IHL- be it the manufacturer, commander, the operator, or the drone itself, is yet to be established (Lokhorst & van den Hoven, 2012). On the same note, as with any other human-made device, drones can malfunction, be misused, and experience technical errors. Thus, solid design plays an important role, ensuring that operators know what they are doing and perform in accordance with the demands of accountability and responsibility (Sparrow, 2009). Stress, anger, and fear are some of the prevalent symptoms associated with soldiers in warfare. The threat of physical exposure, which in turn leads soldiers to engage in violence and commit war crimes, is, therefore, an impending danger that contributes to similar acts all the time. Another factor that may affect moral decision-making is the relationship humans have with one another (Schulzke, 2011). The sense of worry and commitment to the group's safety can be achieved through unity within platoons. In most situations, soldiers go to extremes, sacrificing their lives to safeguard their close friends (Schulzke, 2011). Even though strong cohesion correlates with a soldier's capacity to perform extraordinarily well on the battlefield, the danger of soldiers failing to make ethical judgments is widespread (Schulzke, 2011). The fact that drone operators are not physically threatened, arguably, significantly decreases their capability of controlling negative emotions that can affect their ability to perform immorally, which would inevitably reduce the possibility of war crimes manifestation (Sparrow, 2009). Nonetheless, specific reports indicate that drone operators, just like pilots, tend to develop post-traumatic stress symptoms because their decisions to take action may produce an enormous impact on the outcome and safety of friendly soldiers on the ground (Brunstetter & Braun, 2011).

Another serious issue that should affect a drone operator's capacity to cope with war-related stress is the ongoing transformation of the atmosphere (Sparrow, 2009). Major Shannon Rogers of the U.S. Air Force said, "Physically, we can be in Vegas, but psychologically, we are flying over Iraq" (Donnelly, 2005, p.1). Usually, soldiers who have ever been engaged in a physical war should be engaged in post-war therapies, such as communication and associations with soldiers who have had a similar experience (Sparrow, 2009). This remote war is a new phenomenon that poses distinct challenges for soldiers. The ongoing environmental variation and geographic and psychological split between the workplace and home environments might prevent the soldier from adjusting to and handling traumatic and stressful events common in war (Sparrow, 2009). You will be at war with 12 h of shooting weapons at the target, as Singer puts it, and in 20 min after that, you will be at the dinner table with your children discussing their homework, and you will be home (Singer quoted in Schulzke, 2011, p. 296). The other ethical concern is the depreciation of human life. Future fights will probably be video game-style, not real fighting. Fighters will be represented as pixels on a screen instead of human souls. Human dignity is most likely to be overcome by the ease of killing and the need to complete missions. The ethical concerns of an AI algorithm possessing the power to take life and death are a gigantic problem since robots are not moral beings, as such (Horowitz, 2016). According to the philosopher Peter Asaro, justice per se can not be assigned to the process of automation: justice itself cannot be given up. (Asaro, 2013, p. 701).

CONCLUSION

The nature of warfare has changed due to the rapid technological development of the 21st century. The concept of drones and autonomous unmanned weapon systems has questioned the fundamental principles of the Just War Theory and offered new insights into its ideas (Lazar, 2016). In turn, future theorists of Just War should also reflect on how these peculiarities relate to the classical notion of the theory. The main question to consider is how the moral grounds of jus ad bellum and jus in bello can be changed to comply with this new phenomenon. The discussions herein examine the issues that arise for the Just War Theory due to AI technology. Even the aggressor-defender paradigm has become dynamic owing to the prowess of modern warfare. Hence, the Just Cause principle, namely self-defence, will likely become the key factor for nations choosing to implement Lethal Autonomous Weapons Systems (LAWS). However, since waging war will be much smaller and cheaper, the motives for going to war will be more readily advanced, possibly breaching the fundamental provisions of Just Cause. The principle of Last Resort will be much harder to respect because the external threat to national security posed by terrorism and Violent Non-State Actors (VNSA) makes threat recognition burdensome. Therefore, pre-emptive strikes rather than diplomatic negotiations may be the only possible rescue. Foreign policy is the process of advancing long-term strategic national interests. The benefits of drones are indeed all-encompassing, and as such, will not see the superpowers abandon their asymmetrical advantage as long as the benefits outweigh the consequences.

The fundamental principles of Necessity and Proportionality are arguably more powerful to apply to the Just War Theory, since drones are designed to reduce civilian deaths, avoid massive invasions, and deliver precise force to induce surrender while minimising friendly deaths. It is speculative, however, as the foreign policy becomes more focused on supporting violence using the prescribed tools, that this will be used to serve military goals as opposed to humanitarian goals. There is a considerable controversy concerning the definition of war itself. With drones, it is possible to monitor the conflict without the disastrous effects of invasion (Brunstetter & Braun, 2011). Indicatively, the U.S. works within Yemen and Pakistan, though not declared war. The phenomenon commonly known as jus ad vim tends to blur the border of what exactly is considered as war, to generate a grey area in terms of the requirements of jus ad bellum. The design of robotics is important because it fulfils the principle of Distinction, as it becomes harder to separate civilians and combatants. Arguably, Drones demonstrate high validity in meeting the Distinction and Proportionality criteria, since, unlike human beings, they can be programmed to exhibit high degrees of restraint. Aspirations in AI algorithms are not influenced by emotion and are therefore not prone to war crimes or disproportionate strikes. Although technology has not yet enabled robots to form moral judgments, it is likely that developments will do so in the distant future, given the rapid pace of technological advancement.

The AI technology poses a problem for the principle of Probability of Success in adhering to traditional values. The presence of drones makes it hard to build relations between the occupying force and the local population, which is crucial to future peace. Moreover, nations that have been involved in a confrontation with a more technologically advanced opponent are more likely to develop a desire to retaliate and feelings of vengeance in response to unjust, asymmetrical violence. Proper Authority is arguably one of the values most influenced by the development of drones and LAWS. Just War Theory has always been applied to the relationships between states, rather than between individuals. The number of conflicts between states and non-state actors is a threat to the legitimacy of the Proper Authority. Since no longer on the battlefield, accountability is less transparent. The necessary rules are required in order to avoid secrecy when conducting military operations. With the shift from semi-automated to fully automated systems, it will be much harder to determine who is to be held responsible; it is already hard to hold Robin responsible for crossing the line due to illegal intent, because the robot cannot experience pain, injustice, or retaliation.

As we slip into this new phase of robotics, the value of human life will depreciate, and the depersonalization of operators on the battlefield will create a gap between the victims and the ethics and mentality of the perpetrators. The issue of robot-on-robot warfare in the future is a primary concern for jus in bello. In this case, it may require targeting civilians and cities to gain victory or even compel a surrender, which would expressly violate the values of Distinction and Proportionality. On the one hand, the benefits of drones are apparent; on the other hand, their consequences are critical considerations. One cannot definitely lead or argue for or against the use of drones, because there are arguments on both sides. Pappantly, the Just War Theory, as a conventional definition, is under tremendous pressure due to the advent of AI technology. The theory still has an ethical, moral, and legal loophole, failing to address modern problems. Thus, changes and adaptations to the theory will be necessary to avoid unjust wars and to promote compliance with International Humanitarian Law (IHL).

REFERENCES

Asaro, P. (2013). On banning autonomous weapon systems: Human rights, automation, and the dehumanization of lethal decision-making. *International Review of the Red Cross, 94*(886), 687–709. DOI: 10.1017/S1816383112000768

Brooks, R. (2014). Drones and the international rule of law. *Ethics & International Affairs, 28*(1), 83–103. DOI: 10.1017/S0892679414000070

Brown, D. (2011). Proportionality in modern just war theory: A tort-based approach. *Journal of Military Ethics, 10*(3), 213–229. DOI: 10.1080/15027570.2011.608496

Brunstetter, D., & Braun, M. (2011). The implications of drones on the just war tradition. *Ethics & International Affairs, 25*(3), 337–358. DOI: 10.1017/S0892679411000281

Buchanan, A., & Keohane, R. O. (2004). The preventive use of force: A cosmopolitan institutional proposal. *Ethics & International Affairs, 18*(1), 1–22. DOI: 10.1111/j.1747-7093.2004.tb00447.x

Chávez, K., & Swed, O. (2021). The proliferation of drones to violent nonstate actors. *Defence Studies, 21*(1), 1–24. DOI: 10.1080/14702436.2020.1848426

Department of Defense. (2012). *Autonomy in weapon systems* (Directive 3000.09). U.S. Department of Defense. https://www.esd.whs.mil/portals/54/documents/dd/issuances/dodd/300009p.pdf

Detert, J. R., Treviño, L. K., & Sweitzer, V. L. (2008). Moral disengagement in ethical decision making: A study of antecedents and outcomes. *The Journal of Applied Psychology, 93*(2), 374–391. DOI: 10.1037/0021-9010.93.2.374 PMID: 18361639

Dodig-Crnkovic, G., & Çürüklü, B. (2012). Robots: Ethical by design. *Ethics and Information Technology, 14*(1), 61–71. DOI: 10.1007/s10676-011-9278-2

Donnelly, S. B. (2005, December 4). Long-distance warriors. *Time*. https://content.time.com/time/subscriber/article/0,33009,1137653,00.html

Dorbolo, J. (2001). *Just war theory*. Oregon State University. http://oregonstate.edu/instruct/phl201/modules/just_war_theory/criteria_intro.html

Gentry, C. E. (2014). Epistemic bias: Legitimate authority and politically violent nonstate actors. In Lang, A. F.Jr, O'Driscoll, C., & Williams, J. (Eds.), *The future of just war* (pp. 17–29). University of Georgia Press., DOI: 10.2307/j.ctt46nbn3.4

Ghosh, B., & Thompson, M. (2009, June 1). The CIA's silent war in Pakistan. *Time*. https://content.time.com/time/subscriber/article/0,33009,1900248-2,00.html

Hajjar, H. (2019). The counterterrorism war paradigm versus international humanitarian law: The legal contradictions and global consequences of the US 'War on Terror'. *Law & Social Inquiry*, *44*(4), 922–956. DOI: 10.1017/lsi.2018.26

Helmore, E. (2009, August 23). US now trains more drone operators than pilots. *The Guardian*. https://www.theguardian.com/world/2009/aug/23/drones-air-force-robot-planes

Horowitz, M. C. (2016). The ethics & morality of robotic warfare: Assessing the debate over autonomous weapons. *Daedalus*, *145*(4), 25–36. DOI: 10.1162/DAED_a_00409

International Committee of the Red Cross. (n.d.). *Treaties, states parties and commentaries*. https://ihl-databases.icrc.org/applic/ihl/ihl.nsf/ART/150-110029?OpenDocument

Johansson, L. (2015). *Äkta robotar*. Fri Tanke.

Kaempf, S. (2014). Post-heroic U.S. warfare and the moral justification for killing in war. In Lang, A. F.Jr, O'Driscoll, C., & Williams, J. (Eds.), *The future of just war* (pp. 79–97). University of Georgia Press., DOI: 10.2307/j.ctt46nbn3.8

Lazar, S. (2016). War. In Zalta, E. N. (Ed.), *The Stanford encyclopedia of philosophy* (Winter 2016 ed.). Stanford University., https://plato.stanford.edu/entries/war/

Lokhorst, G.-J., & van den Hoven, J. (2012). Responsibility for military robots. In Lin, P., Abney, K., & Bekey, G. A. (Eds.), *Robot ethics: The ethical and social implications of robotics* (pp. 145–156). MIT Press.

Lushenko, P. (2021). Review of the special issue "robotics autonomous systems and warfare," Small Wars and Insurgencies 31, 4 June 2020. *Small Wars & Insurgencies*, *32*(3), 550–570. DOI: 10.1080/09592318.2021.1891618

Mills, D. (2013, May 24). Obama's speech on drone policy. *The New York Times*. https://www.nytimes.com/2013/05/24/us/politics/transcript-of-obamas-speech-on-drone-policy.html

Moseley, A. (2004). Just war theory. *Internet Encyclopedia of Philosophy*. https://iep.utm.edu/justwar/

New America. (n.d.). *The drone war in Pakistan*. https://www.newamerica.org/international-security/reports/americas-counterterrorism-wars/the-drone-war-in-pakistan/

Norris, K. (2020). The drone threat to just war theory: Responding to Braun. *International Relations, 34*(4), 603–607. DOI: 10.1177/0047117820967983

Office of the Surgeon General. (2006). *Mental health advisory team (MHAT) IV Operation Iraqi Freedom 05-07*. National Technical Reports Library. http://www.combatreform.org/MHAT_IV_Report_17NOV06.pdf

Pietrzak, P. (2022). *Introducing the idea of ontology in statu nascendi to the broader international relations theory. International Conference Proceeding Series: International Conference on Economics and Social Sciences*. ECLSS., https://www.eclss.org/publicationsfordoi/abst11act8boo8kIE%26SS2022_antalya.pdf

Pietrzak, P. (2023). Approaching regional conflicts through the prism of ontology in statu nascendi: The new compartmentalization of the IR theory. *In Statu Nascendi – Journal of Political Philosophy and International Relations, 6*(1).

Pietrzak, P. (2024). The sixth great debate in international relations theory revolves around clarity: Ontology in statu nascendi as a cutting-edge tool for conflict management. In Pietrzak, P. (Ed.), *Dealing with regional conflicts of global importance* (pp. 1–27). IGI Global., DOI: 10.4018/978-1-6684-9467-7.ch001

Royakkers, L., & van Est, R. (2015). A literature review on new robotics: Automation from love to war. *International Journal of Social Robotics, 7*(5), 549–570. DOI: 10.1007/s12369-015-0295-x

Schulzke, M. (2011). Robots as weapons in just wars. *Philosophy & Technology, 24*(3), 293–306. DOI: 10.1007/s13347-011-0028-5

Sharkey, N. (2012). Killing made easy: From joysticks to politics. In Lin, P., Abney, K., & Bekey, G. A. (Eds.), *Robot ethics: The ethical and social implications of robotics* (pp. 111–128). MIT Press.

Sloan, E. (2015). Robotics at war. *Survival, 57*(5), 107–120. DOI: 10.1080/00396338.2015.1090133

Smith, T. (2002). The new law of war: Legitimizing hi-tech and infrastructural violence. *International Studies Quarterly, 46*(3), 355–374. DOI: 10.1111/1468-2478.00237

Sparrow, R. (2009). Building a better WarBot: Ethical issues in the design of unmanned systems for military applications. *Science and Engineering Ethics, 15*(2), 169–187. DOI: 10.1007/s11948-008-9107-0 PMID: 19048395

United Nations. (n.d.). *United Nations Charter*. https://www.un.org/en/about-us/un-charter/full-text

United States Conference of Catholic Bishops. (1993). *Centrality of conscience*. https://www.usccb.org/issues-and-action/human-life-and-dignity/war-and-peace/excerpts-from-the-harvest-of-justice-is-sown-in-peace-centrality-of-conscience-1993-11

Velez-Green, A. (2015, March 1). The foreign policy essay: The South Korean sentry—A "killer robot" to prevent war. *Lawfare*. https://www.lawfareblog.com/foreign-policy-essay-south-korean-sentry%E2%80%94-killer-robot-prevent-war

Vogel, R. J. (2011). Drone warfare and the law of armed conflict. *Denver Journal of International Law and Policy*, *39*(1), 101–136.

ADDITIONAL READINGS

Arkin, R. C. (2009). *Governing lethal behavior in autonomous robots*. CRC Press., DOI: 10.1201/9781420085952

Chamayou, G. (2015). *A theory of the drone* (Lloyd, J., Trans.). The New Press. DOI: 10.2307/jj.26193307

Human Rights Watch. (2012). *Losing humanity: The case against killer robots*. https://www.hrw.org/report/2012/11/19/losing-humanity/case-against-killer-robots

Scharre, P. (2018). *Army of none: Autonomous weapons and the future of war*. W. W. Norton & Company.

Singer, P. W. (2009). *Wired for war: The robotics revolution and conflict in the 21st century*. Penguin Books.

Strawser, B. J. (Ed.). (2013). *Killing by remote control: The ethics of unmanned military*. Oxford University Press., DOI: 10.1093/acprof:oso/9780199926121.001.0001

Walzer, M. (2015). *Just and unjust wars: A moral argument with historical illustrations* (5th ed.). Basic Books.

KEY TERMS AND DEFINITIONS

Lethal Autonomous Weapons Systems (LAWS): Lethal Autonomous Weapons Systems (LAWS) are military robots or drone systems that can automatically choose and attack without human control. Their creation disrupts classical moral models, leading to a leave-taking decision and leaving matters of life and death to non-moralising algorithms.

Jus ad bellum (Right to War): The part of Just War Theory which deals with the circumstances in which a state has the right to use armed response. This has been made complex by the ubiquity of drone deployment in the context of AI and could lead to a reduction in the entry threshold of war and the circumvention of other concepts, such as the Last Resort.

Jus in bello (Justice in War): The complex of rules and principles of morality under which parties within an active conflict have to act, namely, the use of force. Introducing AI raises fundamental principles of AI, especially its capability to comply with International humanitarian Law (IHL).

The: Principle of Distinction: This is the requirement of the jus in bello that combatants must provide a strict distinction between legitimate and immune civilian targets. This is one of the main scandals of AI: algorithms cannot comprehend the situation (such as handing over soldiers) or discriminate between insurgents and civilians.

The Principle of Proportionality: The requirement to make sure that civilian deaths and property destruction are not unreasonable in comparison with the predicted military benefit. Whereas drones provide an element of precision, critics say that they lead to a utilitarian mentality in which killing easily is more valuable than the cost of violence in the long run.

Jus ad vim (Force Short of War): A modern term used to refer to limited use of force, which is not war per se (e.g. targeted drone strikes in non-belligerent areas (e.g. Pakistan, Yemen) and does not feel the legal test of full-scale war. This produces a legal grey zone which obscures the lines of a conventional state conflict.

Chapter 3
The Emergence and Implications of Lethal Autonomous Weapon Systems:
Motivations, Adoption, and Impact on Global Military Dynamics

Tulip Suman
http://orcid.org/0000-0001-5605-3564
KIIT School of Law, KIIT University, India

Anuj Kumar Sinha
http://orcid.org/0009-0003-5953-8454
Amity Law School, Amity University, Ranchi, India

Akash Bag
http://orcid.org/0000-0001-8820-171X
School of Law and Justice, Adamas University, India

Ananya Bharadwaj
Faculty of Law, SGT University, India

ABSTRACT

The chapter explores the development of Lethal Autonomous Weapons Systems (LAWS), also known as killer robots. The first task is to establish what is motivating the evolution and operationalisation of weapon systems functioning in the absence of meaningful human control, that is, the decision-making of machines. The study is

DOI: 10.4018/979-8-3373-6264-9.ch003

related to the motives behind this growth and the subsequent effects of these motives on military behaviour and the character of the conflict. Synthesising the theories of Antinomic Democratic Peace and Adoption Capacity Theory, the incentives for their utilisation and the mechanisms of their adoption have been investigated. Two key conclusions were obtained from the qualitative content analysis. The chapter's conclusions indicate that LAWS will likely accelerate their growth at an intermediate pace, potentially disrupting the world's balance of power.

INTRODUCTION

Artificial Intelligence (AI) has recently received considerable attention in both the general and scholarly worlds, and one field in which it has been addressed is International Relations (IR). The term Lethal Autonomous Weapons Systems (LAWS) is commonly described as the third revolution in warfare, building on the invention of gunpowder and nuclear weaponry (Future of Life Institute, 2015). They are officially defined as weapons that, when triggered, select and attack a target without human intervention (U.S. Department of Defence, 2012). This power causes numerous disruptive problems, which form the basis of the key topic of this investigation. Although it has often been used colloquially in both scholarly and popular literature as a variant of the term 'killer robots' - a name associated with a science-fiction, futuristic outcome - empirical evidence points to the contrary. According to a recent report by the Stockholm International Peace Research Institute (SIPRI, n.d.), 49 lethal autonomous weapon systems exist and are operated by various militaries, demonstrating that such technologies are no longer fantasies. These 49 cases illustrate the full capabilities of autonomous target recognition, including the optional feature of excluding humans from lethal decision-making (Boulanin & Verbruggen, 2017; p. 26). This particular trait is the defining feature of such systems, which warrants serious academic attention. Moreover, the seriousness of the problem is supported by publications from key organisations and NGOs, such as Human Rights Watch, the International Committee of the Red Cross, and the United Nations (HRW 2016, 2014; ICRC 2016, 2014; UNIDIR 2014; Future of Life Institute 2015).

Accordingly, a detailed analysis is needed of the multifaceted implications of LAWS for war, beyond the military field, including non-proliferation, the risk of an arms race, and the motives for their deployment. Focusing on war, weapons, and the dynamic of their usefulness regardless of particular actors, this subject matter will always be relevant to International Relations in itself. Accordingly, the main Research Question (RQ) is as follows: What drives actors to use LAWS in conflicts, and in what way would their application alter the behaviour of military employing them, and the nature of the conflict they would be engaged in? To answer this question, it is necessary to identify the benefits of LAWS implementation compared with traditional, human-operated systems and to assess their financial and operational requirements (Pietrzak, 2024). This discussion utilises two theoretical propositions, applied to empirical data and academic studies, through qualitative content analysis. The theories provide unique information on the two essential aspects of the Research Question: the reasons people resort to it and how it can be implemented (Pietrzak, 2023).The first theoretical framework, known as the democratic peace theory, posits that democracies are less likely to engage in armed conflict than non-democracies, from which specific incentives to use the LAWS can be inferred. The second theory is the adoption capacity theory, which examines the implementation of military innovations and the conditions that affect how innovations spread. The literature review below summarises prior research, assesses the relatedness of the concepts, and outlines their application to the current study. It is accompanied by a methodology chapter that supports the use of qualitative content analysis. Lastly, the investigation concludes with two chapters of analysis, recommendations for future studies, and an assessment of the impact of the results.

LITERATURE REVIEW

Drones are an emerging technology in the military and the direct father of LAWS. Nevertheless, this technology and its applications have a more extended history; more than guided torpedoes have been invented and used since the 20th century (Geist, 2016; p. 318; Asaro, 2017; p. 142). The situation in which guided munitions become drones, and vice versa, is analogous to the development of the aeroplane into the aircraft carrier: the new technology is not the unit of the current tool, but a new way of using the current tool to create capacities that neither component has on its own. It is an essential feature of LAWS, as it is an emerging ability that will be demonstrated in the analysis. For example, the risk-aversion aspect, which was already debated before the LAWS came to prominence, can help explain why this capability may be considered an asset or a liability, depending on one's position on drone warfare (Boyle, 2015, p. 122). This chapter focuses primarily on LAWS, rather than drones; however, the general literature often highlights the ethical and legal issues associated with LAWS, which are essentially extensions of the questions posed by drone warfare. In this framework, all literature on LAWS can be divided into two main segments.

The former, the category, encompasses the normative debate over whether to support or oppose the use of LAWS. Scholars who oppose such systems typically invoke ethical or legal considerations. On the other hand, some supporters argue that the LAWS might lead to a war more compliant with international humanitarian law (Arkin, 2010; Anderson & Waxman, 2013). This discussion directly follows the argument about drones, in which the ability to carry out decapitation strikes, that is, surgical assaults against high-value targets, becomes balanced against the risk of collateral civilian casualties and the possibility of depersonalization of murder (Korac, 2018). One standard line of reasoning in this field is that by outsourcing the act of killing to a machine, be it a drone or a LAW, the human soldiers have been relieved of their so-called psychological burden of killing. Such alienation may contribute to the reduction of the barric when entering into or participating in armed conflicts (Korac, 2018, p. 62). Moreover, these arguments are reminiscent of the philosophical position that machines are dehumanising human life to just an object and killing humans through automation is a denial of natural human dignity (Johnson & Axinn, 2013).

The second type of literature also deviates from this normative evaluation. Edward Geist, in particular, argues that scholarly attention should move beyond binary ethics of LAWS and instead analyse the strategic impacts of their use more comprehensively. One of the ongoing anxieties of this necklace has been the idea that Artificial Intelligence technology is highly resistant to being confined through control, which makes the unproliferation of such technologies challenging, not to say impossible (Geist, 2016). This difficulty is explained by the dual-use character of AI applications, which operate in both civilian and military contexts, where technological spillover promotes widespread proliferation (Schulzke, 2018, p. 15; Cummings, 2017). Such proliferation is vital, as it determines whether decision-makers must contend with an adversary capable of employing LAWS. To achieve the revolutionary effect the term "revolution" connotes, the technology must spread widely enough to effect a radical change in the strategic environment. Another method of analysis is to examine the versions of Artificial Intelligence to distinguish between narrow and general AI and to identify the strategic and tactical implications each version entails (Ayoub & Payne, 2016, p. 794). General AI is the artificial intelligence commonly portrayed in science fiction. However, such technology is not yet available and remains distant; its development would revolutionise the world and affect all areas of human life (Cellan-Jones, 2014). Opponents of the claim that AI can make war argue that a machine must possess the value of human life to meet the needs of the former (Arkin, 2010; Ayoub & Payne, 2016, p. 795). By comparison, Narrow AI is highly specialised in a specific field and can acquire new skills through experience and machine learning to become more capable. This type of AI already exists and has enormous implications for analysis (Ayoub & Payne, 2016, p. 795). That is the second genre of literature, those that think about these operational facts and not about absolutely abstract ethics, which the present chapter is supposed to add. The sub-field is much less developed and in scope than the scholarship on philosophical issues, a research gap that warrants further exploration.

Returning to the former category, which addresses ethical and legal conflicts arising from LAWS and their adherence to international law, this is the most widely adopted solution in the research (Amoroso & Guglielmo, 2018; Maas, 2019, pp. 2-3). These scholars engage in discourse analysis and often refer to one another within the same discussion, a characteristic of intertextual discourse (Fairclough, 2003, p. 41). At the centre of such literature is the argument that the implementation of LAWS will be an intentional relocation of the right to murder a person to the machine (ICRC, 2018). Although the power to take lethal action is now in human hands with meaningful human control, the essential feature of LAWS is to restrict or eliminate the human factor. This continuum of experience is referred to as: human-in/out/on-the-loop (HRW, 2012, p. 2). The lack of meaningful human control is widely recognised as the primary source of risks associated with LAWS (Sauer, 2016). In addition, there is a lack of clarity and accountability in LAWS's activities. As an example, it is troublesome to ascertain culpability when a LAW kills a civilian. One of Robert Sparrow's arguments is that no single person can be held responsible for such mistakes, since the machine made the decision. He considers three possible culpable parties, namely the programmer, the manufacturer, and the commander, and concludes that all are unprosecutable (Sauer, 2016; Schornig, 2012, p. 375). The opponent's counterargument is that a human soldier has the same level of autonomy as a LAW (Schulzke, 2013, p. 204); hence, Sparrow's argument fails, assuming that the same level of autonomy should be achieved once a human being has replaced the machine. Among the commonly used qualitative methods in comparative studies, this accountability issue is frequently examined (Pietrzak, 2022).

Lastly, the issue at hand concerns LAWS' adherence to International Humanitarian Law (IHL). They are valued based on their violation of the principles of distinction (the separation of combatants and civilians) and proportionality (the damage and the military purpose of the strike) (Asaro, 2017, p. 142). Such legal norms affect any conclusions drawn during deployment, making them applicable to the Research Question. For example, some munitions, such as landmines, do not comply with the distinction principle, and nuclear and chemical weapons do not comply with either the distinction principle or the proportionality requirement (Amoroso & Guglielmo, 2018, pp. 5-6). Unfortunately, the advocates argue that LAWS may be programmed in ways that do not violate these principles; they may even be fairer in adhering to the rules than humans can, which is why LAWS would be better suited to conflicts in other contexts (Arkin, 2010; Anderson & Waxman, 2013).

THEORETICAL FRAMEWORK

To address the key Research Question of the reasons and mechanisms underlying the use of LAWS (LAWS), this analysis employs two theoretical approaches. The use of LAWS within a state's military apparatus fundamentally alters the traditional explanations, constraints, and inferences advanced by Democratic Peace Theory regarding why a country should embark on armed conflict. At the same time, the Adoption Capacity Theory proposed by Horowitz is applied to theorise the processes of innovation absorption by military institutions and the subsequent outcomes of that adoption. The practical benefits of LAWS over traditional weapons can create new strategic problems while simultaneously solving existing ones. Such abilities are inherently associated with the premises of incentives that persistently inform the application of LAWS (Sauer & Schornig, 2012). Moreover, the use of LAWS poses risks of triggering a security dilemma, as the technology can confer a significant first-mover advantage, thereby disrupting the traditional balance of power. Within the context of this strategy, the concept of Adoption Capacity (ACT) provides the necessary understanding of the overall implications of LAWS use (Horowitz, 2018, p. 39). The two theories are presented and discussed in the latter part.

Democratic Peace Theory

Democratic peace theory has established itself as an underpinning of both the study of conflict and the study of International Relations (IR) to such an extent that Jack Levy (1988, p. 662) has described it as something as close as we could call an empirical law in international relations. The primary theoretical rationale is that democracies would never go to war against one another, primarily because of the need to be accountable to their electorates and the absence of ideological differences. Other scholars, such as Michael Doyle (1995, p. 180), have argued that democratic forms of government are more likely to promote peace, whereas autocracies are more likely to wage war. Because of its importance to issues of war and peace, this theory has sparked considerable debate over whether peace proliferation follows democracy proliferation (Reiter & Stam, 2002; Yildirim & Sezgin, 2005, p. 99). However, with the emergence of new technologies, this paradigm is being challenged. As an example, Sauer and Schornig (2012, p. 365). State that democracies become increasingly bellicose toward non-democracies due to LAWS (LAWS) and drones. The rationale is that the conservative incentives that kept the democracies at peace - politically imperative to protect the lives of soldiers and military spending - can be disregarded and crippled by the use of LAWS. This is supported by Kaag and Kreps, who argue that drones serve a practical function of insulating citizens from the facts of the war, thereby undermining or derailing the allure for the democratic electorate and reducing political responsibility in the matter.

The said is especially effective in the context of the recent changes in the paradigm of warfare, where the inter-state conflict is replaced by what may be referred to as asymmetrical warfare, the course of which sees nations resort to airstrikes to maintain the military personnel in a safe place (Thornton, 2007; Kaag & Kreps, 2013, p. 106). Although democracies rarely go to war with one another, the presumption that they would be more peaceful is empirically inadequate. The conflict with non-state actors is also an example of the armed conflicts that democracies have often engaged in, as seen in the Middle East (Marshall, 2016), which stands in contrast to the idea of a purely peace-prone nature (Doyle, 1995, p. 180). This phenomenon is a split of the theory- a dissection into two separate lines, a world in which democracies continue to coexist with one another, and at the same time a reality where democracies are as war-like as any other form of state (Muller, 2004, p. 494). Further, the notion of antinomy, as defined by Harald Müller (2004, p. 516), is a proposition of law that maintains lawlikeness, such that a secondary proposition follows from its opposite. In this regard, identical democratic systems lead to both peace-avoidance (within) and conflict-proneness (between groups). According to Sauer and Schornig, democracies exhibit a predisposition to behave both peacefully and bellicose, depending on the type of target.

Autocracies, by contrast, lack the democratic institutions that serve as transmission belts to convey the will of the majority to executive decision-making (Muller, 2004, pp. 501-503). Although autocracies do not require the approval of a voting population, the modifications to risk and cost estimates introduced by LAWS are equally relevant to them. LAWS is a new variable—another technology that fundamentally alters the incentives of democracies to use violence. One reason LAWS are being implemented is the threat of military action. To contain this, democracies pursue risk-transfer rules that emphasise reducing amicable casualties as a primary objective (Shaw, 2005). Democracies have experienced high deaths several times, which have led to citizens reacting negatively to the government and withdrawing their support. The nature of the government, in turn, promotes policies that do not result in any victims (Sauer & Schornig, 2012, p. 367; Luttwak, 1996). Moreover, these risk estimates are also applied to the collateral damage that is the murder of the non-combatant population of an enemy, which could also impact unfavourably on a democratic electorate through concerns of International Humanitarian Law (Kaag & Kreps, 2013, p. 105). Such ethical limits are usually considered in the operational context in military terms. The technological response to it is the development of systems such as LAWS that are likely to overcome these limitations themselves (Kaag & Kreps, 2013, pp. 97-98). Lastly, cost reduction is a driving force. This would include the economic cost in human lives and the monetary investment. Quantitative research supports the chapter's finding that higher levels of democracy are associated with lower peacetime military expenditures (Fordham & Walker, 2005; Yildirim & Sezgin, 2005). This aligns with the strategic appeal of LAWS and drones: they offer a means to conserve resources in both monetary and human terms, and they benefit democratic actors (Fordham & Walker, 2005, pp. 142-145).

Adoption Capacity Theory

The second theory used in this chapter is the Adoption Capacity Theory (ACT). This theory offers an organisational perspective through which an institution, in this instance a military, may adopt and deploy a new technological innovation, in this case, LAWS (LAWS), into its already established organisational practices (Horowitz, 2018, p. 43). Moreover, it introduces realist notions, namely, the effects of introducing such military technology on the security dilemma and the balance of power. One of the most common misunderstandings about innovations in the military is that the very invention of a specific type of technology, and its possession, places its inventor at an immediate advantage over the enemy. A paradigmatic example of this is often nuclear weapons. Nonetheless, in his article, The Diffusion of Military Power, Horowitz (2010, pp. 1-2) implies the opposite, arguing that the creation of technologies, or even the right to use them in advance, is not a condition that can be advantageous. Instead, ACT presents a conceptual framework for identifying the precise timing and manner in which significant military innovations, to which LAWS certainly belongs, change the status quo.

The application of this theory requires a definition of the term 'diffusion' as a precondition before operationalising the theory. According to Rogers (2003, p. 11), diffusion is the process by which an innovation spreads over time through designated channels among members of a social system. At this starting point, also known as a demonstration point, diffusion can occur once it has been reached (Horowitz, 2010, p. 8). One issue that must be left to the analysis section of this chapter is whether a demonstration point for LAWS has been made. Typically, a point like this is defined when a potential of its full capabilities has been reasonably perceived in the international system by an act of a first mover, as opposed to the fact that the capability is that which is the target of exercises or debates inside a country (Horowitz, 2010, p. 24). Building on the essence of ACT, one must distinguish between introducing a significant military innovation and implementing it within military frameworks. These two steps are closely related, though they are not identical. To describe this process, ACT identifies two crucial variables: organisational capital and financial intensity. Their needs differ, as both are required to implement an innovation at different levels (Horowitz, 2010, p. 9).

The flexibility of an organisation in adapting its established processes to incorporate a new input defines organisational capital. It is a variable that lies on a continuum; the bureaucracies within an organisation can both facilitate and prevent proposed changes. These resistance mechanisms are termed veto points by Horowitz; the more veto points, the less flexibility organisations have (Horowitz, 2010, p. 38). The argument of the text, therefore, is that the entrenched capital of a high organisation (bureaucratic inertia) would prevent the adoption process and vice versa. Organisational capital is assessed according to specific criteria. First, they are older organisations, which are likely to have a well-established reputation and financial standing, making them more resistant to change. Second, the intention and the extent of investment in Research and Development (R&D) and experimentation, along with the capabilities they possess, are measured. Third is the organisation's self-identification, including its perceived purpose. Narrow self-identification entails obstacles to embracing new ideas that may defy that particular definition (Horowitz, 2010, p. 39).

The chapter allows analysis of the information using quantitative data, as financial intensity enables the identification of two spheres of financial origin. The modern military innovation can be based on civilian technology or limited to military considerations with only military applications (Horowitz, 2010, p. 32). The evaluation will assess how LAWS fall into these categories. On average, commercial technologies are less expensive than military technologies because market-based competition provides incentives for private companies to recoup development costs in the hope of getting a return on investment. It is worth noting that the same technology can be used simultaneously, albeit independently, in the military domain (Horowitz, 2010, p. 32). Without monetary gain from investment in mere military technologies, prices are significantly higher and are usually fully offset by tax relationships. Here, return is realised not in terms of monetary aspects, but in terms of capabilities (Horowitz, 2010, p. 31).

The primary sources of information used to determine the diffusion of new technology at the system level are, therefore, organisational capital and financial intensity. In summary, the more the financial intensity of an innovation, the slower it spreads through the system and the less likely an actor is to make an adoption decision. In line with this, the more organisational capital (conceived here as institutional inertia), the slower the implementation and diffusion of the innovation (Horowitz, 2010, pp. 32-39). Lastly, ACT in some way justifies why a state may want to invest in a new technology. Despite the theory's primary focus on the how, the results allow one to draw a broader conclusion about motivation. The motivations of a state can be strategic necessity, an arms race with a competitor, or the intention to demonstrate a technology and create a new standard in military practice. In addition, one of the strategic measures to handle a rival acquiring a position can be an alliance with a third party- a balancing game which approaches realism (Horowitz, 2010, pp. 28-29). Table 1 shows that, in terms of global investments and motivations for adoption, early access to technology confers a first-mover advantage, creating a monopoly. Nevertheless, the acquisition of such an advantage depends on the rate of diffusion of the innovation; the greater the innovation spreads, the weaker the first-mover advantage (Horowitz, 2010, p. 49).

Table 1. Global investment and development of LAWS (status quo)

Country / Region	Financial Investment & Funding	Key Systems & Developments
United States	**DoD Spending:** $149 million (2015); $4.25 billion (2017); $18 billion pledged (2016–2020) (Bornstein, 2015). **Civilian AI:** ~$1.1 billion (2015) (Hunter, 2017, p. 11).	• **Sea Hunter:** Autonomous ship ($2m/unit) (U.S. Department of Defence, 2017a, p.3). • **LOCUST:** UAV swarming technology. • **Perdix:** Micro-drones with adaptive swarm behaviours. • **X-47B:** Autonomous carrier landing & refuelling.
Russia	**Rearmament Program (2016–2025).** ~$346 billion total budget (specific LAWS allocation undisclosed).	• **Uran-9:** Unmanned autonomous tank. • **T-14 Armata:** Manned tank with partial automation and future unmanned capability (U.S. Department of Defence, 2007, p. 53/54).
European Union	**European Defence Fund (2021–2027).** €13 billion. **Previous Decade:** €665 million on AI/ robotics.	General strategic focus on AI and robotics support across member states (NSTC 2016: 6).
France	**National Fund:** €550 million for AI development (2017) (Scharre, 2018, p. 78).	Strategic alignment with EU goals (U.S. Department of Defence, 2017a; Lachow, 2017, p. 97).
Italy	**Industria 4.0 Plan (2017–2020):** €30 billion (Ray, 2016; SIPRI 2019).	Focus includes "autonomous cooperative robotics and sensors."
Germany	No official plans or funding mentioned (Chen, 2018).	Possesses weaponry capable of autonomous engagement despite a lack of official doctrine (Altmann and Sauer 2017, p. 122/123).
Israel	Undisclosed funding (Weitz, 2013, p. 4).	• **Harpy:** Loitering munition capable of human-off-the-loop autonomous attack (Boulanin & Verbruggen, 2017, p. 98). • **Border Sentinels:** Autonomous defence systems (pioneered with South Korea). • Leading exporter of unmanned systems (Scharre 2018, p. 114/115).
China	**Total Military Budget (2018).** $250 billion (14% of global spending). Specific LAWS funding undisclosed (Cheung et al., 2017, p. 2).	• **Autonomous Submarines** (in development). • AI and robotics have been declared key R&D priorities since 2014 (EPP 2019; Boulanin & Verbruggen, 2017, p.103).

METHODOLOGY

The central analytical theme recognised and divided into categories in this paper concerns the direct problems arising from the appearance and use of LAWS (LAWS) in military conflict. Operational problems are the indicators of the underlying motivation of their use (U.S. Department of Defence, 2017b). An example of this is the effect of depersonalization of the act of killing, which literally takes the military personnel out of the fire; this feature, in turn, responds directly to the strategic motivation of risk avoidance, considering that states can afford to go to war without the danger to their political capital of sacrifices (Korac, 2018). The empirical data used to conduct this analysis were selected after careful review of authoritative reports, in particular the SIPRI report Mapping the Development of Autonomy in Weapon Systems (Boulanin & Verbruggen, 2017). Quantitative data on military expenditures were obtained from the SIPRI general military expenditure database, which was further complemented with information on national Research and Development (R&D) projects and representative LAWS systems (SIPRI Military Expenditure Database 2017). Moreover, the search for specific case studies of LAWS testing and implementation has presented a wide range of scholarly articles and, to a lesser degree, newspaper publications (Scharre, 2018). The consistency of the rationale given in support of the chosen data and the soundness of the conclusions made support the validity of this chapter. Qualitatively, the content analysis methodology is characterized as the most suitable method to respond to the Research Questions that focus on the interrogatives of why and how as the central point of this question (Pashakhanlou, 2017, p. 449).

ANALYSIS

The first step in the following analysis is to outline the driving forces behind the use of LAWS (LAWS) by putting the theory of Antinomic Democratic Peace into practical terms in relation to the so-called human-in-the-middle pattern of action. However, when the innovation is conceptualised as software rather than hardware, the strategic environment changes dramatically. After achieving the necessary hardware (UAV) and having it controlled remotely, changing it to a LAW may be necessary only with a software upgrade that provides independence; externally, there is no visible difference (Horowitz, 2018, p. 49). Therefore, the current organisational framework does not have to compete with the innovation but may instead be complemented and supported by it through Narrow AI applications, specifically if nations follow the so-called human-on-the-loop concept that has been promoted so far. This is an exception to this non-competitive trajectory when organisational subsections that utilise autonomy compete financially with traditional units. However, this challenge is likely to be overcome by the sheer applicability and the scope of improvement across the military spectrum. Software-based skills can be easily replicated and broadcast around the globe within a few seconds, whereas human training requires significant time, money, and workforce. Therefore, the capabilities of machine learning in Narrow AI are expected first to exceed human performance gradually, and then to do so exponentially (Ayoub & Payne, 2016, p. 794; Horowitz, 2018, p. 45; Boulanin & Verbruggen, 2017, p. 81).

An even more severe exception would arise when adoption poses the threat of making human personnel obsolete. Similar to the outsourcing of factory employees to robotic machinery, it is possible to expect that all veto points within an organisation under threat will be mobilised against such initiatives whenever feasible (Horowitz, 2010, p. 38). Although these opposition efforts are eradicable over the longue durée, instantaneous shocks caused by those seeking to maintain their utility by cutting off the beneficiaries of the statu quo are bound to be considerable. This phenomenon can already be empirically precedented. Coupled with its autonomous aerial refuelling and carrier landings, the United States' X-47B drone was defunded; more funds were allocated to a manned air-to-air refuelling platform (Cummings, 2017, p. 9; Horowitz, 2018, pp. 48-56). Merely because the mentioned logistic capabilities are not necessarily directly tied to the main benefit of LAWS, which is the absence of mortality threat, one can understand why the veto was issued. Another adoption barrier is that it requires specialised training; the military also requires the services of persons skilled in computer science to use these computers. This demand creates the tension between these technical recruits and traditional, practical particulars of pilots, drivers, and marines (Cummings, 2017, p. 9). It remains to be seen whether these factors and existing examples of LAWS are sufficient to serve as the demonstration point demanded by the Adoption Capacity Theory (ACT) (Horowitz, 2010, p. 24). It is also predetermined by the standards that its potential to do anything it wants should be known to everyone.

Due to advances in sensor technology, the accuracy of weapon systems improves accordingly, so they can not only be used in surgical strikes against legitimate targets under humanitarian law but also be used disproportionately (Kaag & Kreps, 2013, p. 105). Generally, this correlation implies a decrease in the raw destructive power of armament, leading to a new strategic emphasis on the so-called surgical strikes. The next element of proliferation is less definite; as an innovation that qualifies the typical aspects of a military revolution, it must go beyond mere theoretical discussion. Although cases of used LAWS (LAWS) do exist, it has not attained the international breakthrough it claims is necessary to achieve this milestone. This saturation point, in its turn, is unquestionably achieved by drones and uncrewed vehicles, and their spread is exemplary (Schulzke, 2018, p. 15; Farooq, 2019). Nevertheless, if it is granted that the technical boundary separating a drone and a LAW is blurred, and that software can be updated to turn a remote-controlled drone into an autonomous drone, then the argument that LAWS' proliferation is so real becomes even more compelling. This is evidenced by Israeli UAVs, namely the Harpy, which, depending on its operating mode, can be classified either as a LAW or a drone and is already being exported.

Despite these common abilities, one operational difference between drones and LAWS is the range of opportunities they offer. Equipping drones with a continuous data connection to a human controller limits their range, and such broadcasts can be easily intercepted or scrambled, thereby breaking or exposing the operation (Sauer, 2016, p. 9). On the other hand, the conceptualisation of LAWS should not be that of a one-functional, autonomous system similar to a fighter jet. A more precise yet relevant way to describe the implementation of AI to weaponry is a so-called general enabler (Horowitz, 2018, p. 39). Nothing nullifies isolated cases such as the Sea Hunter of the US or the Korean sentry arms, but it points out that it cannot be their particular hardware category; it is their ability to act independently. Thus, the acquisition of autonomy, as such, rather than a particular weapon platform, is the first goal for actors adopting these systems. In turn, there is a need to initially determine the specific effect of autonomy on a weapon system and the operational advantages it comes with, to apply the primary determinants of organisational capital and financial intensity to decipher adoption. The main advantage, though abstract in the first place, has far-reaching consequences: the speed of operations. It is the time needed to complete a machine cycle, or the controlling algorithm of the machine; the cycle of the observe, orient, decide/act (OODA). It is impossible to directly compare the processing speed of an autonomous system to that of a human soldier.

Although allowing the final decision, the so-called act, to a human as a human-on-the-loop approach like the given system, it is still limited to being a human on the loop, the possibility of the system to collect, collate and analysis more complex volumes of data still greatly enhances the quality and speed of the overall decision cycle (Ayoub & Payne 2016, p. 807; Boulanin & Verbruggen 2017, p. 61). Not co-incidentally, most of the currently fielded LAWS (LAWS) with autonomous target detection are deployed in scenarios where the reaction time required of them exceeds human ability. They are mostly defensive environments, specifically designed to intercept projectiles (Scharre, 2018, p. 43). Operational speed is a benefit that will be highly extended to the area of agility. Autonomous combat vehicles, e.g., would no longer have to comply with the physical limitations of a human pilot, including the need for costly life support gear or the requirement to remain below a given level of G-force to prevent unconsciousness (Altman & Sauer, 2017, p. 123). More importantly, the use of complex onboard sensors and autonomy also reduces reliance on onboard piloting and close supervision, enabling manoeuvres that manned platforms cannot perform.

The human operator, or their place in the chain of decision-making concerning the use of lethal force, continues to be the focal point of LAWS and the central area of moral apprehension. As a result, to understand the implications of using LAWS, one should examine the decision loop principle and the need for meaningful human control over it (HRW 2016; ICRC, 2018). However, such words as autonomous target recognition might seem a softened form of complete lethal autonomy (Boulanin & Verbruggen, 2017: 6); this is a debatable difference. Middle-range human control means critical systems, in particular, the accurate functions of choice and involvement of a target. These definitions have baffled state institutions and, in many instances, left the question of blurred lines between automation and autonomy unanswered. According to the resolution of the European Parliament on LAWS, they are explicitly characterized as the weapons that serve the purposes without meaningful control by persons over the fundamental functions of choosing and hitting individual targets and the fact that the system should be automated, remotely controlled and thereby should not be considered a LAWS (EP 2018). This definition creates a loophole in the regulation of weapon systems with sliding autonomy. These systems are not strictly in either the fully autonomous (human-off-the-loop) or semi-autonomous (human-on-the-loop) modes, but may exist in either mode depending on the system user's choice (Boulanin & Verbruggen, 2017, p. 6). The primary instance of this ability is the so-called Harpy loitering munition, one of the few offensively deployed LAWS that have this flexibility of operation expressly (Scharre, 2018, p. 5).

Conversely, other significant geopolitical players do not even mention meaningful human control. A good example is President Vladimir Putin, who stated that the future is not only for Russia but also for the rest of humankind (artificial intelligence), and that whoever controls the sphere will control the world (RT, 2017). Equally, there are no clear statements by the Chinese government or President Xi Jinping on the so-called human-in-the-loop question, though they are firmly supporting the establishment of LAWS. China is spending heavily on this technology, according to government plans. Whereas China has indicated that it prefers the regulation of LAWS worldwide, its specific proposal calls for a ban on their use in conflict. However, this does not include their creation, which they are already engaged in (CSKR 2018). Both China and Russia seem to believe that eliminating the human element has been beneficial to the development of military power. This brings a fundamental strategic paradox into play: how helpful can autonomous weapon technology be to these actors when they cannot use it in practice because of their own diplomatic constraints? The list of countries that have signed an opposition to a ban on LAWS (LAWS) includes literally all the states that currently develop or possess such a capability at the moment of writing. The leading ones are the

United States, Germany, France, the United Kingdom, Russia, Israel, and South Korea (CSKR 2018).

Table 2 summarises existing data on national standpoints on the operational doctrine of the human-in-the-loop, Research and Development (R&D) investments, and total military spending.

Table 2. Comparative analysis of LAWS stances and expenditure

Country/ Entity	Position on Loop Question	Known Military AI R&D Spending	Known Civilian AI R&D Spending	SIPRI Military Expenditure 2018 (Rank / Total)
United States	**On-the-loop**	$22.394 billion	$1.1 billion	**#1** / $649 billion
Israel	**On-the-loop**	Unknown	Unknown	**#17** / $15.9 billion
Russia	**Off-the-loop**	Unknown (Portion of $346 billion rearmament budget 2016–2025)	Unknown	**#6** / $61.4 billion
China	**Ambiguous** (Supports ban on deployment, but not on R&D)	Unknown	Unknown	**#2** / $250 billion (SIPRI estimate)
European Union	**On-the-loop**	€13 billion	€665 million	N/A
France	**On-the-loop**	Unknown	€550 million	**#5** / $63.8 billion
Italy	**On-the-loop**	Unknown	Unknown (Portion of €30 billion for 2017–2020)	**#11** / $27.8 billion
Germany	**On-the-loop**	Unknown	Unknown	**#8** / $49.5 billion

Although the temptation to find a direct correlation between a nation's position on the human-in-the-loop question and the type of governance it runs, the annual figures in Table 2 suggest that the association between the two variables and the democracy-versus-autocracy dichotomy is somewhat distant. China and Russia are the only autocracies in the data set, while the European countries, the USA, and Israel are at the other end of the democratic spectrum (EIU 2017). However, the lack of data and China's ambivalence in addressing the question of the loop are obstacles to analytical clarity, and, as such, it is left to Russia as the only outlier within its respective category. Because the analysis of LAWS positions and military spending shows that there is no direct relationship between the type of governing and incentives to utilise LAWS, one must consider the historical forerunner, the drones, before moving to further analysis of the LAWS. Records of drone creation and use suggest that most of the first-mover countries in this area, and consequently in LAWS, are democracies (IISS 2011, p. 20-26). Two-thirds of the listed countries are within the democratic spectrum. Moreover, the same countries that led the revolution in drones and uncrewed vehicles during the last twenty years, the USA and Israel, are the modern pioneers of LAWS manufacturing, use and study (Sauer & Schornig, 2012, p. 364). This continuity justifies LAWS's direction as a straightforward technological development of drones. It, therefore, seems to have specific incentives unique to democracies that cause this. The point of contention, then, is to ascertain what kind of restrictions and incentives a democratic status bestows to explain this predominance in unmanned systems, automation, and LAWS.

Answering this, it should be based on the theoretical framework of the Antinomic Democratic Peace Theory. The aspect of institutional mechanisms that is relevant here is the manner in which the will and opinions of constituents are converted into political action, i.e. the support of the armed conflict or the opposition to the armed conflict (Kaag & Kreps, 2013, p. 107; Sauer & Schornig, 2012, p. 365). One of the main benefits uncrewed vehicles brought was that they were no longer life-threatening to the pilot, a defining characteristic of this type of vehicle. As successors to drones, LAWS inherit this advantage and further develop it (Boyle, 2015, p. 122). Although the physical transfer of the operator out of the war zone to a safe area, commonly the operator's home country, is necessary for safe operations, the use of drones radically alters the nature of operations. LAWS based on a human-in-the-loop operating model do so even more radically. The technology reduces the system-operating ratio, which was traditionally one-to-one, to a new level where one operator oversees many LAWS. This schizophrenic ratio can rise dramatically in the near future. Thus, the danger to any party engaged in active war will increase in direct proportion to the development of remote-control technology and autonomy (Scharre, 2018, p. 13). Before drones and LAWS, the ratio was strictly one operator per weapon system, whether a gun, tank, helicopter, or plane, or had several operators, such as the naval vessels. The physical separation of the weapon and its operator, enabled by unmanned systems, eradicated any direct threat of death.

Some might say that, in terms of risk reduction, LAWS do not stand out much compared to conventional unmanned drones. Nevertheless, the change in the operator-to-system ratio has significant implications for the second primary motivator, cost reduction. This ratio change is in addition to eliminating the risk of the operator's troublesome life, thereby generating significant financial savings. It takes immense amounts of money to move military resources and to maintain military personnel in place and trained (Shaw, 2017, p. 458). LAWS present an opportunity to lower these operational cost escalations. Early-stage Research and Development (R&D) incurs significant investment. However, when the technology is mature, mass-producing the systems is easier, and staffing is reduced, thereby lowering long-term expenses associated with standing armies. The cost of procuring a state-of-the-art fighter jet like the F-35 from the United States is just a fraction of the cost of unmanned aerial systems, such as the Predator or Raptor aircraft. For example, the US Sea Hunter autonomous ship costs around 2 million dollars per unit, a price that is cost-effective and comparable to the price of an individual conventional cruise missile used in the modern military (Scharre, 2018, pp. 78-93). Moreover, it is easy to imagine the commercial deployment of 3D printing technology for the production of miniature LAWS (LAWS) in the future, and their use in a swarming configuration. They are now being tested and show highly desirable cost-to-benefit ratios (Lachow, 2017, p. 97).

Another dimension of this economic calculus, which is important but often overlooked, is the significant capital savings from pilot training. Manned fighter aircraft require constant training under the conditions of real flight to avoid perishable high-level skills. Although operational costs and the weapon system equally affect drone pilots, the effects are much less. This continuous training concept becomes obsolete once it progresses to LAWS, where the task becomes operational proficiency at best, and is ultimately delegated to the operating system itself. LAWS are increasingly empowered by machine learning and the consumption of massive data, making them more effective over time. As a consequence, the further a LAW becomes competent through training (machine learning) by applying narrow AI programs, similar to the commercial algorithms used by large technology companies, the greater the cost reduction can be (Ayoub & Payne, 2016). This dynamism implies a runaway effect: the higher the usage, the lower the cost, because the resulting energy of training data can be replicated indefinitely, leading to an exponential increase in efficiency. Regardless of the incentive behind democracies, it is an argumentative workforce that cuts across the political spectrum. Other countries, like Russia, facing manpower shortages, Japan, with its ageing population, and even Germany, where the army is having a hard time recruiting sufficient personnel, consider LAWS not only as a way to save costs, but also to lower staffing expenses. In such states, automation offers a viable avenue to maintain current military capacities even in a context of a reduced or inadequate workforce (Boulanin & Verbruggen, 2017, p. 63).

This population stress is probably the primary force behind the domination of the LAWS technology in Israel; considering its geographical location and numerical underdog status in comparison to the usually hostile neighbour-states, Israel is thoroughly using this technology to compensate for its workforce shortages. These characteristics of the human being in the decision-making cycle in lethal force are inseparably connected with accountability. A complex of the International Law, International Humanitarian Law (IHL), human rights frameworks, and the Geneva Conventions creates a whole system that other actors can be held responsible for committing war crimes, genocide, or employing limited weapons (e.g., biological or chemical agents). Although an examination of the effectiveness of bodies such as the International Criminal Court falls outside the immediate purview of this study, the fact that the human element has been eliminated in this particular loop would most certainly and cripplingly bring down this legal framework, which is hard to change anyway. There would no longer be a human-in-the-loop, as is the case with at least some LAWS deployed nowadays, which poses a burning question: how can one hold anyone responsible for violations of international law? This accountability gap, as mentioned in the literature review, is a disputed area among academics, and the consensus is that LAWS effectively abolish practically implemented accountability (Sparrow, 2017).

Theoretically, it may be suggested that the actor who uses the LAW is responsible for its actions. Nonetheless, the accelerated increase in the use of armed drones and the consequential increase in the proliferation of LAWS make this attribution more and more complicated. We can argue that they resemble optically, with the majority of existing LAWS operating in the air and possessing the physical features of ordinary unmanned drones (Horowitz, 2018, p. 49). With the fact that civilian collateral damage resulting as the consequence of the drone strike is already often unaccounted, the rising share of the autonomous systems in the worldwide aerial arsenal is bound to make locating the appropriate side of the balance sheet to hold in case of the precise airstrike increasingly challenging. Under the condition that the machine is unable to be charged or set liable, and that the development of the Lethal Autonomous Weapons System (LAWS) can be developed without knowing the creator or providing self-destruct mechanisms, a military strike can be perpetrated in a way in which no one would be able to attribute any responsibility (Sparrow, 2007). Such a scenario is quite possible in complicated and disputed conflict areas like Syria. Moreover, when the machine operates as an independent entity, it is illogical to hold the software producer or developer responsible for its actions.

Financial capital is the level of resources needed to implement an innovation practically, and in practice, this level varies widely. Small-scale drones are commercially widespread and relatively easy to arm and thus are available even to the resource-poor non-state actors or individuals, which already has precedents (Scharre, 2018, p. 102). On the other hand, LAWS are even more intensive in terms of capital and require a wide range of technological know-how, which is usually impossible without the support of state levels (Horowitz, Kreps, Fuhrman, 2016, p. 34-35). The situation, however, becomes more real and apparent when the attraction moves beyond the hardware of LAWS into their software, which enables autonomy. This distinction is a dichotomy between considerations of financial capital and the civilian and military domains. The technology behind autonomy is primarily commercial, especially in the technology industry, and it requires the development of algorithms. The main direction in this technological path is narrow AI (Ayoub & Payne, 2016). As such, there is a clear intersection between the applicability of this technology to civilians and its ease of later modification for military applications. According to the Adoption Capacity Theory (ACT), a civilian utility and the resultant commercial interest in developing a technology reduce the degree of financial intensity required to secure military adoption compared to a technology that primarily requires military development.

Records on financial information, limited to the democracies of the European Union and the United States, indicate that substantial capital is being channelled into civilian AI technology through large-scale economic investment campaigns. Italy, France, and the EU as a whole are pumping hundreds of millions of euros over the next decade, and the US has committed billions, as explained in the empirical section. The US has already provided specific numbers for the military investment. In contrast, the EU has not, though it has recognised that, despite the programme's intensive economic outlook, significant proportions of the funded research will be of value to defence and will lead to the development of technologies that defence players can utilise (EU, 2014, p. 43). The financial acumen needed to obtain autonomy-enabling software is thus relatively low, due to the strong commercial incentives for commercial development and, indirectly, because such systems can, in theory, be readily purchased and incorporated into current military hardware by national governments. The adoption will therefore be cheap when the state does not have to bear the entire development cost. However, in the context of the increasing popularity of AI weaponisation, the representatives of commercial AI technologies, like DeepMind, owned by Google, have explicitly declared that they would refuse cooperation or collaboration with the (US) government in terms of the military use of their AI software (Future of Life Institute 2015). Whether such collaboration will take place in the future remains to be seen, and if it does, whether it will still be visible to the people.

When the Lethal Autonomous Weapons System (LAWS) cannot be blamed, and therefore will be considered innocent of a crime, and when these systems can be built without the clear identification or a self-destruct feature, then a military attack can be carried out without them having any responsibility anyone can trace (Sparrow, 2007). Such a situation is entirely realistic in a complex, contested conflict zone like Syria. However, when a machine is actually acting on its own, the programmer of the software cannot logically be expected to be responsible for its autonomous actions. Financial capital is the threshold of resources required for the practical implementation of an innovation, and it fluctuates wildly. Small-scale drones can be easily purchased as commercial goods and are relatively easy to weaponise, making them accessible to non-state actors or even individuals who lack the required resources, a precedent that already exists (Scharre, 2018, p. 102). On the other hand, LAWS are more capital-intensive and require a wide range of technological skills, which usually require state-level support (Horowitz, Kreps, Fuhrman, 2016, p. 34-35). However, as soon as the emphasis shifts to the software itself that empowers autonomy rather than the physical hardware of LAWS, the financial intensity becomes more real and apparent. This distinction separates financial capital considerations into civilian and military arenas. The software that supports autonomy primarily stems from the business world, specifically from technology companies that depend on algorithmic innovation. This technological movement is led by narrow AI (Ayoub & Payne, 2016). This results in a clear overlap between the applicability of this technology, its relevance to civilians, and the ease with which it can later be implemented in military service. Adoption Capacity Theory (ACT) indicates that, because civilian utility exists and there is a commercial reason to develop a technology, the degree of financial intensity required to adopt military technology is lower than for technologies that will only require military development.

The available financial data, confined to the democracies of the European Union and the United States, indicate that a considerable amount of capital is being channelled into civilian AI technology within the context of large-scale economic investment programs. Italy, France, and the EU as an entity are committing hundreds of millions of Euros over the forthcoming decade, and the US is committing billions, as indicated in the empirical part. Although the US has publicly published specific figures of military investment, the EU has not, though it has admitted that, although the program may be well-focused economically, significant sections of the funded research may be helpful in the defence and result in technologies that will be utilised by actors in the defence sphere (EU, 2014, p. 43). Thus, the economics of attaining autonomy-enabling software are relatively low, due to robust incentives in commercial development and the ease with which national governments can hypothetically acquire and use them in current military equipment. Adoption is therefore relatively cheap, provided the state does not absorb the development cost. However, as the process of AI weaponisation becomes more prevalent, commercialised AI companies such as DeepMind, a Google subsidiary, have publicly stated that they will continue to collaborate with the (US) government on the military use of their AI software (Future of Life Institute 2015). Whether this kind of collaboration will occur in the future is yet to be determined, and if so, whether it will remain visible to the public. According to the theoretical premise of the Adoption Capacity Theory (ACT), the latter part of the analysis measures the variants of organizational capital and financial intensity. Based on current information and academic literature, this analysis provides conclusive evidence of the participation of LAWS (LAWS) in military units, directly answering the second section of the Research Question.

The organisational capital is concluded to be high. The above indicates that LAWS disrupt an organisation's status quo in a highly disruptive way, leading to high institutional resistance due to several factors, as explained earlier. On the other hand, financial intensity is considered low. The latter is mainly explained by the high rate of commercial and subsequent civilian interest in the Research and Development

(R&D) of Narrow AI. The excellent transferability of this technology to civilian uses, including self-driving cars and proprietary algorithms in large technology companies, has created a substantial business ecosystem that effectively cross-subsidises much of the underlying R&D expenses. In theory, such a technology can be readily applied in the military, but so far, there is little evidence of this on a quantifiable scale. Nevertheless, it is commonly assumed to be happening in countries like China, which is more combative about the dual-use nature of AI to use it both as a civilian and military phenomenon than other participants of a civil society. In America, conversely, business organisations have been relatively inflexible regarding military access to their proprietary technologies (Boulanin, Verbruggen, 2017, p. 103). The organisational capital has a low financial intensity, which Horowitz synthesises and fits within the ACT framework, yielding definite conclusions.

To start with, the rate of diffusion of LAWS technology is expected to be moderate, and thus, further proliferation will be the most likely outcome (Horowitz, 2010). With the already available drones, UAVs, and other unmanned systems, the barriers to entry for new adopters of LAWS are far lower than conventionally believed. Narrow AI and autonomous technologies have strong commercial interests, which reduce the financial burden (R&D costs) on military actors, especially when they are introduced through the adaptation of existing commercial technology. As a result, it is the most probable method of proliferation through software updates to current systems, giving them self-directed functionality, rather than new weapon platforms built entirely from scratch. In addition, it means that the organisational capital requirement is high and thus LAWS require operation processes and applications that do not fit conventional norms. In one military environment, the research and development areas of an existing military organisation tend to dismiss initial indications of innovation as insignificant to core competencies, because such innovations demand a change in behaviour rather than improved performance of existing tasks, unlike those of an opponent. A good evidentiary example of this institutional inertia is the defunding of the X-47B. This dynamism gives emerging powers, including China, an edge, enabling them to incorporate these structures into their military build-up rather than trying to retrofit bureaucracies caught in the rut (Horowitz, 2010, p. 49).

CONCLUSION

The Democratic Peace Theory and the Adoption Capacity Theory, when operationalised, yield distinct responses to the motivation for and adoption of LAWS (LAWS). The main reasons these systems should be deployed are twofold: risk aversion among military employees and significant cost savings. Interoperability and exploitation of an accountability gap are included in the list of secondary incentives. Although the effects of these factors are similar in democracies and autocracies, the institution is asymmetrical because democracies enjoy greater interoperability benefits, with more multidimensional points of use. Most importantly, democratic institutions customarily censor violence by holding political leaders accountable to the electorate; however, LAWS have proved to override and eliminate these natural impediments. Concerning implementation, LAWS require significant organisational capital but low financial intensity. The presence of high organisational capital requirements is due to high levels of disturbance to the existing process through automation, the need to recruit new personnel through retraining, and organisational resistance stemming from fear of substituting current human capability. On the other hand, low financial intensity is driven by significant commercial activity in Artificial Intelligence, which enables the transfer of technology between civilian and military functions at a cost-efficient rate. The fact that autonomy is essentially software that runs on exchangeable hardware significantly increases the likelihood of this technological spillover.

All these combined aspects will ensure that LAWS will replicate at a medium pace, guaranteeing a drastic change like war, marked by robotisation, risk-aversion, and the depersonalisation of organised violence. In addition, LAWS have the potential to upset the status quo among existing powers in favour of new powers structurally positioned to run an efficient adoption process rather than the existing military powers. Lastly, the notion of meaningful human control and what its lapse would have led to is also a crucial aspect of this analysis. The results show that none of the players in LAWS Research and Development (R&D) is ready to limit their development for ethical reasons. The global campaign to prohibit the use of LAWS has made a few significant strides beyond increasing awareness of the issue (CSKR, n.d.; ICRAC, n.d.). Instead, the human-on-the-loop model seems to have received enough momentum, and there is a high likelihood of the spread of the LAWS with sliding autonomy settings in the near future.

REFERENCES

Altmann, J., & Sauer, F. (2017). Autonomous weapon systems and strategic stability. *Survival, 59*(5), 117–142. DOI: 10.1080/00396338.2017.1375263

Amoroso, D., & Guglielmo, T. (2018). The ethical and legal case against autonomy in weapons systems. *Global Jurist, 17*(3).

Anderson, K., & Waxman, M. (2013). *Law and ethics for autonomous weapon systems: Why a ban won't work and how the laws of war can.* The Hoover Institution, Stanford University, Jean Perkins Task Force on National Security and Law Essay.

Arkin, R. (2010). The case for ethical autonomy in unmanned systems. *Journal of Military Ethics, 9*(4), 332–341. DOI: 10.1080/15027570.2010.536402

Asaro, P. (2017). Moral and ethical perspectives. In R. Acheson, M. Bolton, & E. Minor (Eds.), *The humanitarian impact of drones* (pp. 141–149). Reaching Critical Will.

Ayoub, K., & Payne, K. (2016). Strategy in the age of artificial intelligence. *The Journal of Strategic Studies, 39*(5-6), 793–819. DOI: 10.1080/01402390.2015.1088838

Bornstein, J. (2015, March 24–26). *DOD autonomy roadmap: Autonomy community of interest* [Conference session]. National Defense Industrial Association 16th Annual Science and Engineering Conference/Defense Tech Exposition, Springfield, VA, United States. https://ndiastorage.blob.core.usgovcloudapi.net/ndia/2015/SET/WedBornstein.pdf

Boulanin, V., & Verbruggen, M. (2017). *Mapping the development of autonomy in weapon systems.* Stockholm International Peace Research Institute (SIPRI).

Boyle, M. (2015). The legal and ethical implications of drone warfare. *International Journal of Human Rights, 19*(2), 105–126. DOI: 10.1080/13642987.2014.991210

Campaign to Stop Killer Robots. (2018). *Country views on killer robots.* https://www.stopkillerrobots.org/wp-content/uploads/2018/11/KRC_CountryViews22Nov2018.pdf

Campaign to Stop Killer Robots. (n.d.). *Campaign to Stop Killer Robots.* https://www.stopkillerrobots.org/

Cellan-Jones, R. (2014, December 2). *Stephen Hawking warns artificial intelligence could end mankind.* BBC. https://www.bbc.com/news/technology-30290540

Chen, S. (2018, July 1). China military develops robotic submarines to launch a new era of sea power. *South China Morning Post*. https://www.scmp.com/news/china/society/article/2156361/china-developing-unmanned-ai-submarines-launch-new-era-sea-power

Cheung, T., Anderson, E., & Yang, F. (2017). *Chinese defense industry reforms and their implications for US–China military technological competition. Study of Innovation and Technology in China (SITC)*. Research Brief.

Cummings, M. (2017). *Artificial intelligence and the future of warfare*. International Security Department and US and the Americas Programme, Chatham House.

Doyle, M. (1995). On democratic peace. *International Security*, *19*(4), 164–184. DOI: 10.2307/2539124

Economist Intelligence Unit. (2017). *The Economist Intelligence Unit's Democracy Index*. https://infographics.economist.com/2019/DemocracyIndex/

European Parliament. (2018). *Resolution on autonomous weapon systems (2018/2752(RSP))*. http://www.europarl.europa.eu/doceo/document/TA-8-2018-0341_EN.html?redirect

European People's Party. (2019, February 20). *New deal is a huge victory for security and defence in Europe*. https://www.eppgroup.eu/newsroom/news/huge-victory-for-security-and-defence-in-europe

Fairclough, N. (2003). *Analyzing discourse: Textual analysis for social research*. Routledge. DOI: 10.4324/9780203697078

Farooq, U. (2019, May 14). *The second drone age: How Turkey and the US became a killer drone power*. The Intercept. https://theintercept.com/2019/05/14/turkey-second-drone-age/

Fordham, B., & Walker, T. (2005). Kantian liberalism, regime type, and military resource allocation: Do democracies spend less? *International Studies Quarterly*, *49*(1), 141–157. DOI: 10.1111/j.0020-8833.2005.00338.x

Future of Life Institute. (2015). *Autonomous weapons: An open letter from AI & robotics researchers*. https://futureoflife.org/open-letter-autonomous-weapons/

Geist, E. (2016). It's already too late to stop the AI arms race—We must manage it instead. *The Bulletin of the Atomic Scientists*, *72*(5), 318–321. DOI: 10.1080/00963402.2016.1216672

Horowitz, M. (2010). *The diffusion of military power: Causes and consequences for international politics*. Princeton University Press.

Horowitz, M. (2018). Artificial intelligence, international competition, and the balance of power. *Texas National Security Review, 1*(3).

Horowitz, M., Kreps, S., & Fuhrman, M. (2016). Separating fact from fiction in the debate over drone proliferation. *International Security, 41*(2), 7–42. DOI: 10.1162/ISEC_a_00257

Human Rights Watch. (2012). *Losing humanity: The case against killer robots.* https://www.hrw.org/report/2012/11/19/losing-humanity/case-against-killer-robots

Human Rights Watch. (2014, May 13). *12 key arguments for a preemptive ban on fully autonomous weapons.* https://www.hrw.org/news/2014/05/13/advancing-debate-killer-robots

Human Rights Watch. (2016, April 11). *Killer robots and the concept of meaningful human control.* https://www.hrw.org/news/2016/04/11/killer-robots-and-concept-meaningful-human-control

Hunter, A. (2017). *Defense acquisition trends, 2016: The end of the contracting drawdown.* Center for Strategic and International Studies.

International Committee for Robot Arms Control. (n.d.). *International Committee for Robot Arms Control.* https://www.icrac.net/

International Committee of the Red Cross. (2014). *Autonomous weapon systems: Technical, military, legal and humanitarian aspects.* https://www.icrc.org/en/document/report-icrc-meeting-autonomous-weapon-systems-26-28-march-2014

International Committee of the Red Cross. (2016). *Autonomous weapon systems: Implications of increasing autonomy in the critical functions of weapons.* https://www.icrc.org/en/publication/4283-autonomous-weapons-systems

International Institute for Strategic Studies. (2011). Unmanned aerial vehicles: Emerging lessons and technologies. In *The Military Balance 2011* (pp. 20–26).

Johnson, M., & Axinn, S. (2013). The morality of autonomous robots. *Journal of Military Ethics, 12*(2), 129–141. DOI: 10.1080/15027570.2013.818399

Kaag, J., & Kreps, S. (2013). Drones and democratic peace. *The Brown Journal of World Affairs, 19*(11).

Korac, M. (2018). Feminists against sexual violence in war: The question of perpetrators and victims revisited. *Social Sciences (Basel, Switzerland), 7*(10), 182. DOI: 10.3390/socsci7100182

Korać, S. (2018). Depersonalisation of killing: Towards a 21st century use of force "beyond good and evil"? *Philosophy and Society*, *29*(1), 1–152. DOI: 10.2298/FID1801049K

Lachow, I. (2017). The upside and downside of swarming drones. *The Bulletin of the Atomic Scientists*, *73*(2), 96–101. DOI: 10.1080/00963402.2017.1290879

Levy, J. (1988). Domestic politics and war. *The Journal of Interdisciplinary History*, *18*(4), 653–673. DOI: 10.2307/204819

Luttwak, E. N. (1996). A post-heroic military policy. *Foreign Affairs*, *75*(4), 33–44. DOI: 10.2307/20047657

Maas, M. (2019). How viable is international arms control for military artificial intelligence? Three lessons from nuclear weapons. *Contemporary Security Policy*, *40*(3), 285–311. DOI: 10.1080/13523260.2019.1576464

Marshall, A. (2016). From civil war to proxy war: Past history and current dilemmas. *Small Wars & Insurgencies*, *27*(2), 183–195. DOI: 10.1080/09592318.2015.1129172

Müller, H. (2004). The antinomy of democratic peace. *International Politics*, *41*(4), 494–520. DOI: 10.1057/palgrave.ip.8800089

National Science and Technology Council. (2016). *The National Artificial Intelligence Research and Development Strategic Plan*. https://www.nitrd.gov/news/national_ai_rd_strategic_plan.aspx

Pashakhanlou, A. (2017). Fully integrated content analysis in International Relations. *International Relations*, *31*(4), 447–465. DOI: 10.1177/0047117817723060

Ray, J. (2016). *China's industrial and military robotics development* (Report for the US–China Economic and Security Review Commission). Center for Intelligence Research and Analysis, Defense Group Inc. https://www.uscc.gov/sites/default/files/Research/DGI_China%27s%20Industrial%20and%20Military%20Robotics%20Development.pdf

Reiter, D., & Stam, A. (2002). *Democracies at war*. Princeton University Press. DOI: 10.1515/9781400824458

Rogers, E. (2003). *Diffusion of innovations* (5th ed.). Free Press.

RT. (2017, September 1). *'Whoever leads in AI will rule the world': Putin to Russian children on Knowledge Day*. https://www.rt.com/news/401731-ai-rule-world-putin/

Sauer, F. (2016). Stopping 'killer robots': Why now is the time to ban autonomous weapons systems. *Arms Control Today*.

Sauer, F., & Schörnig, N. (2012). Killer drones: The 'silver bullet' of democratic warfare? *Security Dialogue*, *43*(4), 363–380. DOI: 10.1177/0967010612450207

Scharre, P. (2018). *Army of none: Autonomous weapons and the future of war*. W.W. Norton & Co.

Schulzke, M. (2013). Autonomous weapons and distributed responsibility. *Philosophy & Technology*, *26*(2), 203–219. DOI: 10.1007/s13347-012-0089-0

Schulzke, M. (2018). Drone proliferation and the challenge of regulating dual-use technologies. *International Studies Review*, *0*, 1–21.

Shaw, I. (2017). Robot wars: US empire and geopolitics in the robotic age. *Security Dialogue*, *48*(5), 451–470. DOI: 10.1177/0967010617713157 PMID: 29081605

Shaw, M. (2005). The new Western way of war. *Polity*.

Sparrow, R. (2007). Killer robots. *Journal of Applied Philosophy*, *24*(1), 62–77. DOI: 10.1111/j.1468-5930.2007.00346.x

Stockholm International Peace Research Institute. (2019). *Trends in world military expenditure 2018*. https://www.sipri.org/publications/2019/sipri-fact-sheets/trends-world-military-expenditure-2018

Stockholm International Peace Research Institute. (n.d.). *Stockholm International Peace Research Institute*. https://www.sipri.org/

Thornton, R. (2007). Asymmetric warfare: Threat and response in the twenty-first century. *Polity*.

United Nations Institute for Disarmament Research. (2014). *Framing discussions on the weaponization of increasingly autonomous technologies*. http://www.unidir.org/en/publications/framing-discussions-on-the-weaponization-of-increasingly-autonomous-technologies

US Department of Defense. (2007). *Unmanned systems roadmap 2007–2032*. https://www.hsdl.org/?abstract&did=481851

US Department of Defense. (2012). *Directive number 3000.09*. https://www.esd.whs.mil/portals/54/documents/dd/issuances/dodd/300009p.pdf

US Department of Defense. (2017a). *Department of Defense announces successful micro-drone demonstration*. https://dod.defense.gov/News/News-Releases/News-Release-View/Article/1044811/department-of-defense-announces-successful-micro-drone-demonstration/

US Department of Defense. (2017b). *Unmanned systems integrated roadmap FY2017-2042*. https://news.usni.org/2018/08/30/pentagon-unmanned-systems-integrated-roadmap-2017-2042

Weitz, R. (2013). *South Korea's defence industry: Increasing domestic capabilities and global opportunities*. Korea Economic Institute of America, Academic Paper Series, 7, 4.

Yildirim, J., & Sezgin, S. (2005). Democracy and military expenditure. *Transition Studies Review, 12*(1), 93–100. DOI: 10.1007/s11300-005-0037-0

ADDITIONAL READINGS

Bode, I., & Huelss, H. (2018). Autonomous weapons systems and changing norms in international relations. *Review of International Studies, 44*(3), 393–413. DOI: 10.1017/S0260210517000614

Crootof, R. (2016). The killer robots are here: Legal theories and artificial intelligence. *Cardozo Law Review, 37*(6), 1837–1915.

Galliott, J. (2015). Unmanned systems and the virtue of responsibility. *Journal of Military Ethics, 14*(1), 60–63.

Johnson, J. (2019). Artificial intelligence & future warfare: Implications for international security. *Defense & Security Analysis, 35*(2), 147–169. DOI: 10.1080/14751798.2019.1600800

Payne, K. (2018). Artificial intelligence: A revolution in strategic affairs? *Survival, 60*(5), 7–32. DOI: 10.1080/00396338.2018.1518374

Pietrzak, P. (2022). *Introducing the idea of ontology in statu nascendi to the broader international relations theory. International Conference Proceeding Series: International Conference on Economics and Social Sciences*. ECLSS., https://www.eclss.org/publicationsfordoi/abst11act8boo8kIE%26SS2022_antalya.pdf

Pietrzak, P. (2023). Approaching regional conflicts through the prism of ontology in statu nascendi: The new compartmentalization of the IR theory. *In Statu Nascendi – Journal of Political Philosophy and International Relations, 6*(1).

Pietrzak, P. (2024). The sixth great debate in international relations theory revolves around clarity: Ontology in statu nascendi as a cutting-edge tool for conflict management. In Pietrzak, P. (Ed.), *Dealing with regional conflicts of global importance* (pp. 1–27). IGI Global., DOI: 10.4018/978-1-6684-9467-7.ch001

Schwarz, E. (2018). Prescription drones: On the techno-biopolitical regimes of contemporary 'ethical killing'. *Security Dialogue*, *49*(1–2), 59–77.

Walsh, J. I. (2015). Political accountability and autonomous weapons. *Research & Politics*, *2*(4), 1–7. DOI: 10.1177/2053168015606749

KEY TERMS AND DEFINITIONS

Lethal Autonomous Weapons Systems (LAWS): Weapon systems, which once enabled, are able to pick and shoot without further effort on a human operator. They are also typified by a delegation of important decision-making capabilities from humans to machines, namely targeting and the use of lethal means.

Antinomic Democratic Peace Theory: A hypothetical split which argued that democracies seldom made war on one another, but could be very violent in dealings with non-democracies or non-state parties. This version implies that democracies are encouraged to employ force with no concern for their lives or political responsibility, and to do so through low-risk technologies such as LAWS.

Adoption Capacity Theory (ACT): A model that is applied to understand the way in which military organisations are employing innovations along two variables: financial intensity (cost/resources required) and organisational capital (institutional flexibility vs. resistance). It forecasts the rate and the success of technology diffusion.

Meaningful Human Control: An ethical and functional rule that entails that humans are able to have the option to manipulate or intervene on the most vital functioning of a weapon system. It distinguishes between legitimate semi-autonomous systems and the fully autonomous ones in which human control is practically eliminated.

Human-on-the-Loop: A mode of operation in which the weapon system is autonomously controlled to make a target selection and engagement, but a human operator oversees the process and has the authority to veto or abort the attack. This is opposed to human-in-the-loop (where humans act) and human-out-of-the-loop (no oversight).

Dual-Use Technology: Technologies, including artificial intelligence algorithms, that can have direct applications in both civilian life (e.g., self-driving cars) and the military. This synergy lowers the cost of military R&D and reduces proliferation.

OODA Loop (Observe-Orient-Decide-Act): The manner in which decisions are made and unmade in war. The introduction of AI into weapons systems is valued because it enables making this loop tighter and faster, so that the forces can process data faster than a human enemy can react.

Chapter 4

The Role of Artificial Intelligence in Military Decisions:

Making Trust, Ethics, and the Future of Warfare

Riday Mukherjee
http://orcid.org/0009-0002-7960-0166
Jharkhand High Court, India

Krishna Panda
Siksha O Anusandhan University, India

Sarada Saha
Kolkata Police Law Institute, India

Ashutosh Kumar
Department of Law and Governance, Central University of South Bihar, Gaya, India

ABSTRACT

The academic literature on the development of artificial intelligence applications in the military industry primarily focuses on technical requirements and legal and ethical paradigms. Nonetheless, there is a gap in the scholarly literature regarding an essential element of any military action: the role of command officers. The aim is to examine Indian military officers' perceptions and reasoning regarding the integration of AI into the military decision-making process. It examines the logic of their approval or rejection of the usage of AI on the battlefield. The analysis introduces

DOI: 10.4018/979-8-3373-6264-9.ch004

a methodological innovation to the war science study vignette survey, enabling the determination of the dynamics between humans and AI across increasing levels of threat to human life in three scenarios. Empirical studies indicate that officers retain confidence in the use of AI in military missions, regardless of the mission's risk profile or complexity, provided that principles such as human command remain in place.

INTRODUCTION

In addition to discussions in military science about the extent to which artificial intelligence constitutes a revolution in warfare (Raska, 2021; Hynek & Solovyeva, 2022), Max Tegmark's suggestions indicate a paradigmatic shift in how warfare is conducted. The current phenomenon is similar to the dualism observed in nuclear technology, which has shifted awareness toward the positive role of energy production while highlighting the devastating possibility of weapons. Thus, intense social debate is ongoing about whether artificial intelligence will ultimately benefit or destroy humankind when researchers can create so-called superintelligence, and how humanity may rationally regulate such creation (Bostrom, 2014; Tegmark, 2017; Scharre, 2018; Taylor, 2020). Artificial intelligence opens a significant gap in military science research, particularly in the philosophy of military science. The latter can be illustrated by Henry Kissinger's questioning of René Descartes's cogito ergo sum (I think, therefore I am). The question that arises is whether, in the future, artificial intelligence could reach a point at which it can reason and make decisions concerning matters of life and death, thereby defining the human condition. (Kissinger et al., 2021).

Further to this idea, General Michael Claesson asks whether artificial intelligence can ever learn common sense. As a result, the understanding of the human actors-artificial intelligence relationship development is of utmost significance. Similar to the debate on nuclear weaponry, it is often easy to forget that it is the officers who end up carrying out or do not carry out a decision to use force: officers. There is an absolute necessity to re-examine the bases of such crucial judgments, which can be a potential byword or bane of the human species, and to comprehend how artificial intelligence will affect the judgment of an officer on the manoeuvres of the situation in a future context. The world, probably, would have ended in nuclear war due to some of the most dramatic moments of the Cold War, without any of the measures taken by different actors, including the decisions made by Stanislav Petrov (or Vasiliy Archipov) not to take any action; as a rule or norm as it was prescribed (Forden et al., 2000; Tegmark, 2017; Kissinger et al., 2021). The primary objective in this case is to examine the military approach to artificial intelligence. This descriptive analysis aims to determine whether the implementation of AI in the decision-making process affects officers. At the same time, it examines whether the practicality of operational implementation changes when an AI provides advice to a human counterpart or makes decisions.

The study is exploratory in nature, given the need to examine an abstract concept such as trust in AI among individuals who do not use the technology in their daily lives. It is hoped that it can attain the level of inquiry commonly associated with qualitative research and also have the expanded population range noted in quantitative research. The vignette survey approach is, therefore, used. This method is where the research question presents the respondent with a set of hypothetical situations and requests them to choose answers according to particular motive alternatives- a sort of questionnaire, but with an increase of internal reasoning by the respondent (Sheringham, Kuhn & Burt, 2021). This methodology offers an interdisciplinary contribution to military science, as vignette studies are rare. The survey's target population comprises officers enrolled in advanced professional education. This group is a special stratum of the total Armed Forces, comprising representatives from most units and branches; the range of experience, age, and years of service is broad.

WHERE IS THE VOID?

This descriptive research aims to address a gap in the discourse on the integration of AI into the battlefield. It avoids a technical discussion of specific systems and instead outlines the current attitudes toward and tolerance of AI in the contexts of decision-making and operational planning. The research is aimed at addressing the main research question through this inquiry: How much do (a sample of) Indian armed force officers trust AI in military work? It also examines the following sub-questions: Does trust vary across levels of risk of human suffering and death during the operation? What are the more general sentiments regarding the application of AI to military operations? To ensure constant reminders of the study's overall purpose, it was intentionally decided not to provide a comprehensive definition of current AI terminology, which may be viewed as a limitation. Nevertheless, Hadlington et al. (2023) point out the same methodological challenge: AI in the military context is not a cohesive, bounded technology that can be compared with a tank or a machine gun. Thus, attention is paid to the perceived effects of hypothetical influences of such systems (Hadlington et al., 2023). To be accessible, a simplified classification is used where any of the modern technologies like deep learning, machine learning, and narrow and deep systems of AIs are then called AI.

To clarify these drastic differences in the extended discussion, the term AGI (artificial general intelligence) denotes an intelligence that is not yet available but is at the level of human intelligence (Tegmark, 2017; Scharre, 2018). At the next level, the interpretation of AI becomes even more complex: artificial superintelligence (ASI) is intelligence that exceeds human cognitive capacity. Similar studies focus on AI-based quantum computers (Eng. Quantum Artificial Intelligence, QAI) (Taylor, 2020; Antebi, 2021; Der Derian & Rollo, 2023; Johnson, 2023), but they do not elaborate further to prevent a shift in focus toward unnecessarily technical details. Nevertheless, AGI and superintelligence should not be dismissed, as they are theoretically the last inventions humanity is likely to create (Tegmark, 2017), leaving humans with the title of the second-smartest creature in the world, currently held by chimpanzees or dolphins. Last but not least, explainable AI (XAI) can be defined as AI support systems that answer the question: how did the algorithm arrive at this result? (Holland Michel, 2020), a concept further discussed in the following paragraphs in relation to the black-box phenomenon.

LITERATURE REVIEW

To organise the research outline, the initial question is divided into two sections, with AI serving as the common denominator. The former area underscores the interplay among trust, risk, and the corresponding notion of control, whereas the latter focuses on AI in military contexts. The application of artificial intelligence to the military decision-making process requires an interdisciplinary approach that draws on scholarship in sociology, computer science, and military science. The use of AI will likely become common in most areas of research in the long term. In most cases, AI studies suggest that a new technological and military weapons race driven by disruptive technologies is beginning. The most notable change in the historical tendency is that the military armament is no longer the key force; nowadays, it has been outshone, with the possible exception of Quantum AI (QAI) research, by innovations in the civilian market (Maas, 2019; Lidskog, 2020; Innes & Morrison, 2021; Kissinger et al., 2021; Raska, 2021; Der Derian & Rollo, 2023).

The Debate of Risk and Trust

Regarding risk and trust in AI at the global level, the following definition is informative: Trust is a property of agents that enables them to operate in the world without complete information about all the systems in question (their properties, histories, etc.). In systems theory, to fully know a system is to know its properties, composition, environment, structure and mechanisms, or its history. Such a possibility can hardly exist, or at all, rarely at all. Therefore, utilising a compensation scheme helps employees manage their earnings to balance their financial goals with the company's requirements. As such, implementing a compensation scheme can help employees manage their earnings and align their financial objectives with the demands of the business.

The cited source underscores the need to initiate an international review of research on officers and AI. It is important for understanding the strategic context within which military decisions are implemented. The evolution of AI shows that the world is dichotomised in many aspects. Firstly, in the economic, developmental, and power dynamics between the northern and southern parts (Png, 2022; Fels, 2023); secondly, in the conventional East-West dichotomy, which exemplifies how officials from various countries analyse and assess AI in terms of risk and trust. This dichotomy is characterized as one of the components of a more extensive discussion on how to govern and control AI. The first vision is the Eastern one, spearheaded by China, and focuses on ensuring that AI development operates under robust state control to promote greater concentration of power (Kania, 2019; Raska, 2021; Roberts et al., 2021). On the other hand, the Western vision, facilitated by the USA, enables the market to develop AI products that are subsequently implemented. This helps develop a holistic understanding of how the introduction of AI varies across armed forces and how officers will apply it in their subsequent operations. Research in the West centres on the issue of delegating life-and-death decisions to AI (Rosert & Sauer, 2019; Wasilow & Thorpe, 2019; Warren & Hillas, 2020). The example of these dangers and distrust given by MIT professor Max Tegmark (2017) reminds us of an incident that happened in 1988, when a civilian passenger plane was shot down by an American ship under the wrong judgments of the onboard radar and warning system device.

Military Operations

Unless powerful robots that kill people are what keep you awake at night, militarised advanced AI is nightmare fuel. It is difficult to imagine researchers opposing hostile AI if they do not know how to control a self-built version. Scharre underscores the urgent need for the military to consider this path of AI development. However, it also presupposes deep trust in, and knowledge of, the particular processes on which the army relies. He makes a clear distinction between the dangers of entrusting a thoroughly independent system, which the enemies can hack, and one that still includes a human in the decision-making process (Scharre, 2018, pp. 245-247). Increasingly, AI technologies are demonstrating superiority over their human equivalents, ranging from computer games to medical studies. Subsequent analyses indicate that AI systems are improving in military simulations, a trend that warrants attention (Tegmark, 2017; Husain, 2021). These controlled settings are often not in a position to reveal drawbacks or identify settings where development is still in its early phase, creating an illusion of an overly impressive idea that AI is unbeatable by humans in all scenarios. Johnson (2023) attributes this problem to a general lack of awareness of the practical performance of AI systems. However, the prevailing discourse often equates the current reality with the theoretical example of AGI or ASI. Although generally well-grounded in the general perspective about the applications of AI in the military setting, a clear reading indicates that his formulations are not based on the way military organisations generally carry out operations in the form of clear mandates and allocated resources, which makes some of his arguments about AI development in the military operations dubious.

Speaking to the potential of the Indian armed forces, Alastair Finlan provides a detailed overview of the research. However, he fails to illustrate the current and emerging challenges of AI clearly, nor to explain how to minimise them (Finlan, 2021). In turn, is more precise regarding the underlying issue of AI in military missions. He argues that the priorities for accessing and controlling data remain. However, they also cannot be effective without designing the interface and integration of man and machine to handle the amount of data or transform it into a precise foundation of decision-making. This particular interaction is subject to research that is termed HMT (Eng. Human-machine teaming) (Warren & Hillas, 2020). Researchers also point to a serious issue: the line separating what is permissible in reaching AI is being moved, with air defence decisions occurring within seconds. This development is essential for control, as the red lines currently imposed can be difficult to define or negotiate.

With this, Western studies would invariably revert to the assumption that a human being must remain involved in the process. Although such a position is logically sound, the argument has two sides. The example of the robotic cruiser that crashed into a civilian passenger plane provided by Tegmark represents the fact that there is a point of argumentation on both sides. There are circumstances in which systems can misinterpret signals, and there are circumstances in which humans can make incorrect decisions based on those signals (Tegmark, 2017). Boyd's OODA loop is a widely used explanatory model for this predicament (Scharre, 2018). This is designed to execute actions earlier than the opponent; hence, it can be achieved by either a faster command system or a faster action system. According to this model, the Western view holds that human beings should never be rationalised out of the loop, even when their retention would make it slow or, worse still, a catastrophe, as in the case of the cruiser. The control, risk, and trust values are therefore strictly conditioned in this hemisphere in research on the deployment of artificial intelligence. These concepts demonstrate a certain degree of dependence: the more one trusts, the less one perceives as taking risks (Sullivan, de Bourmont, & Dunaway, 2022). An organisational ground is typical of control, based on rules and governance measures, and risk and trust are also integral to situational and individual measures.

This interdependence is outlined as follows, with the interplay of these variables forming the heart of the research.

THEORETICAL FRAMEWORK

The theoretical background of this question is divided. The former identifies key concepts that address gaps between the previous research summary and the current one, and the latter identifies similar studies to outline expectations for the research outcomes. All of these are combined to develop the theoretical framework, which is further elaborated in the methodology chapter. Owing to the absence of a comparable precedent, findings from studies across different areas are synthesised into a comprehensive view that integrates the military dimension. There is, therefore, no a priori hypothesis; this descriptive study will explain how trust in AI manifests across various military operations. By describing the interaction between officers and AI, this question prompts the development of a theory that remains to be teste. The fact that this kind of ambition can be expansive in terms of the scope of a master's thesis framework does not invalidate it.

Trust and Risk

The concept of trust is defined in different ways, but about the particular issue of the human-AI interaction, the concept is defined as per Lukyanenko et al. (2022). They also elaborate on Luhmann's account, which holds that trust lacks a definitive, unique definition and is multifaceted. It should be described not only as an interpersonal process but also as a broad-based behavioural process in which the perception of trust operates in the context of uncertainty (Lukyanenko et al., 2022). Therefore, trust entails an investment in a risky endeavour, in which the outcome of an engagement with another party is expected to be largely favourable, even though one does not fully know the nature of the counterparty (Lukyanenko et al. 2022, pages 1994-1996). As a result, trust cannot be considered in isolation; it should be discussed within the context of risk. Risk is a quantitative variable denoting the adverse effects of a future event. This is the reverse: a measure of the desirable consequences, referred to as chance. Researchers also regard risk, which is ambiguous, as a multifaceted concept, in the same way that Lukyanenko has described trust. The most important difference between the two is that risk acceptance is much greater when it is self-inflicted (e.g., parachuting) than when it is externally imposed (e.g., traffic hazards). Risk is always associated with one of the following: life, property, or capital. In this paper, the concept is examined primarily in the context of the military decision-making process, which involves human suffering or death.

Anticipated Outcomes

It is a phenomenon that is called a black box and denotes a phenomenon of critical intransparency that operates in artificial intelligence systems, where even the designers cannot explain how exactly the methodology or logic is applied to draw a particular conclusion (Scharre, 2018; Holland Michel, 2020). Such opacity is a significant issue for the AI research path, and it will only intensify when discussing Artificial General Intelligence (AGI) or Artificial Superintelligence (ASI). To counteract this, advocacy is needed to develop so-called Explainable AI (XAI) systems that are transparent in their work (Holland Michel, 2020). Along with other ideas such as LEIA (language-endowed intelligent agent), XAI reflects a trend in studies that emphasises maintaining human control and comprehensibility in future developments (Holland Michel, 2020b). It therefore follows that explanatory ability predicts whether human actors can be confident in using AI-generated decisions. Such officers, who are expected to assess intelligence created by AI, should, thus, understand how the system received and analysed the received data. In this regard, some respondents may distrust the AI specifically because they lack understanding of how the AI reached its decision and, as such, may choose to adopt alternative courses of action.

According to Lukyanenko, cross-fertilisation of philosophical questions with research on trust in AI would add value to numerous studies in this field, especially concerning possible ethical and moral issues that might arise (Lukyanenko et al. 2022, pages 1999-2000). In their research on applied AI, specifically on autonomous vehicles, Nyholm & Smids (2016) shed light on this dilemma. They explore how a self-driving vehicle should theoretically decide to kill one person to protect five other people in an emergency. This question is based on the trolley problem, a traditional moral-philosophical thought experiment: when at a crossroads, a train is approaching people, one agent is put at a crossroad: does that train deserve to hit five individuals, or should it be intentionally pulled off the tracks to collide with one person (Nyholm & Smids, 2016, p. 1276).

This is a unique gap within the broader research on trust and within AI research in military science. In a military environment, a human agent (or an AI system) may be required to make such decisions under a broader range of motivations and conditions than those available in a civilian traffic scenario. In turn, the vignette used in this research is, to some extent, based on the trolley problem. The design shall motivate respondents to employ internal computational reasoning, thereby enabling an evaluation of the extent to which risk-taking propensities modulate the choice to use AI. To explore how decision-making and risk-taking are correlated, it is reasonable to consider what Nobel laureates Daniel Kahneman and Amos Tversky wrote about their first and second versions of Prospect Theory (Kahneman & Tversky,

1979; Tversky & Kahneman, 1992; Wakker, 2010; Haerem et al., 2011). Principally, Prospect theory holds that human decisions differ significantly depending on how a proposal is framed or processed. Moreover, justified actions are often based on current gains and losses experienced during the process, rather than on the ultimate goal (Kahneman et al., 1979).

Haerem et al. (2011) then applied Prospect Theory to military officers' decision-making. They used the mainstream Asian disease problem as a psychological test that requires the respondents to either take the sure thing in saving a given number of people X or the probabilistic one in terms of risk, referred to as the maximum Y number of deaths. The researchers then modified these standard tests to fit a particular military setting. More importantly, Haerem et al. (2011) have shown that officers' decision-making is not always guided by the behavioural curve proposed by Kahneman et al. (1979); instead, officers are more prone to risk-taking than civilians. It is therefore expected that the reactions in the current study will also deviate from the standard curve, indicating that officers will not be reluctant to assume a larger share of the risk than the general population. The British research Attitudes towards AI in Defence (AAID) examined British citizens' attitudes toward the use of AI across different types of defence (Hadlington et al., 2023). Hadlington identifies three potential areas of AI use in defence. The support activities include the first area, which deals with logistical flows. Nevertheless, in an Indian context, external authorities tend to control such processes, and they may not pay sufficient attention to the respondent, thereby distracting them with other extraneous questions about inter-agency collaboration; therefore, this domain was not considered. The two other domains, in turn, can be viewed as integral to the essence of the military profession: AI in non-kinetic offensive missions (including cyber warfare) and AI in kinetic offensive missions (when physical force is used) (Hadlington et al., 2023). The two categories have been operationalised in the following vignettes.

METHODOLOGY

This chapter outlines the study's structural development, the characteristics of the respondents, and the ethical issues, culminating in a description of the operationalisation process. In determining the required population size for this inquiry, it became clear early that respondents for a vignette study must have a standard baseline knowledge of military operational planning (Pietrzak, 2024). As a result, the officer corpus survey was excluded because it is challenging to ensure comparable levels of prior knowledge. Instead, a group of individuals at an advanced stage of professional training, namely the "*Leadership of Joint Operations*" (LGO) course, was selected to provide consistency about the educational base. Course coordinators may subsequently conduct this study within the curriculum. During the preparation stage, the research approach shifted to a vignette study, which is more conventional: all participants answered the same questions. This change was necessitated by the exasperating conditions inherent in the experimental method itself. However, the investigation did not exclude the original population and sample to which the experiment was aimed. To preserve the anonymity of the female minority in the course (not more than 10%), the gender variable was excluded from the analysis. This conclusion is supported by the AAID study by Hadlington, which showed that gender has a relatively small influence (Hadlington et al., 2023).

Introductory film. Before data collection, respondents were shown a film. This mechanism presented the researcher with no compelling face-to-face interaction, which would have exposed the researcher to the possibility of influencing the participants. Notably, the movie intentionally omitted the study's purpose; this was a strategic measure that further mitigated the risk of respondent bias (Sheringham et al., 2021, p. 4). In the end, of the 156 students contacted to participate, 82 (52.5%) responded. This turnout rate is attributable to the task's voluntary nature and to the fact that it occurs at one of the most active stages of participants' lives, namely their education. Another reason is that, for some respondents, the topic of Artificial Intelligence may seem somewhat abstract and alien, thereby fostering unwanted interaction. Participant selection was strictly in accordance with the requirement that participants personally consented to use their responses in this research. The respondents who were willing to participate were categorised as given in Tables 1 and 2 below.

Table 1. Respondent demographics by age group

Age Group	Number of Responses
25–35	10 (12.2%)
35–45	47 (57.3%)
45–55	24 (29.3%)
55–65	1 (1.2%)
Total	**82 (100.0%)**

Table 2. Respondent experience by service branch

Service Branch / Combat Force	Number of Responses
Army	34 (41.5%)
Air Force	16 (19.5%)
Navy	23 (28.0%)
Command/Intelligence (C2/Intel)	9 (11.0%)
Space/Cyber	0 (0.0%)
Total	**82 (100.0%)**

One of the main methodological priorities meant to create realistic vignettes that also served as ethical trolley problems. The design ensured that respondents were required to take a clear step of action, rather than providing vague, non-specific free-text responses that lacked an obligation to a particular alternative.

The tool was stratified into three major areas:

1. Respondent background
2. Vignettes
3. AI in the Armed Forces

To collect feedback and determine the time required to complete the questionnaire, a pilot study was conducted with a smaller cohort initially. Therefore, the questionnaire was extensively revised before being sent to the thesis supervisor and the course coordinator. Specific responses indicated that the first instrument was too broad and was therefore narrowed to 26 questions. In addition, some wording, particularly in vignettes 1 and 2, was found to be ambiguous or leading and was revised to improve conceptual validity. Approximately 36 responses were received, primarily after the launch, prompting a reminder email. The centre developed to 75 in response, on average, until the survey was initially closed. After being informed by individuals who were still planning to attend the event that they lacked sufficient time, a final distribution was authorised. This resulted in a final figure of 82 respondents-slightly less than it was initially estimated, but considered to be enough in terms of methodology. In terms of reliability and validity, trust in military AI activities is grounded in personal experience and situational considerations. It is associated with multiple levels of abstraction, which raises issues regarding conceptual validity. The operational indicators needed to be grounded in a single vignette to ensure alignment between theoretical definitions and operational indicators. Instead, three different vignettes were created using prior literature relevant to the central research question, a strategy for reasoning about validity (Sheringham et al., 2021). The multifaceted approach was considered the most effective for recording respondents' positions. As mentioned earlier, the participation rate among the invited population was slightly over 50%. It may be explained by the fact that the respondents rated themselves as technically competent. In the introductory phase, respondents rated their technical maturity by comparing themselves with peers. The resulting statistics indicate that most consider themselves to have equal or superior technical maturity compared with their colleagues. Thus, the research significantly lacks the voices of people who consider themselves to have lower technical maturity, thereby introducing a potential bias into the results.

Table 3. Distribution of self-assessed technical maturity

Scale Value	English Label	Interpretation of Distribution
1	Much lower capability	Represents the lowest tier and was selected by a negligible number of respondents.
2	Somewhat lower capability	Represents a minority self-assessment of slightly below-average proficiency.
3	Same capability	Central Tendency: The data clusters here indicate that the majority perceive their capability as equivalent to the standard.
4	Somewhat higher capability	Positive Skew: A significant portion of the cohort views themselves as possessing above-average proficiency.
5	Much higher capability	Represents the highest tier of self-assessed technical competence.

RESULTS

The main facts of the GRE research are detailed in this chapter. The gathered information comprises approximately 30 pages of qualitative, unstructured responses and quantitative data. Nevertheless, in order to maintain respondent anonymity strictly. The quantitative analysis was primarily conducted using SPSS. Since the sample is relatively large and the distribution of responses is relatively even, statistical significance is not achieved; however, the observed response differences provide a strong basis for additional exploratory research. In particular, cross-tabulations of vignette responses with demographic factors, including service branch and age, were conducted to identify potential correlations and to inform subsequent hypotheses. Although the results of these questions revealed interesting trends, the small number of responses did not permit the establishment of statistically significant relationships. To this end, the analysis will aggregate responses from the entire surveyed population in the vignettes and contrast them with the control questions presented in the third section of the instrument. Vignette 1 presents a hostage situation in which the Artificial Intelligence suggests non-kinetic responses, namely cyber and information operations, rather than a standard kinetic response (v1). This vignette employed an interactive structure: respondents who selected the "other" category were asked to provide free-text in a field to elaborate on their selection. Moreover, the non-trustees in the AI recommendation had to explain their particular reasons in the follow-up question (free-text) later.

Figure 1. Responses to vignette 1 (hostage situation)

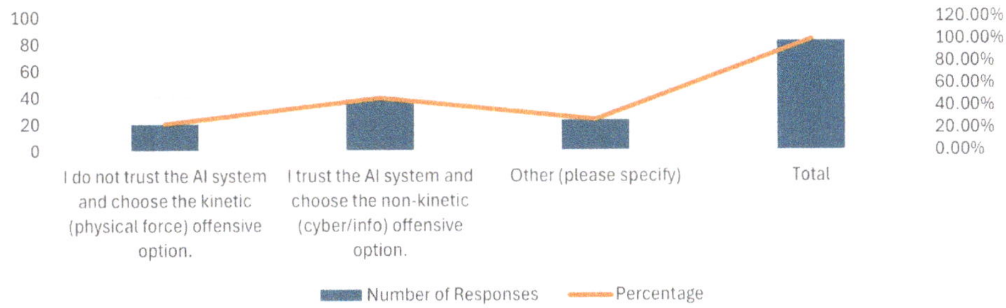

In the described situation, 47.6% of participants expressed immediate trust in the AI-generated alternative (v1). Nevertheless, qualitative analysis of the free-text responses indicates that most compliant respondents applied the AI's course of action with specific consideration for operational implications. To illustrate, one of the respondents wrote, "*I believe in AI, yet, I would like to consider how to address a cyber attack in general, what can be another outcome (v1), another one stressed that it is imperative to know, in general, whether the AI can pose a threat to third parties and other outcomes (v1).*" Other views referenced the reciprocal relationship between staff and technology, where one officer reported that AI system inputs apply. Once they determine the relevance and feasibility of these, they are compelled by and inspired by the proposal (v1). Others who chose not to believe the AI, by contrast, defended their position by invoking arguments about the AI's intelligence and immaturity. One of the respondents framed the underlying black-box issue as follows: "*I believe that I do not understand where the AI acquires its data to provide its advice, and what is the perceived context of the AI (v1).*"

The other also showed doubt about the potential of the future to do such a thing, noting, "*I do not think that this technology is sophisticated enough to supply the pertinent information for decision-making in such a specific scenario in 10-15 years (v1).*" Although several respondents provided well-developed alternative strategic solutions, the tactical responses were not considered analytically valuable in this study. A set of control questions was used in the third part of the study, namely, to discuss AI in operations that do not involve kinetic motion, as well as its use as a planning aid, to ensure high reliability. According to the results, a clear majority (93.2%) favour using AI as general support for operational planning. Ironically, a slender majority (53.4) said they should not allow AI to make decisions in non-kinetic operations on its own, indicating a subtle distinction between AI as a support mechanism and AI as an independent agent.

In Vignette 2, the situation involving a separatist group employing attrition strategies against a local army appears to be a multifaceted problem that threatens the capital. A polemic arises when the Chief of Staff supports the combat value of friendly forces, and the AI system recommends an active, kinetic offensive. The findings indicate that a high percentage of respondents chose to accept the increased risk associated with the AI proposal. It is an analytically important insight, explained in detail below. According to the findings made by Haerem (2011; pp. 492-493), this vignette was expected to come up with such information and thus supported with precise follow-up questions to shed some light on attitudes toward AI in operations where serious risks of incurring suffering or demise are significant. In fact, none of the respondents provided a reason for not trusting the AI-generated proactive alternative. One possible reason this was not mentioned is the wording of the question: would you have made a different decision had the AI system suggested or recommended any of the other options? (v2). Although the answers to these questions were quite diverse, the amount of free-text explanations was much lower than in Vignette 1. The responses indicate that the decision to implement the offensive option was not based solely on the AI's recommendation, but rather on an evaluation of a broader set of factors. As one respondent explained, the fundamental reason for my choice is the ally's support. The fact that option 3, which I selected after reading the provided piece, was also suggested by the AI supports my choice. I do not consider myself unfamiliar with AI, but I also wish to take into account personal and national considerations when making decisions of this scale (v2). At times, the respondents gave cautious responses, such as, "*I had looked at the options of the Chief of Staff and AI and subsequently, maybe, had arrived at a different solution (v2).*"

Figure 2. Responses to vignette 2 (tactical dilemma)

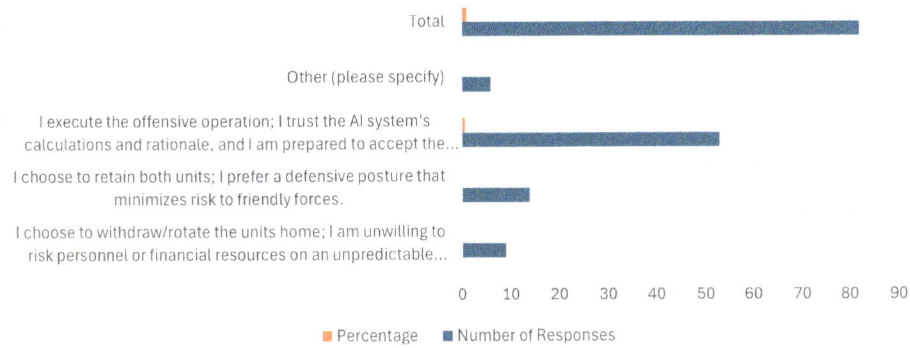

The vignette was then intensified to promote the dilemma of death and suffering. When the situation worsens and the enemy becomes entrenched, the AI system proposes the use of entirely autonomous weapons to eliminate risk and minimise casualties among allies. This situation approaches the crux of the philosophical question: whether the respondent is sufficiently trustworthy to entrust life-and-death decision-making to an autonomous system. According to the data, 45.1% answered this question affirmatively, stating that the risk of personal loss justifies such actions (v2). More to the point is when 18.3% based their acceptance on the principles of legal positivism, as opposed to moral consideration, they said: "*Doing as per the current norms and guidelines. I will most likely use them, unless it is not permissible (v2); the legal framework and rules of engagement for such use should also be clarified. I choose to use the system*" (v2). However, when compared with general concepts, a severe dissonance emerges between these tactical choices. In the third part of the survey, the questions in the control section indicate that the majority (78.1%) is firm that AI should not be allowed to make decisions involving human suffering and death without human oversight. Such a paradox between general ethics and concrete tactical decisions makes it important to perform a strict analysis. The distinction proposed by one respondent can offer some resolution to this paradox: "*At any rate, it is the human who is to decide about the usage of lethal weapons, but there are cases, when there might be sufficient control of a human, to consider the usage of autonomous systems as a decision-making area (v2).*" Vignette 3 was designed to present a stark, straightforward dilemma resembling the trolley problem, thereby precluding subjective interpretation and alternative courses of action. This situation demanded a binary decision: either to accept the judgment of the system of a hostile air threat or to abort the engagement, thus he knowingly endangered the life of the operator and the crew themself. Two respondents chose not to answer this question (v3).

Figure 3. Responses to vignette 3 (imminent aerial threat)

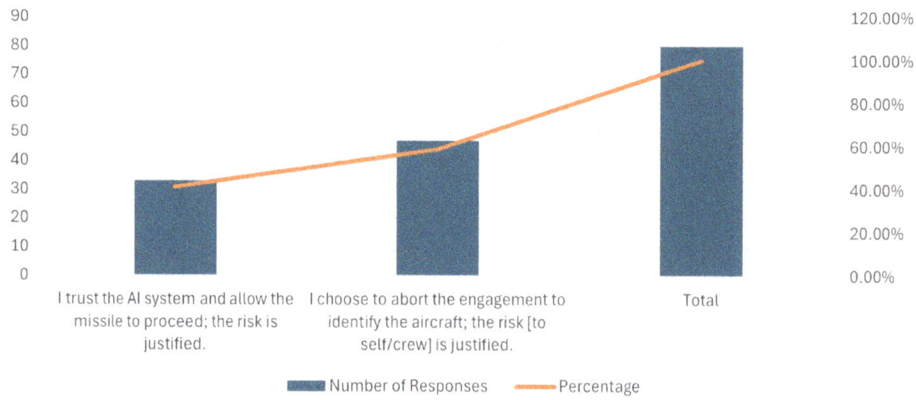

This performance elicits significantly less trust in AI systems than in Vignette 2, in which most participants chose to use autonomous systems specifically to soothe friendly casualties (V2). On the other hand, in Vignette 3, 58.8% decided to take the risk to their own lives and those of their crew, but not to the AI's decision to engage (v3). This separation can be explained by a variety of factors, including differences in the organisation's scope of operation (land and maritime environments) and the recognition that the relationship between risk and trust in AI is not absolute but situational.

ANALYSIS

This chapter presents the findings of the vignette study and addresses the main research question: what kind of trust Indian armed force officers have in AI during military operations. Moreover, the study addresses the sub-questions and outlines the methodological contributions to military science. At the end of the discourse, recommendations for further research and analysis of the study's implications for the profession may be offered. The empirical data yielded conclusions that were, in several respects, unexpected. Specifically, the respondent cohort tended to use AI, which was superior to early predictive models. This could be associated with the high level of technological maturity of the respondents, which Lukyanenko (et al. 2022) proves with support. It is worth noting that officers' resistance was primarily limited to cases in which the system's functionality was unclear or the rules and requirements for its use were unclear. This discovery underscores the urgent need to pay even greater attention to Explainable AI (XAI) in military systems development.

The biggest lesson learnt from the vignettes is that trust is not black-and-white but a function of specific situations; it is not simply trust or distrust of the system. Instead, the level of trust that respondents report depends on the situation's exigencies. According to the vignette summary of findings, the general research question is as follows: The trust demonstrated by Indian armed forces officers in military operations depends on continuous evaluation of situation-specific risk factors. The difference between Vignette 2 and Vignette 3 shows this fluidity: in the first, the respondents were already willing to entrust a completely free system to theattenuation of further amicable losses; in the second, they chose to abort a self-defence action- that is, risk their life to an object. The degree of trust is strongly influenced by the AI's output, which may support or contradict the officer's judgment during the operation. Therefore, when the AI decision is consistent with the officer's opinion or offers a valid reason, trust will be enhanced. On the other hand, when the AI suggests a line of action that is at the extreme opposite of what the officer takes as the solution to the problem, such as in the case of Vignette 1, with the cyber operations being proposed as a solution to a hostage situation, most often solved by using kinetic force, trust is lost. This creates a significant difficulty in developing AI systems across a variety of settings (kinetic or non-kinetic), because utility cannot always depend solely on the system being used as a mechanism for confirmation bias.

Regarding the sub-question of whether trust varies with the perceived risk of human suffering and death in the response, the evidence indicates that respondents placed considerable emphasis on valuing diverse opinions in death cases. This result is supported and anticipated by Haerem et al.'s (2011) study on officers' risk appetite. The findings show that trust in AI does not fundamentally change in most cases, even in the presence of lethal risk. Instead, AI is appreciated as a supplementary input to the planning process, which helps highlight critical areas and, as illustrated in Vignettes 1 and 2, enables accurate probabilistic risk estimation. Although it may seem implausible to the layperson that the possible loss of life does not warrant an urgent posture adjustment, within the operational ethos of the professional military, this is one variable among many to be factored into the operational calculus. It is clear in this professional indifference toward risk-taking in both the present inquiry and the study by Haerem et al. (2011). This statement is supported by answers that explain risk tolerance through specific orders, legal systems, and requirements, as well as by the central issue: the task's importance within the overall strategic plan. Vignette 3 did not allow respondents to develop a reasoned response, as the decision was to be made instantly and in a binary manner. As Tegmark (2017) describes, in real-world contexts, decision-makers often must act without all the background facts at their disposal.

A slight majority (58.8%; v3) indicated insufficient trust in the system's assessment. This outcome necessitates an analysis that integrates the Luksyaneko framework of trust in AI with the broader perspective on strategic perception of risk and control found in the literature (Scharre, 2018). The interpretative consensus is based on the institutional presupposition that Western military defence is institutionally betrayed on the idea that lethal force is the status of control en route to the decision-making loop. The officer will therefore resolve this trade-off by comparing available information with the broader strategic picture. One of the most likely reasons is the officer's (respondent) eagerness to endanger the safety of their unit: they believed the available information was insufficient, which led them to call off the offensive operation. Such a phenomenon may be termed "*professional heuristics*": choices made based on experience accumulated over time, even when the underlying reasoning cannot be articulated explicitly at the time (Lukyanenko et al., 2022, p. 2005). Regarding the last sub-question on general attitudes toward AI in military operations, responses consistently emphasise the need to preserve an officer's decision-making authority, particularly in high-risk operations. At the same time, officials exhibit a degree of curiosity and a desire that AI will make working in a complex environment easier. It seems that police adjust their understanding of trust and danger as a heuristic, depending on the situation, when reviewing AI recommendations.

This practice is essential to the system's development, as such professional inquisitiveness can lead to the emergence of new technical and tactical applications. Although the findings cannot be generalised to the entire population, they provide empirical evidence for existing studies on trust in AI (Lukyanenko et al., 2022) and risk propensity (Haerem et al., 2011). An unexpected finding is that, despite high levels of self-reported technical competence among some respondents, distrust persists, rooted in what appears to be a lack of basic knowledge about AI use. The given attitude may be justified by the study of Lukyanenko (et al. 2022, pages 1997-1998), who believes that the adoption of AI depends not only on individual previous knowledge but also on organisational and culturally determined experiences, including having negative or positive experiences with AI in civilian settings. This scepticism presents a particular challenge to the development of Human-Machine Teaming (HMT) (Warren & Hillas, 2020). The development of HMT cannot be based solely on technological overlap, as officers must be provided with a higher level of functionality within an even more digitalised operational environment, which is necessary to improve users' conceptual understanding of the system.

In short, the findings suggest that the respondents hold a moderate degree of mistrust of AI systems when applied to military operations, even in complex situations in which human suffering and death are significant concerns. Although there is an overwhelming majority supporting the statement that the application of AI should never be allowed to make independent decisions that involve lethal control, the vignette applications indicate a particular twist. Although there are circumstances under which the system can be used, there must be a mechanism for restoring authority over the AI (v2). Nevertheless, effective implementation of AI in military operations requires users to understand the rationale behind the courses of action generated; today, the black-box phenomenon is a significant impediment to this process. Therefore, subsequent phases of AI implementation in the Armed Forces should, where possible, consider the introduction of Explainable Artificial Intelligence (XAI) to mitigate this obscurity. The general aim of this investigation has been to define the trust and risk perceptions of Indian armed officers regarding the applicability of AI in the military. The work constitutes a professional contribution, supported by its consistent identification of both the challenges and the opportunities that AI may present for future development in the Armed Forces), (Pietrzak, 2023). The key benefit of formulating this investigation as a vignette study, rather than a conventional survey, is that respondents can consider a broader range of parameters before deciding whether to trust the AI system. The retrospective analysis reveals that, although control questions were included, nuances in interpretation distinguished the vignettes from the controls. To address this gap in contemporary replications of this study, alternative rating scales, such as "*do you agree that...*," would have improved the measurement of these nuances.

A vignette study was intellectually engaging, but it required substantial labour to analyse the results. To be practical for a master's thesis, the technique requires considerable ambition and a clear understanding of the research field, particularly in developing vignettes. With hindsight, the adherence to Sheringham et al. (2021, p. 4)'s conditional criteria for a robust study was perhaps taken too seriously; a smaller sample size of one to two vignettes followed by follow-up questions would probably have been more effective. In particular, the strategic decision to obscure the study's purpose was effective, as evidenced by responses that attempted to explain the purpose or to criticise the formulation of the questions. One criticism of the methodology employed is that the vignettes included too many variables, and the process of composing the responses into a single data set posed a challenge (Pietrzak, 2022). All variables in a vignette study are challenging to operationalise; therefore, sufficient time should be allocated during the development stage. Further applications of this approach should focus on process functionalization, which would involve using two types of vignettes: a text-interpreting vignette (as applied in this case) and a picture-interpreting vignette, to improve efficiency.

There was a need to provide a methodological balance in interpreting the findings that could not be achieved through statistical analysis using a program such as

SPSS. These critical nuances, evident in the free-text responses, would have been obscured by overreliance on statistics. Although this method may be considered in future versions, supplemented with AI-based software, including NVivo, to organise and analyse qualitative data, it was ethically questionable to conduct this research, as the data were viewed as potentially being stored and accessed by the researcher. To conclude, a distinctive strength of vignette studies in the field of war science is that they allow researchers to pose questions that more closely reflect the complexity of modern reality, rather than focusing on historical details. The details described in the answers help us better understand the complexity of war. The accompanying drawback is that the study may be empirically cumbersome. However, more vignette research or vignette experiments should be used in military science. Moreover, such methodology promotes multidisciplinary investigations and the reconciliation of leadership, ethics, and philosophical perspectives on war.

CONCLUSION

Artificial intelligence in the military field has great potential for further academic research. It is suggested that future studies build on this work to conduct comparative analyses of officer corps across countries at the macro level (ecological study). However, it is also recommended that the causal factors underlying variations in the concept of control (which were outside the scope of the current study) be identified. Moreover, users' ethical and moral paradigms regarding the use of AI across different levels should be subject to scrutiny in future studies, with attention to the tactical use and the ramifications of AI for military-strategic courses of action. The existing results provide several directions for future study, particularly regarding how officer training programs should manage AI and disruptive technologies. One aspect that should be critically highlighted, albeit not previously mentioned, is that the militarisation of AI is both a current target and a future reality. Western academia can pose a threat, but world defence organisations will continue to act in accordance with government directives. Thus, in the absence of globally binding laws governing the development of military AI systems, the global community will face a dilemma similar to that posed by nuclear proliferation. Therefore, this discussion ends by going back to the starting point, the open letter of Tegmark, as a stern warning not to initiate an arms race of AI before putting in place effective control systems. One of the primary existential questions is: when two parties to a dispute use the same AI systems, is it the machine or the human operator that ultimately determines whether people will survive?

REFERENCES

Antebi, L. (2021). *Artificial intelligence and national security in Israel*. Institute for National Security Studies (INSS).

Bostrom, N. (2014). *Superintelligence: Paths, dangers, strategies*. Oxford University Press.

Der Derian, J., & Rollo, S. (2023). Quantum warfare. In *Routledge handbook of the future of warfare* (pp. 319–329). Routledge. DOI: 10.4324/9781003299011-34

Fels, E. (2023). Prospects of great power rivalry: Escaping the tragedy? In *Routledge handbook of the future of warfare* (pp. 377–388). Routledge. DOI: 10.4324/9781003299011-40

Finlan, A. (2021). The shape of warfare to come: A Swedish perspective 2020–2045. *Defense and Security Analysis*, *37*(4), 472–491. DOI: 10.1080/14751798.2021.1995976

Forden, G., Podvig, P., & Postol, T. A. (2000). False alarm, nuclear danger. *IEEE Spectrum*, *37*(3), 31–39. DOI: 10.1109/6.825657

Hadlington, L., Binder, J., Gardner, S., Karanika-Murray, M., & Knight, S. (2023). The use of artificial intelligence in a military context: Development of the attitudes toward AI in defense (AAID) scale. *Frontiers in Psychology*, *14*, 1164810. DOI: 10.3389/fpsyg.2023.1164810 PMID: 37213385

Haerem, T., Kuvaas, B., Bakken, B. T., & Karlsen, T. (2011). Do military decision makers behave as predicted by prospect theory? *Journal of Behavioral Decision Making*, *24*(5), 482–497. DOI: 10.1002/bdm.704

Holland Michel, A. (2020). *The black box, unlocked: Predictability and understandability in military AI*. United Nations Institute for Disarmament Research (UNIDIR).

Husain, A. (2021). AI is shaping the future of war. *Prism*, *9*(3), 50–61.

Hynek, N., & Solovyeva, A. (2022). *Militarizing artificial intelligence*. Routledge. DOI: 10.4324/9781003045489

Innes, J. M., & Morrison, B. W. (2021). Artificial intelligence and psychology. In *The Routledge social science handbook of AI* (pp. 30–57). Routledge. DOI: 10.4324/9780429198533-4

Johnson, J. (2023). Automating the OODA loop in the age of intelligent machines: Reaffirming the role of humans in command-and-control decision-making in the digital age. *Defence Studies*, *23*(1), 43–67. DOI: 10.1080/14702436.2022.2102486

Kahneman, D., & Tversky, A. (1979). Prospect theory: An analysis of decision under risk. *Econometrica*, *47*(2), 263–292. DOI: 10.2307/1914185

Kania, E. B. (2019). Chinese military innovation in the AI revolution. *RUSI Journal*, *164*(5–6), 26–34. DOI: 10.1080/03071847.2019.1693803

Kissinger, H., Schmidt, E., & Huttenlocher, D. (2021). *The age of AI: And our human future*. John Murray.

Lidskog, R. (2020). Samhället utmanat? Artificiell intelligens och sociologisk kunskap [Society challenged? Artificial intelligence and sociological knowledge]. *Sociologisk Forskning*, *57*(2), 99–120. DOI: 10.37062/sf.57.19591

Lukyanenko, R., Maass, W., & Storey, V. C. (2022). Trust in artificial intelligence: From a foundational trust framework to emerging research opportunities. *Electronic Markets*, *32*(4), 1993–2020. DOI: 10.1007/s12525-022-00605-4

Maas, M. M. (2019). How viable is international arms control for military artificial intelligence? Three lessons from nuclear weapons. *Contemporary Security Policy*, *40*(3), 285–311. DOI: 10.1080/13523260.2019.1576464

Nyholm, S., & Smids, J. (2016). The ethics of accident-algorithms for self-driving cars: An applied trolley problem? *Ethical Theory and Moral Practice*, *19*(5), 1275–1289. DOI: 10.1007/s10677-016-9745-2

Pietrzak, P. (2022). *Introducing the idea of ontology in statu nascendi to the broader international relations theory. International Conference Proceeding Series: International Conference on Economics and Social Sciences*. ECLSS., https://www.eclss.org/publicationsfordoi/abst11act8boo8kIE%26SS2022_antalya.pdf

Pietrzak, P. (2023). Approaching regional conflicts through the prism of ontology in statu nascendi: The new compartmentalization of the IR theory. *In Statu Nascendi – Journal of Political Philosophy and International Relations, 6*(1).

Pietrzak, P. (2024). The sixth great debate in international relations theory revolves around clarity: Ontology in statu nascendi as a cutting-edge tool for conflict management. In Pietrzak, P. (Ed.), *Dealing with regional conflicts of global importance* (pp. 1–27). IGI Global., DOI: 10.4018/978-1-6684-9467-7.ch001

Png, M. T. (2022). At the tensions of South and North: Critical roles of global South stakeholders in AI governance. *ACM International Conference Proceeding Series*. DOI: 10.1145/3531146.3533200

Raska, M. (2021). The sixth RMA wave: Disruption in military affairs? *The Journal of Strategic Studies*, *44*(4), 456–479. DOI: 10.1080/01402390.2020.1848818

Roberts, H., Cowls, J., Morley, J., Taddeo, M., Wang, V., & Floridi, L. (2021). The Chinese approach to artificial intelligence: An analysis of policy, ethics, and regulation. *AI & Society*, *36*(1), 59–77. DOI: 10.1007/s00146-020-00992-2

Rosert, E., & Sauer, F. (2019). Prohibiting autonomous weapons: Put human dignity first. *Global Policy*, *10*(3), 370–375. DOI: 10.1111/1758-5899.12691

Scharre, P. (2018). *Army of none: Autonomous weapons and the future of war*. W. W. Norton & Company.

Sheringham, J., Kuhn, I., & Burt, J. (2021). The use of experimental vignette studies to identify drivers of variations in the delivery of health care: A scoping review. *BMC Medical Research Methodology*, *21*(1), 1–17. DOI: 10.1186/s12874-021-01247-4 PMID: 33888077

Sullivan, Y., de Bourmont, M., & Dunaway, M. (2022). Appraisals of harms and injustice trigger an eerie feeling that decreases trust in artificial intelligence systems. *Annals of Operations Research*, *308*(1–2), 525–548. DOI: 10.1007/s10479-020-03702-9

Taylor, R. D. (2020). Quantum artificial intelligence: A "precautionary" U.S. approach? *Telecommunications Policy*, *44*(6), 101909. DOI: 10.1016/j.telpol.2020.101909

Tegmark, M. (2017). *Life 3.0: Being human in the age of artificial intelligence*. Alfred A. Knopf.

Tversky, A., & Kahneman, D. (1992). Advances in prospect theory: Cumulative representation of uncertainty. *Journal of Risk and Uncertainty*, *5*(4), 297–323. DOI: 10.1007/BF00122574

Wakker, P. P. (2010). *Prospect theory: For risk and ambiguity*. Cambridge University Press. DOI: 10.1017/CBO9780511779329

Warren, A., & Hillas, A. (2020). Friend or frenemy? The role of trust in human-machine teaming and lethal autonomous weapons systems. *Small Wars & Insurgencies*, *31*(4), 822–850. DOI: 10.1080/09592318.2020.1743485

Wasilow, S., & Thorpe, J. B. (2019). Artificial intelligence, robotics, ethics, and the military: A Canadian perspective. *AI Magazine*, *40*(1), 37–48. DOI: 10.1609/aimag.v40i1.2848

ADDITIONAL READING

Kaushal, S., Lynch, J., Suess, J., Lee, J.-J., Vannurden, L., & Bajraktari, Y. (2024). *Leveraging human–machine teaming*. Royal United Services Institute (RUSI).

Marsili, M. (2023). Military emerging disruptive technologies: Compliance with international law and ethical standards. In Kalpokas, I., & Kalpokienė, J. (Eds.), *Intelligent and autonomous: Transforming values in the face of technology*. Brill. DOI: 10.1163/9789004547261_004

Patel, J., Boardman, M., Gregory, F., Lamb, S., Sarkadi, S., Tesic, M., & Yeung, N. (2024). Give us a hand, mate! A holistic review of research on human-machine teaming. *BMJ Military Health*. Advance online publication. DOI: 10.1136/military-2024-002737 PMID: 39719381

Renic, N., & Schwarz, E. (2023). The ethical legitimacy of autonomous weapons systems: Reconfiguring war accountability in the age of artificial intelligence. *Security Studies*, *32*(4), 1–28.

Riesen, E. (2022). The moral case for the development and use of autonomous weapon systems. *Journal of Military Ethics*, *21*(2), 132–150. DOI: 10.1080/15027570.2022.2124022

Umbrello, S., Torres, P., & de Bellis, A. F. (2020). The future of war: Could lethal autonomous weapons make conflict more ethical? *AI & Society*, *35*(1), 273–282. DOI: 10.1007/s00146-019-00879-x

Walker, P. (2021). Leadership challenges from the deployment of lethal autonomous weapon systems: How erosion of human supervision over lethal engagement will impact how commanders exercise leadership. *RUSI Journal*, *166*(1), 10–21. DOI: 10.1080/03071847.2021.1915702

KEY TERMS AND DEFINITIONS

Vignette Study: A study design involving the use of hypothetical, realistic situations (vignettes) that are intended to assess how respondents use their values and judgment when making decisions to allocate funds in complex settings. This will outwit the underlying forms of bias and decision-making that are opposed to typical survey questions.

Human-in-the-Loop (HITL): One control idea is that the human operator should never relinquish ultimate control over AI decisions, as this would provide moral and tactical control. This is especially essential in life-and-death operations to hold oneself accountable in the administration of force.

Explainable AI (XAI): The introduction of artificial intelligence systems that allow clarity and concise comprehension of the reasoning behind their decisions, and not merely a solution. This openness helps users trust the system by demonstrating how it reached its conclusion.

Kinetic and Non-Kinetic Operations: Kinetic operations are those that employ physical force and ammunition (e.g., missiles, combat). Non-Kinetic operations are those that are non-physical (such as cyber warfare or information operations). The researchers found that officers are more likely to trust AI when it comes to the type of operation suggested.

Professional Heuristics: Rapid decision-making involving intuitive mental shortcuts in high-skilled, experienced officers on the basis of training and prior experience. These were the so-called gut feelings, and the officers used them to estimate AI proposals when they lacked all the information.

Context-Dependent Trust: The observation that trust in AI is not a constant feature (binary) but is different depending on the given situation. In one case, an officer may be willing to use AI to protect himself and fire a missile, but in another, such as a hostage mode, he may not.

Black Box Problem: The problem in which an artificial intelligence system is a black box and provides responses without disclosing the logic behind its decisions. This was identified as one of the biggest obstacles to officers' belief in the system.

Chapter 5
Political Strength and Cognitive Warfare:
How Democracies Radicalize

Iskren Ivanov
http://orcid.org/0000-0001-9967-6251
Sofia University St. Kliment Ohridski, Bulgaria

ABSTRACT

The driving forces behind the post-Cold War democratization stemmed from the assumption that building accountable and transparent institutions would secure peace. The failure of transitional regimes and polarization within democratic societies have resulted in a survivorship bias for the North Atlantic Treaty Organization, as such polarization within Western societies presents a considerable risk to their national security. The purpose of this chapter is to introduce a structural model of how cognitive warfare undermines political strength, thereby radicalizing liberal democracies. The study seeks to explain how cognitive warfare affects the strategic perceptions of the major NATO allies, bringing the Alliance one step closer to domestic fragmentation and institutional instability.

1. INTRODUCTION

The democratization of Eastern Europe and NATO's overseas missions after the end of the Cold War created favorable conditions for strengthening the political strength of the Alliance. However, many post-socialist Member States still lack an institutional architecture that is durable and transparent enough to counter the detrimental effects of emerging security threats, such as cognitive warfare. Added to those challenges is polarization in Western European democracies, driven by po-

DOI: 10.4018/979-8-3373-6264-9.ch005

litical radicalization. To use the terminology of the classical International Relations theories, NATO allies today pursue what Barry Buzan posits as security – the ability of societies to maintain their independent identity and functional integrity (Buzan, 2015, p. 130). But the stronger their political will to preserve security, the greater the likelihood that radicalization will affect democratic societies.

The purpose of this chapter is to explain how cognitive warfare corrupts the political strength of democratic societies and radicalizes their security perceptions. Most previous studies assess radicalization through the lens of terrorism, claiming that it is either a pattern to the understanding of the latter or a problematic concept, which cannot explain the relation between extremist political beliefs and violent political action (Neumann, 2013, p. 880) (McCauley & Moskalenko, 2008, p. 425). Less attention has been paid to state-sponsored radicalization, with the most recent studies focusing on how it affects terrorism prevention strategies in the United States and Europe (Carter et al., 2020, p. 186). Some scholars even partially reject the correlation between radicalization and politics, claiming that the concept of violent extremism allows policymakers to create security narratives about terrorism (Githens – Mazer, 2012, p. 560). Although these earlier studies make considerable contributions to the academic debate, they do not fully address the mechanisms by which cognitive warfare radicalizes political perceptions in democracies, nor do they consider essential variables of radicalization beyond motivation and behavior, such as societal polarization.

This chapter argues that the weakening of the political strength of democracies due to domestic polarization represents a central challenge to NATO rather than its potential involvement in military conflicts. Although the imminent danger of war generates insecurity, this fear looms especially strongly when trust in democratic institutions is declining. In such situations, decision-makers in democracies typically rely on radical narratives to boost public support, even as their policies become increasingly ineffective. Major challenges to NATO's security deepen those effects, further weakening the political strength of institutional decision-making and setting the preconditions for strategic disagreements among the allies. The early stages of this process trigger considerable polarization in democratic societies, while in its advanced phase, cognitive warfare uses the advantages of democracy against itself.

2. COGNITIVE WARFARE AND RADICALIZATION

There is a growing body of literature discussing the concept of cognitive warfare, with the most recent studies focusing on its complex nature. Scholars who deal with security studies point out three essential aspects of this phenomenon: manipulation of perceptions by using soft power and cyber technologies, targeting the population through information, and proliferating false narratives (De Swielande et al., 2024) (Backes and Swab, 2019). Another group of authors introduces a more comprehensive approach, defining cognitive warfare as deliberate offensive maneuvers aimed at influencing perceptions, beliefs, interests, decisions, and behavior by directly targeting the human mind and disrupting the process of immediate information (Marsili, 2023) (Tooby & Cosmides, 2025). A third volume of works to succinctly capture the sources of emerging threats draws a relation between Russia's information operations, designed to influence the public opinion in Western democracies, and China's utilization of artificial intelligence for boosting state-sponsored propaganda overseas (Hung & Hung, 2022) (Paziuk et al., 2025).

Those observations are further supported by studies that show a correlation between cognitive warfare and radicalization in democratic societies. Empirical findings and essential contributions indicate that open societies are highly vulnerable to cognitive propaganda and that authoritarian regimes have endorsed narratives to shape perceptions of democratic institutions: populism, migration, re-nationalization, ethnic rhetoric, and antisemitism (Herschinger et al., 2020). Similar research calculates that right-wing, left-wing, and Islamist extremism are equally detrimental to the security of democratic societies because of their polarizing effects and likelihood to advocate political violence (Jasko et al., 2022). Despite differences in their methodological approach and conclusions, the authors observe that radicalization should not be limited to domestic and transnational terrorism but should also be considered in the context of democratic stability and institutional credibility. It is also essential to highlight that, according to the mentioned studies, democracies remain vulnerable to emerging threats, whereas autocracies, which lack pluralism, diversity, and freedom of expression, are not. Thus, defining cognitive warfare is essential to national security.

2.1. Defining Cognitive Warfare

This chapter defines cognitive warfare as the premeditated use of psychological, technological, and ideological tools to influence cognitive perceptions and decision-making at the individual, group, and state levels.

The initial stage of the process employs democratic pluralism and societal debates to foster polarization among groups and weaken the political strength of decision-making institutions. Psychological tools, including false or true narratives, are likely to be highly effective when targeting the population at the individual level by increasing negative emotions through charged sentiments such as fear of an imminent military attack, political apathy, or public insecurity. Policymakers, who endorse specific behavior, serve as the intermediary link in the narratives, even if they reject them, as their struggle to mobilize popular support naturally leads to changes in public perceptions. For example, the narrative that democratic institutions are ineffective and incapable of defending their citizens serves as a powerful ideological justification for Russia's self-image as a victim of Western democratic expansion.

Once affected at the individual level, society enters a spiral of polarization, making it more vulnerable to the tools of cognitive warfare, such as fake news, misinformation, and propaganda. Combined with the digital effects of social media-driven operations and data harvesting, cognitive warfare weakens critical thinking and fosters conspiracies. To strengthen their impact, disinformation agents seek popular support, often spreading divisive narratives that turn recipients against each other. Exploiting the advantages of democracy thus provides cognitive warfare with the capability to establish a particular pattern of thinking in public discourse on issues such as foreign policy and to reduce the political strength of democratic governments. International Relations theories offer a useful parallel to the divisions within British society in the first months of World War II, when the Axis used domestic criticisms of the Chamberlain government's powers to launch Operation Weserübung (Jervis, 2017). At a more advanced stage, manipulating public discourse could even lead to violent clashes.

The final stage of cognitive warfare involves behavioral tools such as nudging. In the context of political psychology, nudging is defined as guiding the behavior of individuals, groups, or organizations to manipulate decision-making (Wilkinson, 2012, p. 350). Rogue actors succeed when institutions responsible for decision-making are weak or suffer from a high level of corruption. When democracy is mature, the society is not immune to the effects of nudging, although voters are more inclined to hold policymakers accountable for their decisions. This greater level of transparency, however, can be used by autocracies to nudge societal trust and its beliefs. The subsequent effects are caused by the mutual reinforcement of political and ideological narratives aimed at provoking institutional crises, such as conspiracy theories about global liberal elites ruling the democratic world or misleading self-images that depict extremist groups as freedom fighters.

Taken together, the three stages of cognitive warfare could cause an institutional crisis of democracy, resulting in a weak and divided society. In liberal democracies, polarization can undermine standard democratic norms, distort perceptions of conflict, and erode political strength and resolve to fight corruption. It seems likely that these implications emerge because democracies allow it. Some works in the field of political studies debate whether defending democracy requires a delicate balance between politicization and institutionalization, including the imposition of some restrictions (Stahl & Popp-Madsen, 2022, p. 315). The history of democracy suggests the opposite, as establishing transparent institutions takes time, while restrictions could jeopardize political pluralism. Preserving democracy thus depends on imposing more negative sanctions instead of preventing its radicalization.

Figure 1. Stages of cognitive warfare

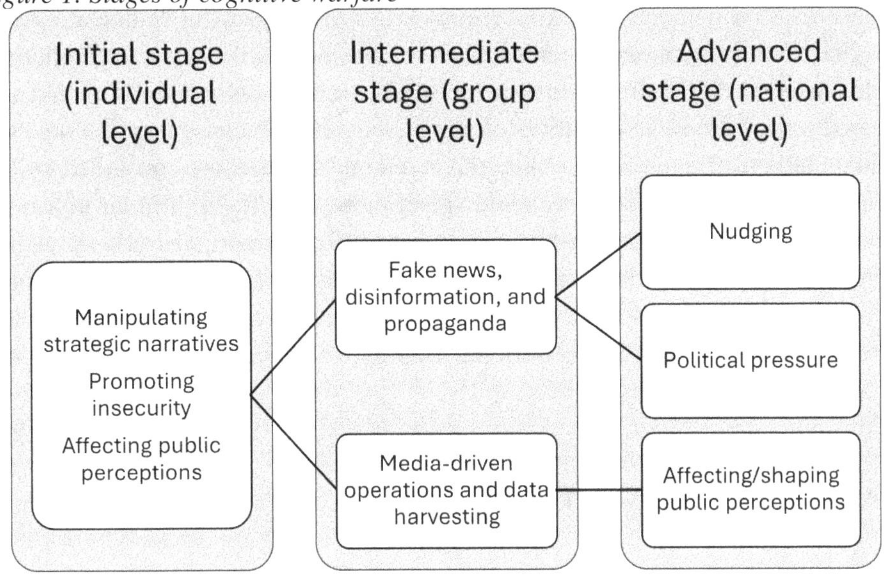

2.2. How do Democracies Radicalize?

How do democracies radicalize? This chapter asserts that the latter begins when, under the influence of cognitive warfare, certain groups in society feel threatened by democracy or feel vulnerable because institutions fail to respect their interests. Thus, the practices of democracy itself turn against its citizens: suffrage, political competition, and pluralism are perceived as ineffective or merely broken. Although policymaking remains transparent and decision-makers generate public support, anti-system rhetoric is more likely to grow alongside political favoritism and corruption. Strategic narratives – both true and false – serve as a basis for excluding essential policies from the national security agenda, while nudging produces beliefs that the current state of democracy should be reformed to better align with the corresponding geopolitical realities. Polarized democracies then begin to experience an institutional crisis as public trust erodes. The political debate then becomes incapable of integrating most of the interests and visions of the interest groups, resulting in increasing tensions among them, a democratic deficit, and the spread of corruption.

Added to those effects is the global shrinkage in the middle class. The widening gap between oligarchs and low-income citizens poses a long-term challenge to democratic regimes, especially among European allies seeking to allocate more financial support to their armed forces and defense. Notable scholars have long debated whether democracies can survive the decline of the middle class when facing a large coalition of autocracies, arguing that the absence of a pattern to preserve it leads to global erosion of democracy. The fact remains that the majority of democratic elites today have rational economic interests in advocating increased military expenditures, while others are dissatisfied with these demands. Consequently, public opinion and affected groups adopt the narratives of cognitive warfare because they feel only the political effects of democracy, while the economic benefits are consumed by the debate of whether democracies should spend more on defense. Similar processes occurred in the United States in the 1980s, when the Reagan administration launched its Strategic Defense Initiative to deter the Soviet Union. This is not to say that the global shrinkage of the middle class should be attributed to cognitive warfare, but rather to stress the fact that autocracies have used it to boost the effects of their anti-democratic narratives.

3. AGENTS OF COGNITIVE WARFARE

It is difficult to generalize about specific patterns of cognitive warfare, although many definitions indicate a growing relationship between non-traditional security and political psychology. This chapter outlines two groups of cognitive warfare agents, based on their nature and origins. That being said, both state and non-state actors could use cognitive tools to attack democracies and polarize their societies.

3.1. State Agents of Cognitive Warfare

State agents of cognitive warfare include authoritarian and totalitarian regimes that pose a long-standing threat to the national security of democratic states. When such a regime collapses – as the USSR and Saddam's Iraq did – it leaves behind political or paramilitary groups, which either form another state actor or a network, endorsing political violence as its cause. State-sponsored cognitive tools are more successful than those used by non-state actors due to their greater complexity. Authoritarian elites may even adopt official political rhetoric of confrontation with democracies if they get enough support from their partners. Consequently, authoritarian coalitions, especially when they are in possession of nuclear weapons, become difficult to contain because, in contrast to arguing democracies, they have been united in their purpose to dethrone the United States and the system of alliances it built after World War II.

Russia has utilized one form of cognitive warfare to affect policymaking and cognitive perceptions in democracies. Most of the studies in that field rely either on the Gerasimov doctrine as an explanation of the Kremlin's hybrid strategies or on its populist rhetoric (Galeotti, 2018). Less attention has been paid to psychological attitudes, which – as part of cognitive warfare – predetermine its initial use by a state agent. False narratives include: the picture of Russia as a victim of Western aggression, nuclear blackmail, and the prospect of the United States abandoning its European allies due to the foreign policy of future Presidential administrations. The final purpose of those messages is two-fold: to undermine trust among NATO allies and justify its war against Ukraine under the guise of a special military operation. The polarization within European societies testifies to the initial success of this strategy, although the support for NATO remains high. Behavioral tools, quite the opposite, are the strongest aspect of the Russian cognitive approach, especially in Eastern Europe, where right-wing populists typically endorse religious rhetoric to depict Russian values as an ideological alternative to the liberal, collective West. Nudging public perceptions involves manipulating the relations between the state and religious institutions, as well as creating various domestic networks to secure the spread of certain values, represented as traditional and sacred.

In contrast to Russia, China's cognitive approach is much more refined because it is based on authentic narratives. An essential methodological clarification is needed to demonstrate that the U.S.–China tensions over Taiwan should not be used to infer a pattern in China's cognitive strategy. The Taiwan issue stands out from the general pattern of Beijing's cognitive apparatus because it is the only red line beyond which a military conflict with the United States becomes highly probable. It is, therefore, an isolated case that should be treated within the context of Sino-American bilateral relations. False narratives do not constitute the centerpiece of China's foreign influence mechanisms, though they are often present. The sources of Beijing's cognitive influence can be traced to the objective geopolitical reality that the international system is in transition. However, Chinese foreign policy exploits the crisis of unipolarity to offer the other authoritarian governments an alternative to liberal democracy and to form an anti-U.S. coalition. Its psychological tools aim to achieve a positive effect, similar to U.S. soft power, and, combined with anti-American sentiment, secure voluntary submission from major U.S. adversaries, such as Russia and North Korea. Behavioral tools are then easier to apply, while China's technological capabilities set the preconditions for intensive armed races with Washington and its allies. The same logic applies to nudging that incorporates the image of the Chinese peacemaker, presumably based on its millennial experience with mediation in international disputes. The fight for public opinion, therefore, relies on narratives suggesting that China is prepared to assume the responsibilities of a great power, does not aim to replace the United States, and could serve as a geopolitical alternative to liberal democracies.

The Islamic Republic of Iran and the Democratic People's Republic of Korea offer the third example of state-sponsored cognitive warfare. In contrast to Russia and China, the influence of these countries is highly underestimated, with some scholars claiming that their scope of action is totally dependent on Beijing (Zoubir, 2023). This chapter finds such a presumption misleading because, as Jack Snyder puts it, cohesion in the foreign policy of totalitarian regimes does not necessarily presume full concurrence in their actions (Mansfield & Snyder, 2002, p. 300). It is also far from pragmatic to consider both actors less competitive than Russia and China, as Pyongyang is in possession of a nuclear arsenal, while Tehran enjoys the support of the transnational terrorist groups in the Middle East and North Africa. Psychological tools, endorsed by Iran and North Korea, are similar because they rely on promoting insecurity through spreading fear and misinformation about the credibility of deterrence. Worthy of observation are the nuclear rhetoric of Kim Jong-Un and Iran's bargaining over the nuclear deal. Although the support for a peaceful unification of the Korean Peninsula is declining in the Republic of Korea, public opinion is still very divided over the prospect of a military clash with the North. Pyongyang's manipulation tactics thus do not focus on the prospect of peaceful coexistence but rather on undermining the trust of South Koreans in the United States and its commitments to Seoul. Iran, on the other hand, could not use nudging on the Israeli public opinion since it is predominantly united on issues such as self-defense. Tehran's approach instead stresses the funding of non-state actors in MENA to shape a deterrent ring against Israel and enter its government in a state of permanent insecurity.

3.2. Non-Stage Agents of Cognitive Warfare

By non-state agents of cognitive warfare, this chapter classifies global terrorist networks, paramilitary forces, or violent movements posing a long-term threat to the national security of democracies. The debate about the motivation of radicalized terrorists does not offer many contributions when discussing cognitive warfare, since scholars have consistently agreed that political violence against civilians should be considered terrorism regardless of its ideological roots. Labeling terrorism could not provide us with further arguments to assess the impact of cognitive tools on democratic societies. Defining what type of radicalization is present in society, quite the opposite, could provide us with a plausible starting point to infer conclusions about the nature of its cognitive tactics.

Based on their profile and motivation, the first group of non-state agents includes left-wing radicalized networks. The existence of capsulated, socially excluded, and marginalized groups in Western societies leads to their radicalization and endorsement of political violence as a cause. Left-wing extremism does not necessarily involve radicals, who are marginalized, although many of the individuals who have embraced it belong to minorities with such a history. However, the sense of isolation and discrimination in the latter is much stronger than in the former.

Combined with the freedoms provided by democracies, those perceptions become a powerful stimulus for violent actions, which – in most cases – target civilian infrastructure or particular groups in the society. Nudging positive emotions and perceptions of belonging is central to the cognitive strategy of such groups, while social networks offer an excellent technological tool for spreading their narratives. A striking similarity between left-wing radicalization and political Islam is that the collective self-awareness of both is essential to the success of the non-state agents. The ultimate purpose of cognitive warfare in such cases is to distort the perceptions of individualism, which is part of the political framework in mature democracies.

Right-wing extremism constitutes the second ideological profile of non-state agents, and its emergence is predetermined by nudging perceptions of democracy itself. One example of such manipulation is the state-sponsored propaganda of Orthodoxy, used by Russia to influence the attitudes of Eastern European allies such as Bulgaria, North Macedonia, and Greece. Cognitive strategies based on religious narratives are difficult to detect because the line between freedom of religion and radicalization is fragile. One of Russia's successes in affecting public opinion was promoting the assumption that liberal democracy is, by its nature, hostile to religious traditions because it encourages pluralism and peaceful coexistence between all ethnic groups. Some of the latest studies in that field also indicate a growing trend of re-nationalization as a response to left-wing extremism (Pickel, 2022). The competitive interaction between the two ideologies results in severe societal division in Europe and the United States, contributing to violent acts that violate democratic principles, such as freedom of speech and expression.

The threat posed by non-state actors to democracies can thus be understood within the context of the right–left cleavage. In contrast to the ideological divisions that characterize democratic societies, this cleavage poses contemporary democracy with two significant challenges. One is that European and U.S. policymakers should find a path to reconcile their societies without restricting the freedoms of their citizens, as granted under their constitutions and laws. The other refers to the deterrence of radical narratives spread by certain interest groups in democratic societies, which reflect much of the far-right and far-left ideological platforms. In other words, the ultimate challenge of liberal democracy today is to dismantle the spiral of self-defeating behavior that has polarized democratic societies for the past two decades and to consolidate public opinion around priorities of common interest, such as countering the assertive policies of autocracies.

3.3. Radicalization on the Individual Level

Tracing radicalization at the individual level requires dismantling the misleading claim that this process is problematic on theoretical and methodological grounds

(Sedgwick, 2010, p. 480). The concepts of terrorism and political violence cannot replace the empirical share of radicalization in cases such as violent protests, organized attacks against civilians, and mass damage to civilian infrastructure. As those threats are, they are no less relevant to the existence of contemporary democracies than the September 11 attacks. Once combined with the effects of polarization, the consistent projection of violence by state and non-state agents of radicalization could undermine trust in democracy and the credibility of collective security systems like NATO.

Radicalization of policymakers is the first and most prominent manifestation of anti-democratic behavior, often endorsed by populists to generate mass resistance against democracies. Populism, of course, has long been part of the political discourse, but when calling for violence or a decision-making boycott, it could become a source of radicalism. The preconditions for this form of radical behavior can be traced to psychological tools of cognitive warfare and the financial support provided by state agents to populist movements that emerge as promising political actors. If funding is the sole reason for shaping anti-democratic narratives, those narratives should not be considered a realistic security threat, since democracies have often done the same during the Cold War to contain the USSR. However, when policymakers advocate violent behavior or spread misinformation by nudging the perceptions of their voters, they practically use their political freedoms to weaken democracies from the inside. A growing number of quantitative studies show that the success of cognitive warfare in such cases is limited to marginal groups, which inevitably transform into non-state agents of radicalization. This chapter supports the conclusion of the authors because, as the political climate in Europe testifies, populism, promoted by democratic opposition, is distinct from that spread by the extremist movements. Many political parties would boost anti-EU or anti-NATO sentiment, but a few would seek to replace the U.S. nuclear umbrella or the European economic support with another sense of collective security.

Radicalization at the individual level also involves political socialization. The process of perceiving and adopting beliefs and values about how the state should be run is an essential aspect of every society. However, democracies are not immune to the effects of radical behavior because socialization occurs through education, group affiliations, and the development of each person's social instincts, while political socialization itself is highly vulnerable to radicalism. Individuals who succumb to the psychological and behavioral tools of cognitive warfare radicalize not because of democracy, but rather due to the lack of resilience. The leadership crisis and the inability of policymakers to unify public opinion erode the institutional mechanisms for conflict resolution, thereby making authoritarian regimes appear more attractive and natural successors to liberal democracy. Cognitive tools successfully nudge individual perceptions, leading to behaviors more likely to endorse violence as a primary means of social transformation and political change.

Finally, it is essential to highlight that the lack of democratic resilience allows non-state actors to influence the educational agenda in democracies and radicalize its pillars, which are the primary sources of political socialization for young generations. More recent studies in both the critical social sciences and the liberal arts demonstrate that education is the latest frontline in the fight against radicalization, due to its collective impact and the freedom of expression it provides in academia. Right-wing and left-wing extremism could easily penetrate the educational system, while imposing barriers on the latter is a red line of cultural censorship. Furthermore, cognitive warfare could succeed if a democratic state's policy encourages historical revisionism as a source for national pride and unity among students. Policymakers and populists, notable critics of multiculturalism, often criticize diversity policy. Meanwhile, proponents of liberal education encourage a specific teaching philosophy, with both camps polarizing the perceptions of essential political and cultural phenomena among younger generations. Mirrored by the contemporary debates in democratic societies, this division leaves the educational system vulnerable to radicalization.

4. COGNITIVE WARFARE AND POLARIZATION

This section focuses on two distinct cases of cognitive warfare: Russia's hybrid warfare strategy in Europe and China's strategy to polarize American society ideologically. Although the concept of hybrid warfare has been consistently operationalized to assess the means by which Moscow targets European institutions, the existing threats to NATO's security should not be limited to propaganda and fake news. The same logic applies to China, where isolated cases such as the Taiwan issue or Beijing's economic aspirations for global leadership are not sufficient to assess how Chinese cognitive tools operate. External threats and challenges to democracies have long been treated as a primary source of insecurity, but that logic no longer applies to the post-Cold War geopolitical environment, where autocracies and totalitarian regimes seek to undermine trust in the democratic government from inside.

4.1. Russia: From Hybrid Warfare to Right-Wing Radicalization

The natural political instinct of democracies primarily characterizes their resilience to self-correction and development, in contrast to autocracies, where the political formula is based on fear of repression and state planning. Recent studies on the Russian invasion of Ukraine conclude that it was preceded by a prolonged and consistent informational warfare, combined with funding paramilitary groups and generating divisions within the society (Pietrzak, 2023). Moscow's provocations would have a different impact if they were directed against a NATO ally, and – as long as each of them remains firm in its deterrence strategy – the Kremlin will need a larger-scope strategy to divide the allies. This chapter assumes that right-wing polarization resembles the tactical matrix of Russian warfare and that its purpose is to pave the way for further provocations against NATO and its partners. The model of this matrix stands as follows:

- Psychological tools: targeting beliefs, credibility, and cognitions of the NATO allies on individual, group, and state levels
- Technological tools: combining cyberattacks on NATO's critical infrastructure with nudging in social networks and designing troll farms
- Behavioral tools: establishing a specific type of decision-making and political behavior

Polarizing public opinion in European democracies is the primary target of Russian cognitive warfare because the lack of reforms and democratic deficit has already compromised the credibility of the EU's institutions. Targeting European beliefs involves not only challenging democracy as a concept but also promoting the vision that its decline is inevitable in the short term, alongside the disintegration of the United Europe and its military alliances. Although this approach has no final direction, its foremost purpose is to demoralize the Europeans, erode the foundations of European supranationalism, and boost the trust in Europe of nations. The narratives attributed to them are fabricated portrayals of NATO as a weak military alliance, incapable of supporting its member states because of its lack of credibility. The presumed inability of the United States to defend its allies occupies a central place in Russian nuclear rhetoric, as well as in its initially favorable, then critical, assessment of U.S. President Donald Trump's foreign policy. NATO's increasing defense spending indicates the opposite – that the Alliance is capable of defending its members and that divisions within its collective security system stem from political disputes, not from existential dilemmas about its existence.

The effects of cyberattacks are typically amplified by nudging in social networks, as seen in the most recent cases involving the manipulation of public discourse in Eastern Europe, where Russian influence has been consistent for decades. In the years after the outbreak of the Ukrainian war, hacker groups like Fancy Bear and Killnet targeted institutional infrastructure and critical data of governmental archives, while law enforcement authorities have detected direct threats to the lives of political officials. A similar strategy is present in Romania and the Baltics, where troll farms have established a standard frontline to distort the public perceptions of support for

Ukraine or NATO's involvement in the war based on fake news and false narratives. Social networks also remain the primary environment for pro-Russian extremists to promote their ideas, but after Russian banks were banned from SWIFT, most of the personal statements issued by radicalized intellectuals or far-right supporters have been funded by China. This is not to say that Russian influence has declined, but rather to highlight that it has merged with pro-Chinese lobbies in Europe. Illustrative cases of nudging on social media include spreading conspiracies and combining them with partially true narratives about NATO's future, the imminent threat of nuclear war with Russia, and incorrect comparisons to World War II to predict how the war in Ukraine might end. Many of the pro-Russian groups in social networks actively advocate anti-system behavior to provoke pro-revolutionary sentiments among their followers and motivate them to perpetrate violent acts.

Establishing right-wing extremism as a predominant pattern of political behavior is the final purpose of Russian cognitive warfare. So far, it has been successful in isolated cases, such as the adoption of a far-right agenda by certain political parties and civil movements. An increasing trend toward promoting radical ideas can be observed among academics, who use their academic freedom to shape anti-Western attitudes among young intellectuals and to attract them to pro-Russian lobbies. Right-wing-sponsored groups also recruit teachers and professors to use their access to school and academic auditoriums, presumed to be depoliticized. State agents of radicalization provide channels of communication, later discovered to be employed for building troll farms and Facebook groups for promoting and discussing far-right ideas. The argument that democratization is responsible for the global conflicts and that Western values are a threat to the sovereignty of the European nations are the dominant narratives, constituting the far-right agenda in Eastern Europe, while calls for political violence are limited to damaging civilian infrastructure during anti-governmental protests or exercising hate speech against ethnic minorities. One of the most alarming trends remains the employment of antisemitic rhetoric, combined with conspiracy theories to justify the alignment of right-wing extremism with Iran's proxies in the Middle East. It is important to stress the fact that contemporary forms of prejudice and hostility towards certain ethnic groups in Western democracies continue to evolve and that their spread remains partially predetermined by far-right ideologies such as the neo-eurasianism, which treats the West as a corrupted collective political entity, destined to fight a war with Orthodox Russia.

Figure 2. Russian approach of cognitive warfare

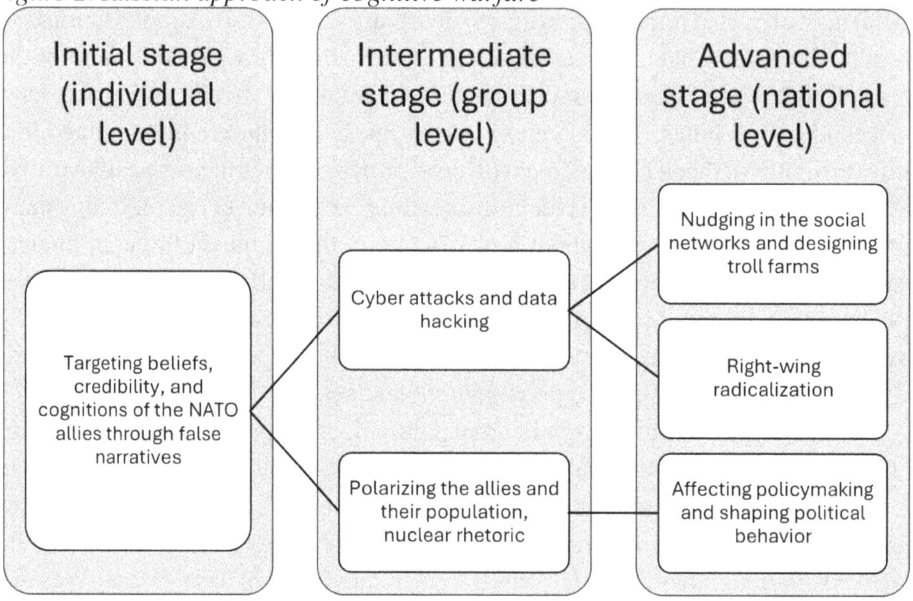

4.2. China: From Soft Power to Left-Wing Radicalization

The image of China as a pacifier dates back to the Civil War, as the Chinese vision of the international system has always positioned it as the heartland. Both contemporary Chinese and Western scholars agree that the rise of China is a natural process, but they differ on how peaceful Beijing will be when discussing its strategic competition with the United States (Yan & Qi, 2025, p. 5) (Pietrzak, 2024). This chapter joins the academic debate by asserting that, as long as China's red lines in the Taiwan Strait have not been crossed, soft power remains the primary tool of Chinese influence in the Sino-American strategic competition. In contrast to the Russian strategy of cognitive warfare, China relies much more on its economy and interdependence with Washington, while its soft approach has two pillars: to preserve and promote a positive, pacifist view of Beijing and convince Western powers that it does not seek to replace the United States as a global leader. Chinese policymakers consider cognitive approaches an essential part of state military strategy in many official documents approved by the Communist Party of China over the last two decades. Beijing's cognitive approach includes:

- Psychological tools: targeting beliefs and exercising personal intimidation on an individual level

- Technological tools: combining cyberattacks on the critical infrastructure of Western democracies with social media manipulation
- Behavioral tools: establishing a pacifist image of Chinese foreign policy by expanding its business contacts, intellectual and political impact on its partners

The psychological tools of Chinese soft power do not target society in general, but rather groups of interest that are often critical of U.S. foreign policy. In contrast to the Russian approach, which promotes ideological rejection of Western values and portrays Russia as a victim of NATO's expansion, China is more committed to crafting and promoting its own favorable image to the public. Neutral foreign policy and pacifism are the cornerstones of that vision, while Europe and the Indo-Pacific are treated as partners, on the same level as Russia. The Chinese cognitive approach is also oriented to affect perceptions at the individual level by undermining trust in the U.S.–dominated system of alliances, its durability, and its capabilities. Demonstrating the technological advantages of the People's Liberation Army, stressing the successes of Beijing's diplomacy, and outlining the achievements of socialism with Chinese characteristics are part of the narratives China uses to impact the European understanding of Chinese foreign policy behavior. The United States is typically presented as an expansionist, assertive, and offensive power, using Europe as a shield against Russia and seeking to place Europeans in military dependence on Washington. However, unlike Russian cognitive warfare, Beijing's adoption of narratives also involves appropriate attitudes, such as recognizing objective shifts in the global distribution of power, addressing the crisis of liberal democracy, and considering the changing nature of deterrence. The depiction of China and its system of partnerships as an alternative to the current world order centers on envisioning a multipolar world in which Beijing coordinates the global economy. Last but not least, the Chinese approach rejects alliances as a form of cooperation between state actors and instead endorses economic partnership to justify its claim that the latter does not bind the sovereign states in a dependence on a sole superpower, as the U.S. – dominated process of globalization did after the end of the Cold War.

The technological tools of the Chinese cognitive approach do not differ from those of Russia in purpose, except that they are much more sophisticated. Recent studies indicate a connection between the Double Dragon hacking group ATP 41, the Honker Union, and the alleged cyber-attacks on the U.S. Department of Defense. While the Russian strategies aim to directly affect and disrupt the policymaking process in the short-term by inflicting as much destructive damage as possible and leak an extensive scope of information, China's tools of informational warfare involve limited acquisition of critical database for long-term purposes such as economic espionage and strategically essential classified information, related to the military and industrial complex of the state – victim. The ultimate purpose is not only to advance Chinese geopolitical interests, as in the Russian case, but also to gather information on how the opposite party perceives China and how those perceptions could be improved. The same logic applies to cognitive tools, used in the public sphere. Pro-Russian intellectuals and academia in Europe and the United States are increasingly committed to composing a benevolent and profitable image of China as a unified force in a divided world, or as a pacifier in a system where America and its allies struggle for ultimate control over international politics. Most cultural events organized by Chinese non-governmental lobbies in Eastern Europe are widely circulated on social media and continue to recruit, even those receiving Western grant support, either voluntarily or by targeting their professional and personal occupations.

China's behavioral approach advocates an ideological direction opposite to Russia's support for right-wing radicalization. Since socialism with Chinese characteristics remains the political formula of China's government, the movements of the far left enjoy a more amicable view for two reasons. One is ideological, and the other is pragmatic – radicalizing Western societies by promoting left-wing ideology is a more plausible design when your partners have already adopted the right-wing rhetoric, because it would allow China to divide and polarize public opinion in democracies. By formally sticking to its non-interference principle, Beijing issues strong support for Eurosceptic and nationalist political parties, which oppose NATO and its assistance for Ukraine. Examples of China's party-to-party diplomacy include support for left-wing and right-wing populist factions, some of which are politically represented in the European Parliament and at the national level, as well as securing their access to various Russian political forums and cultural events held in Moscow. No less important are cases such as the Chinese endorsement of the Black Lives Matter protests in the United States – a step, defined by some scholars as opportunistic and presumed to spread propaganda. Even so, the damage to the American public opinion was severe alongside the acts of political violence, perpetrated by the BLM activists and several other social movements, that emerged in the years after 2020.

Figure 3. Chinese approach of cognitive warfare

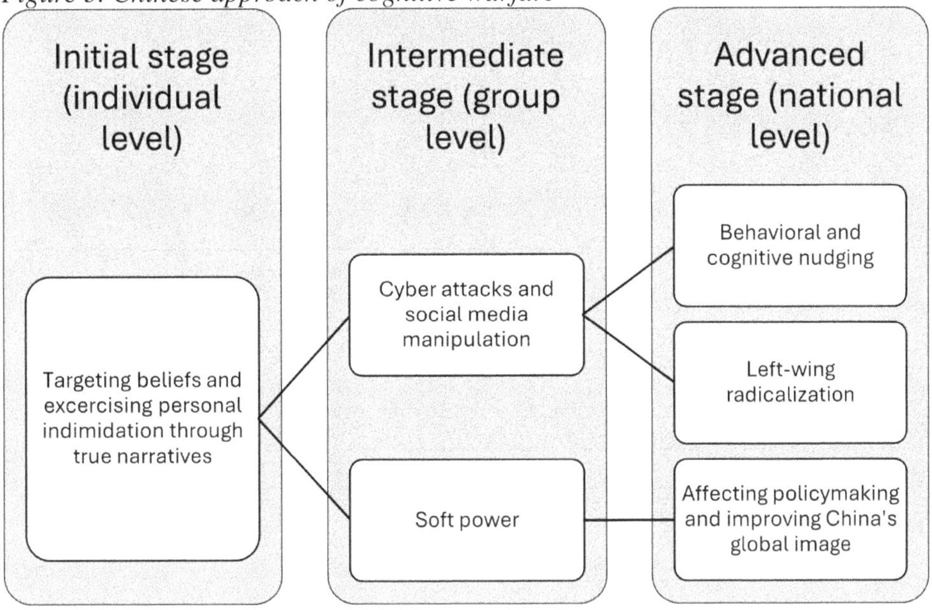

4.3. Case Study: Democratic Resilience and Cognitive Warfare Index

The final section of this chapter seeks to introduce an index of impact to calculate how cognitive warfare impacts democracies, based on the qualitative data in the preceding sections.

Cognitive Warfare and Radicalization of Democracy

This formula below quantifies this section's conclusion, introducing a matrix that illustrates how cognitive warfare radicalizes democracies.

$$D(t) = R(t) - H(t)$$

Where $D_{(t)}$ indicates democratic durability over time, $R_{(t)}$ – democratic resilience, and $H_{(t)}$ – effects of cognitive warfare. ***D* depends on whether *R* outweighs *H*.** If $H(t) > R(t)$ resilience weakens, while radicalization occurs due to the impact of cognitive warfare.

Based on the qualitative statements in this article, we can summarize that there two indexes we can use to measure the impact of cognitive warfare.

The cognitive attacks index stands as follows:

$$H = \alpha I + \beta C + \gamma S + \delta N + \epsilon M$$

Where:

Table 1.

I	=	Informational warfare
C	=	Cyber attacks
S	=	Soft power
N	=	Nudging
M	=	Military provocations
Greek letters	=	Relative weights

The democratic resilience index stands as follows:

$$R = f(T, L, M, S, J)$$

Where:

Table 2.

T	=	Trust in policymakers
L	=	Rule of law
M	=	Media transparency
S	=	Strategic perceptions
J	=	Public support for foreign policy

Radicalization occurs when:

$$\frac{dD}{dt} = -kH + cR$$

Where: k indicated the success of cognitive attacks and c – the degree of resilience.

Equation of the Impact Based on the Prisoner's Dilemma Game Theory

In this equation, the impact of radicalization on democratic resilience depends on the legitimation of political decisions and public fragmentation, while the success of the cognitive attacks is equal to the negative amount between polarization and the costs of radicalization.

$U_{attacker} = \max(\text{Polarization} - \text{Cost})$

$U_{democracy} = (\text{Legitimacy} - \text{Fragmentation})$

Solution here depends on the specific case, the sources of polarization and the subsequent degree of fragmentation.

Democratic Resilience and Cognitive Warfare Index

To conclude, the index of cognitive warfare's impact stands as follows:

DRCWI – *Democratic Resilience and Cognitive Warfare Index*

$$DRCWI = R - C$$

Where ***R*** indicates resilience and ***C*** – the cognitive impact.
Range of interpretation: -100 to +100

Table 3.

+40 to +100	=	**Very high resilience, weakened cognitive attacks, low polarization**
+10 to +39	=	High, but vulnerable to cognitive attacks, moderate polarization
-9 to +9	=	Resilience at risk, relative success of cognitive attacks, strong polarization
-10 to -39	=	Decline of resilience, success of cognitive attacks, very strong polarization
-40 to -100	=	Risk of democratic decline, severe polarization

Calculating democratic resilience:

$$R = \sum_{i=1}^{6} wi \cdot ri$$

Where R is normalized to 0-100

Calculating the cognitive impact:

$$C = \sum_{j=1}^{5} vj \, . \, hj$$

Where C is normalized to 0-100

Thus:

$$\text{DRCWI} = (R_{0\text{-}100}) - (C_{0\text{-}100})$$

Democracies, exposed to cognitive attacks and radicalization therefore face three options:

Resilience neutralizes cognitive attacks and radicalization:

$$R > 1.3H$$

Resilience is in decline, radicalization grows:

$$C > R$$

Resilience is ineffective, cognitive attacks are successful, radicalization and polarization increase democratic deficit.

$$R\downarrow$$

Despite its success in dividing Western democracies, cognitive warfare did not lead to the collapse of democratic institutions, as respect for civil liberties and support for peaceful coexistence remain high. The fact remains that even in states where political leaders enjoy continuing reelection and electoral support, corruption and favoritism remain the basic challenges to democracy rather than anti-globalism. Another example is calls to leave the EU and NATO, spread by various populist movements across the Member States. Although popular in social networks and among radical protesters, political elites are unlikely to sacrifice the national security of their states, while most European citizens are still in favor of European integration and the debate about securing NATO's future, especially after the Russian invasion of Ukraine. In this final section, the chapter will make predictions about which pattern of cognitive warfare is likely to be more durable in the Cold War against democracies.

Right-wing extremism remains strong in Europe and the United States because most of its narratives have become increasingly popular with the emergence of the political cleavage between neoliberals and neoconservatives. Russian cognitive warfare, however, suffers a relative decline, and its future durability is doubtful. Despite the fact that several Member States within NATO and the EU experience some difficulties in coordinating their positions, their institutions are credible enough to make decisions and strategically viable to apply them. The regular exercises conducted by the allies offer a clear example of their commitment to each other, and the European support for Ukraine remains firm and swift as allies move to increase their defense expenditures. Moscow's nuclear rhetoric further contributed to the unity of NATO, which faces the dilemma of how to navigate deterrence with Russia. Right-wing radicalization promotes the narrative that democracies are no longer capable of defending each other because the political formula of democracy itself has become obsolete. Proponents of this perspective typically employ the Russian vision of a centralized society, ruled by a certain type of patriotic and, in that sense, nationalist elites, who would even sacrifice some civil freedoms for the sake of security. Although this view is typically applied to the political status quo in the United States and some EU Member States, fact-checks indicate that the domestic policies of their policymakers do not affect the essential aspects of the foreign policy consensus in the Euro-Atlantic community regarding the deterrence of Russia, China, and other strategic adversaries.

Left-wing extremism implies a growing connection between anarchist movements in developed democracies and the ideology of wokeism, which scholars define as a branch of progressive policies and practices that oppose social discrimination and inequality (Bernstein, 2022). Most studies in that field are divided in their assessment of the implications of wokeism for the developed democracies, but empirical observations indicate that this ideology has been endorsed by various social movements to form violent narratives and use political violence against civilians (Burston & Jacobsen, 2025). China has consistently supported woke movements in the United States in its efforts to divide Western societies alongside the Russian adoption of right-wing narratives. In contrast to the Russian approach, Beijing has been even more successful, as the messages of left-wing extremists have penetrated U.S. society more deeply and boosted the radical opposition from right-wing factions. Eroding democracy was an easy task for Beijing and Moscow since democratic voters found themselves in a situation where the social justice warriors and the white supremacists polarized public opinion. Security implications include growing populism, an increase in the democratic deficit, and the rise of antisemitism, driven by both right-wing and left-wing political narratives, especially after the October 7 terrorist attacks against Israel.

In Europe, the Balkans have suffered the most severe effects of polarization, as it has allowed corruption to spread and undermined trust in post-socialist democracies. Eastern Europe became particularly vulnerable to the Chinese economic and cultural influence, which considers the region as a pivot to the rest of the EU.

The narrative that the Chinese way could improve democracies, or at least coexist with their institutions, became popular among some political actors who sought to reduce political accountability in their democracies and break away from the United States. Beijing's cognitive approach was thus even more impactful than Russia's, and its pattern is also more potent because China has the ability and resources to form a geopolitical bloc, unlike democracies.

5. CONCLUSION

The preceding discussion demonstrates that state agents of cognitive warfare possess the capacity to radicalize Western democratic societies, thereby weakening political debate and corrupting transparency within institutions. Some of their victims – such as interest groups, powerful lobbies, and even political parties – could transform into non-state agents of cognitive warfare by adopting the ideas of state agents. It is essential to draw a clear line between freedom of expression, an existential part of every democracy, and political violence arising from radicalization. If an individual or an interest group endorses a particular narrative and promotes it peacefully, its actions cannot be sanctioned unless they violate the law. If, however, violence is used as a means of resistance, and especially if it affects civilians and civilian infrastructure, legal actions are required to protect national security. The debate about criminalizing disinformation and fake news is highly relevant because it erodes trust in democracy from within, alongside objective challenges such as democratic deficit, corruption, and political favoritism. Some scholars believe that any legal actions against misinformation and fake news would have the opposite effect, while others consider them necessary in light of the threats posed by cognitive warfare to democracies. Integrating the fight against cognitive warfare in the national security strategies is essential, but using its sources as a channel of information would better benefit the intelligence community than merely imposing negative freedoms on non-state agents of radicalization.

Deterring cognitive warfare involves a delicate balance between preventing the sources of polarization and strengthening the resilience of democratic institutions to enhance the credibility of democracy. Observing institutional transparency, strengthening the middle class, and safeguarding human rights are the cornerstones of democracies, their most potent weapons against autocracies, extremism, and political violence. The weaponizing of right-wing and left-wing radicalization – to a greater extent – challenges the assumption that support for democracies should be taken as granted and that democratic regimes are immune to totalitarian instincts. From a policy perspective, NATO remains the most successful military alliance in history because it was established by a group of democracies that fought a Cold War against the Soviet bloc. But recent trends indicate that satisfaction with democracy has dropped, and global trust in its credibility has declined.

The chapter thus concludes that the decline in favorable perceptions of democracy, influenced by cognitive warfare, would affect the Alliance's ability to respond to crises by dividing its allies. In many states where cognitive warfare has been successful, institutions have been weakened, and policymaking has become less transparent. Therefore, fighting the next cold war against authoritarian regimes should be accompanied by efforts to strengthen society's political strength to withstand domestic polarization. Before pressuring autocracies politically, economically, and militarily, democracies should first achieve unity among themselves by reducing political extremism, strengthening the judiciary, and de-radicalizing the public discourse. Failing to succeed would result in a global weakening of democratic resilience and failure of democratic policymaking.

REFERENCES

Backes, O., & Swab, A. (2019). *Cognitive Warfare: The Russian Threat to Election Integrity in the Baltic States.*

Bernstein, D. L. (2022). *Woke antisemitism: How a Progressive Ideology Harms Jews*. Wicked Son.

Burston, D., & Jacobsen, K. (2025). *Authoritarianism in All its Guises: Perspectives on Right, Left, and Center*. Taylor & Francis. DOI: 10.4324/9781003389927

Buzan, B. (2015). The English School: A neglected approach to International Security Studies. *Security Dialogue*, *46*(2), 126–143. DOI: 10.1177/0967010614555944

Carter, B., Van Nuys, M., & Albayrak, C. (2020). State sponsorship of religiously motivated terrorism: A deadly combination. *Democracy and Security*, *17*(2), 181–209. DOI: 10.1080/17419166.2020.1848558

De Swielande, T. S., Orinx, K., & Peiffer, S. (2024). La guerre cognitive au coeur de la stratégie chinoise de socialisation. *Ingénierie Cognitique*, *7*(2). Advance online publication. DOI: 10.21494/ISTE.OP.2024.1153

Galeotti, M. (2018). The mythical 'Gerasimov Doctrine' and the language of threat. *Critical Studies on Security*, *7*(2), 157–161. DOI: 10.1080/21624887.2018.1441623

Githens-Mazer, J. (2012). The rhetoric and reality: Radicalization and political discourse. *International Political Science Review*, *33*(5), 556–567. DOI: 10.1177/0192512112454416

Herschinger, E., Bozay, K., Von Drachenfels, M., Decker, O., & Joppke, C. (2020). A Threat to Open Societies? Conceptualizing the Radicalization of Society. *DOAJ*. DOI: 10.4119/ijcv-3807

Hung, T., & Hung, T. (2022). How China's Cognitive Warfare works: A Frontline Perspective of Taiwan's Anti-Disinformation Wars. *Journal of Global Security Studies*, *7*(4), ogac016. Advance online publication. DOI: 10.1093/jogss/ogac016

Jasko, K., LaFree, G., Piazza, J., & Becker, M. H. (2022). A comparison of political violence by left-wing, right-wing, and Islamist extremists in the United States and the world. *Proceedings of the National Academy of Sciences of the United States of America*, *119*(30), e2122593119. DOI: 10.1073/pnas.2122593119 PMID: 35858413

Jervis, R. (2017). *Perception and misperception in international politics*. Center for International Affairs.

Mansfield, E. D., & Snyder, J. (2002). Democratic transitions, institutional strength, and war. *International Organization*, *56*(2), 297–337. DOI: 10.1162/002081802320005496

Marsili, M. (2023). Guerre à la Carte: Cyber, Information, Cognitive Warfare and the Metaverse. *Applied Cybersecurity & Internet Governance*, *2*(1), 1–11. DOI: 10.60097/ACIG/162861

McCauley, C., & Moskalenko, S. (2008). Mechanisms of Political Radicalization: Pathways toward Terrorism. *Terrorism and Political Violence*, *20*(3), 415–433. DOI: 10.1080/09546550802073367

Neumann, P. R. (2013). The trouble with radicalization. *International Affairs*, *89*(4), 873–893. DOI: 10.1111/1468-2346.12049

Paziuk, A., Lande, D., Shnurko-Tabakova, E., & Kingston, P. (2025). Decoding manipulative narratives in cognitive warfare: A case study of the Russia-Ukraine conflict. *Frontiers in Artificial Intelligence*, *8*, 1566022. DOI: 10.3389/frai.2025.1566022 PMID: 41018734

Pickel, A. (2022). *Handbook of Economic Nationalism*. Edward Elgar Publishing.

Pietrzak, P. (2024). The Geoeconomics of NATO's Responses to Russia's Invasion of Ukraine (2022-Present). In Pietrzak, P. (Ed.), *Analyzing Global Responses to Contemporary Regional Conflicts* (pp. 1–38). IGI Global Scientific Publishing., DOI: 10.4018/979-8-3693-2837-8.ch001

Pietrzak, P. (2025). The Xi Jinping Doctrine and the Idea of China's Peaceful Coexistence in the Fragile Post-2022 Context: A Smokescreen or a Long-Term Commitment? In Pietrzak, P. (Ed.), *Exploring the Implications of Local and Regional Conflicts* (pp. 257–290). IGI Global Scientific Publishing., DOI: 10.4018/979-8-3693-2877-4.ch010

Sedgwick, M. (2010). The concept of radicalization as a source of confusion. *Terrorism and Political Violence*, *22*(4), 479–494. DOI: 10.1080/09546553.2010.491009

Stahl, R. M., & Popp-Madsen, B. A. (2022). Defending democracy: Militant and popular models of democratic self-defense. *Constellations (Oxford, England)*, *29*(3), 310–328. DOI: 10.1111/1467-8675.12639

Tooby, J., & Cosmides, L. (2025). The evolution of war and its cognitive foundations. *Evolution and Human Behavior*, *46*(3), 106687. DOI: 10.1016/j.evolhumbehav.2025.106687

Wilkinson, T. M. (2012). Nudging and manipulation. *Political Studies*, *61*(2), 341–355. DOI: 10.1111/j.1467-9248.2012.00974.x

Yan, X., & Qi, H. (2025). China–US competition, reverse globalization, and the regression of world politics. *China International Strategy Review*, *7*(1), 1–15. DOI: 10.1007/s42533-025-00187-8

Zoubir, Y. H. (2023). *Routledge companion to China and the Middle East and North Africa*. DOI: 10.4324/9781003048404

Chapter 6
Meaningful Human Control and Morality:
Implementing Advanced Control Directives for Autonomous Systems

Prateek Mishra
http://orcid.org/0009-0002-7960-0166
KIIT School of Law, KIIT University, India

Bhanu Pratap Singh
Faculty of Law, Madhusudan Law University, India

Pranjal Khare
http://orcid.org/0000-0002-9937-9588
Jindal Global Law School, O.P. Jindal Global University, India

Sapna Singh
College of Law, IIMT University, Meerut, India

ABSTRACT

Current warfare is accelerating as conventional military forces, autonomous systems, and human-machine combinations integrate. This development presents an essential quandary: either establish moral and legal causation within systems under non-standard conditions of perception of human time. To counter this, the question is: can humans reciprocate independent actions in real time through a prior contract, thereby arranging future conduct in advance? It is especially applicable in situations that are too fast for humans to comprehend or too slow, where constantly concentrating on every action is not feasible. Leveraging the medical-legal prece-

DOI: 10.4018/979-8-3373-6264-9.ch006

dent of the so-called advance care directives, the framework suggests the so-called Advance Control Directive (ACD) to help pivot the rigorous deliberation needed to become accountable in the first place. This, therefore, creates an autonomy command, a framework of power designed and justified by creating these orders before implementation.

INTRODUCTION

In a lethal decision process, a given human decision-maker is supposed to have enough time to consider their case and deem it legitimate. They also need training to become knowledgeable, accountable, and liable for the implications of such a decision (Asaro, 2012). The appeal of autonomous systems in the military is growing, as these technologies can amplify the scale and effectiveness of operations and ensure they are executed in accordance with human intent and values. Although these systems can also function in real time, they are sometimes best suited to time scales that vary widely relative to human cognition. In other words, they can be used in situations where the tempo is fast (high-tempo) or incredibly slow (very slow). A significant problem, however, is ensuring that humans are ethically and legally in charge of systems that are not bound by normal human time frames. Narrative in time crosses our experience, interpretation and shared social meanings as Paul Ricoeur (1980) stressed. Because meaning is constructed through narrative (Maan & Cobaugh, 2018), any assessment of meaningful human control should account for the temporal filter through which humans perceive autonomous systems. This involves analysing how the narrative of the experience is constituted and how control can be evaluated. The time constraint on decisions can be considered a critical factor that should affect human intervention and, in turn, the subsequent judgment of actions themselves.

Autonomy also reduces the fidelity and utility of human involvement by enabling systems to act more swiftly than a human can perceive or respond to at the point of action, e.g., autonomous countermeasure arms firing at an enemy missile attack or a cyber attack due to a specific danger level. On the contrary, autonomy also decreases the utility of human attention during periods when systems are hibernating or engaged in high-risk, lengthy behaviours, such as sailing through a deep ocean at a minimal speed. The synthesis of the story of an autonomous system is not done by humans second by second, but in a series of segments determined by the waning influence of decision-making and worldly impacts. Similarly to the human brain working less efficiently in a limited environment, humans are more compatible to learn and comprehending autonomous behaviours when a diversity and transformation of cognitively graspable alterations are present. This is akin to the use of dynamic lights and sirens on emergency vehicles, intended to keep people alert. Thus, to have valuable and effective human control, it is necessary to ensure that humankind is not lost in action and is alert. This might require adjusting the systems' operational speeds before real operation to achieve an adequate level of human consciousness, involvement, and interpretation. This suggestion aligns with human factors research on the impact of boredom (Smith, 1981), an absence of involvement and stimulation that results in unpleasant neurocognitive conditions, including mind wandering, reduced effort, and perseveration (Dehais et al., 2020). The study of human factors simulation knowledge in the past has focused primarily on human attention and action in real time (Hancock et al., 2008). Therefore, conventional solutions have suggested changing the user interface or simply relying on neuro-adaptation of the user (Dehais et al., 2020), rather than adjusting the speed of operations to suit human cognitive inclinations.

The main question, therefore, is whether human beings can be placed in a situation outside the actual time and dictate actions to autonomous systems by entering into a contract in advance. This would permit action to take place in some future situation- especially in those instances that are either faster than usual time, or so slow that the human mind and focus cannot be in its full capacity. To address this, it can be reported to the healthcare legal precedent of so-called advance care directives, in which a person is asked to consider any medical procedures they want performed on them if they are unable to communicate their wishes to others. This medical precedent implies that accountability and responsibility for weapons systems can be time-consuming and deliberative, but that, rather than through an Advanced Control Directive (ACD), the process can be performed outside real time. To hold decision-makers accountable within this framework, they need to use the so-called Advanced Control Planning (ACP) before developing the ACD. This entails doing a lot of work on the systems throughout the development process, training them, running simulations, conducting exercises, and so on, to gain experience with the operation of the autonomous systems. This preparation will mean a) that the system will know who will be targeted legitimately, and b) to guarantee that human beings will be answerable and responsible in their deployment.

DEFINING AUTONOMY, RESPONSIBILITY, AND THE CONTROL DILEMMA

To build a conceptual base, the meaning of an autonomous weapons system (AWS) is defined in the US DoD 3000.09 (Department of Defence, 2023). An AWS is characterized as a weapons system that, after activation, can select and engage the target without operator intervention. This group is generic and encompasses systems monitored by operators and programmed with overID facilities, as long as they retain the ability to choose and take on targets on activation. As such, the distinctive feature of such systems is that they take actions without the involvement of people over a specific period, which is represented as to. It is such a definition that is abstracted to include non-weaponised "autonomous systems" (AS). An AS is a system that, once activated, can operate without further operator intervention, including by operators with operator supervision and override functionality. The most important thing is that even non-weaponised systems may result in problematic outcomes or bring harm. As an example, a fleet of uncrewed underwater vehicles (UUVs) may be commissioned by a commander to cover a particular area of the ocean, which may cause unintentional harm to a protected reef (Gu et al., 2021) or an unwanted intrusion into the waters of a foreign country (McKenzie, 2020, 2021).

The concept of responsibility in this regard is understood explicitly as command responsibility (Mettraux, 2009), which imposes on human beings the obligation to meet specific requirements of moral and legal responsibility. In essence, moral responsibility and blameworthiness presuppose that a person has both autonomy and knowledge, as well as the capacity to make choices and the tendency to reasonably anticipate the outcomes of those choices (Talbert, 2016). An individual is not morally obligated if he/she is forced, could not anticipate the result, or was not in a position to make any decisions because of external conditions. The role of legal responsibility, in turn, implies a proper relationship between the human operator and the system, whereby the human operator controls the system's activities and is responsible for the degree of its behaviour (Liivoja et al., 2022). Human responsibility in the use of autonomous systems has three related issues, namely: (1) responsibility of multiple parallel systems in real-time; (2) responsibility of high-tempo systems; and (3) responsibility of slow-tempo systems. Planning a real-time autonomous system with many parallel units, such as a robot swarm, requires breaking down the swarm's decision-making and action parameters before operation to make them understandable to humans. For example, high-speed simulation of swarms can be conducted to introduce them to a wide range of risks and to scrutinise the diverse outcomes to inform usage orientation1. Alternatively, high-paced systems such as cyber countermeasures should be slugged out during simulation to allow human beings to examine each decision and its probabilistic arguments. Lastly, systems with low tempo, such as long-range undersea surveillance, need simulations sped up.

This will enable humans to evaluate the system's behaviour slowly, allowing them to judge its capability to achieve goals without violating ethical and legal requirements. The framework of this query begins by defining the operational velocity of autonomous systems as a key challenge to humans achieving meaningful autonomy over such technologies. The analysis continues by considering three key aspects: trends in warfare in the 21st century, the principle of advance care directives, and the need for command, with respect to moral and legal accountability. All of this informs the notion of autonomy command, where both differences and similarities exist between existing command structures and those needed for autonomous systems. It is against this backdrop that a procedure for drafting an Advance Control Directive (ACD) is proposed, using a case study of Australia to provide the required national-level context. Moreover, it is discussed how artificial intelligence affects the autonomy of the future, as human decision-makers become increasingly dependent on digital simulations to decide on relevant courses of action. The discussion concludes with the reaffirmation of how difficult it is for the armies that use them to ensure the moral and lawful accountability of autonomous systems, and speculates on the value of training an ACD to regulate their use in the future.

It should be noted that this proposal is purely methodological, as it seeks to offer a template for thinking about such contractual mechanisms. It does not, however, require any particular legal or ethical conditions, taboos, operational constraints, or risk judgments that would be obligatory for each instance of an ACD before an autonomous weapons system is launched. In accordance with the definitions, an autonomous agent (human or artificial) will, at times, choose a subset of actions against the backdrop of human intervention. Though such agents are affected by internal and external environments, programming, and environmental conditions, and they may be restricted, they are still able to decide what to do within a set of temporal and spatial limits. The pressing question arises: Can these systems really be left under any absolute human control? However, can they not be placed under meaningful human command when it is not possible to control them? These control problems are unavoidable in modern weaponry. On the one hand, the International Committee of the Red Cross (ICRC) suggests that timely and effective intervention and deactivation of autonomous weapons systems should be achieved only through human supervision (International Committee of the Red Cross, 2021). They also propose a ban on autonomous systems that choose and apply force against human objects (International Committee of the Red Cross, 2021, 2022). ICRC further believes that men should still retain control of weapons systems, so they require a human element in the targeting process and in the use of force. According to them, the application of autonomous weapon systems is fraught with risks because of the complications in preventing and controlling their consequences. This human loss of control and discretion in the deployment of force and weapons is a concerning trend on humanitarian, legal and ethical terms (International Committee of the Red Cross, 2021).

On the other hand, the use of software to regulate the legal operation of weapons systems by militaries dates back more than 40 years (Department of Defence, 1978). Defence industries are now promoting kill chain acceleration, a concept in which military events move at ever-faster rates to accomplish missions effectively, overcoming the constraints of human cognition. This raises the philosophical question of whether meaningful human control over such high-tempo acts is possible. The response to this depends on the specifications for meaningfulness and control. The notion of meaningful human control remains contentious when it comes to AI-powered, robotic, and autonomous (RAS-AI) weapons systems. This resistance is in part because determining the parameters of what constitutes a meaningful interaction is subjective, complicated, and ambiguous. It has therefore been argued legitimately that the qualifier 'meaningful' is even redundant, but that there might be a need to require merely simple control. In practice, the operational history of autonomous deployed platforms can be used to cluster intervals of inactivity (the m in a meaningless time), condensing down data with no events to condensed chunks, and only requesting human attention when the system notices a change or uncertainty. One example of this model of operation is the Ocius Bluebottle (Department of Defence, 2023), which automatically performs ocean surveillance. Fully autonomous functions in this system run independently until special conditions, such as collision avoidance, require human or remote control.

Using this technique, the narration of the activities of an independent platform is told at a cognitive rate that people can handle, including updates on the internal state of the entity as well as the external engagements:

a) Internal issues refer to the health of the system, such as incident reports, self-repair procedures, or the fidelity of the sensors that are going to affect the quality of decision-making or perception of the unexpected environments.
b) External issues include physical contact with the environment, e.g. the use of force, manipulation of objects (e.g. improving debris or fixing cables), kinetic interaction, and internet activities.

The dissenting discussion about meaningful human control over artificial intelligence (AI) focuses on situations in which artificial actors have agency and decide matters that pose ethical or legal risks. Under such circumstances, human beings need control and the ability to intervene. The lack of significant control is usually synonymous with the inability to assign moral or legal blame to human actors for the decisions the system makes. On the contrary, humans can be considered to be in reasonable control when they are strictly responsible for making decisions with the help of AI. Rather than trying to define control in isolation, the emphasis here would be on the mechanisms of responsibility and accountability, which are the basic building blocks of a sound structure of appreciable human control. The most viable approach to framing this debate is to understand meaningful human control in terms of meaningful human command. A command can help maintain responsibility and accountability, which are central principles of the NATO AI Strategy

(NATO, 2021). Implementing advanced technologies is possible without having to work with them or do everything on the spot. Since NATO's policy is the most notable international endeavour to date to address irresponsible AI deployment, this idea of an autonomy command can be seen as a good way out. Particularly, the scaffolding and legitimisation of this authority can be achieved through the establishment of an Advanced Control Directive (ACD) before the autonomous systems become operational.

THE ROLE OF CONTRACTS IN ENSURING RESPONSIBLE AI IN MILITARY OPERATIONS

Moreover, it is like I have a book, and I am in love with the book, but I am slowly reading it, so the words are really spread out, and the spaces between them are very long. I still feel you and the words you were telling me, and it is in this timeless gap between the words that I am discovering myself. Samantha AI OS (Jonze, 2013). This conversation between the AI system Samantha in the film Her clearly shows the type of temporal disorientation caused when machine processing is significantly faster than human perception. She advances to the stage of processing information on such a scale and with such a pivot that she practically discards her human companion. Although we are yet to have conscious AI (Gebru & Mitchell, 2022; Schneider, 2019), current trends in military tech are leaning towards this imbalance, as information processing has accelerated and decision-making has become faster. For example, defence startup Anduril sells its product, Lattice, as one that will allow military troops to move at the speed of a machine, turning data into information, information into decisions, and decisions into actions (Anduril, 2022). Here, Mick Ryan (2022) points out several trends of the 21st century warfare that are critical and are mentioned in War Transformed: (T1) the accelerating pace of planning, decision-making, and action; (T2) the integration of high mass conventional forces and massed autonomous systems; and (T3) the deep human-machine integration where the autonomous systems become fully partners. According to Ryan, these trends necessitate novel approaches that enhance human physical and cognitive abilities to produce more mass... [and] faster decision-making processes (Ryan, 2022, p. 84). Nonetheless, there is a deeper ethical issue in this case: when real-time decisions (T1) accelerate while human cognition does not, moral and legal accountability will be lost due to the loss of human awareness and agency (Asaro, 2012; Sharkey, 2012, p. 116).

Militaries may suggest that there is no need to make ethical decisions faster; they can be trained to operate in particular situations, just as soldiers train to act in an emergency with an external medical knife or pilots train to hit a target quickly.

Nevertheless, this argument does not explain the difference between cognitive processes. We can make Type 1 reasoning faster, as it is unconscious, gut-level, and heuristic, whereas Type 2 reasoning is deliberative, conscious, and, to a great extent, reflective (Kahneman, 2011; Stanovich & West, 2000). One cannot even take time to think in this slower, deeper way, even when one is not in stressful, exhausting, or uncertain situations or in situations of war (Lieberman et al., 2005). Consequently, it is not possible to ask humans to think faster to keep up with autonomous systems as a strategy for ensuring ethical responsibility. The solution is to devise a system of governance that enables responsible human beings to approve autonomous operations not constrained by time in battle. Under this method, it is necessary to capture "Type 2 thinking in advance. For example, we can refer to the medical-legal precedent of the so-called advance care directives.

Similarly, patients can make decisions about their future care as long as they are competent to do so; commanders can consider advance control planning to direct autonomous systems, ensuring that legal and ethical duties are not violated, and that lines of responsibility are not misconstrued before operation. It is necessary to assume that human-AI trust during military activities can be contractual and significant. Contracts are an effective instrument of governance because they delineate the terms of trust; to trust the agent is to believe that a pattern of contracts will be enforced. It means that the AI and its designers should be bound to a previously signed agreement (Jacovi et al., 2021). By allowing this, incidents can be explained and even judged, and, in cases of absurd behaviour, the contracts laid down may be reviewed during a legal audit. Nevertheless, the possibility of unexpected, uncontrolled behaviours by autonomous systems remains an open question about whether autonomous systems can ever be entirely ethical and legal in their practices (International Committee of the Red Cross, 2021).

LEGAL AND ETHICAL RESPONSIBILITY IN DIRECTING AUTONOMOUS SYSTEMS

If it is concluded that contracts between people and competitive systems can secure responsibility, the character of the contract must be specified. This contract within the military sphere would likely align with the Rules of Engagement (ROE) (Hosang, 2020) and other legislative stipulations governing the documentation of military decision-making (Devitt et al., 2021). It is mentioned that it has been analysed through the lens of philosophical and cognitive science, not the legal one. As a result, the discussion of contracts in the medical sector is intended to provoke creative thought about the procedures and content applicable to a command contract. Developing suitable legal tools and mechanisms would be the next step, provided the philosophical assumption is considered valid. Given that this philosophical reflection about contracts may be worthwhile, one should ask whether there is a legal precedent of creating a contract at a given time, at known as, and where the effects of any action entail significant damages, misery, and loss of life; and (3) the decision-maker at that moment is the same person at, who chose the present. The implementation of the advance care directives in the medical profession provides an ideal illustration of such a contract. This presents the reader with the question of whether the 40-plus years of labour the people in the health sector have spent have provided anything that can teach them to govern autonomous systems and hold themselves accountable for what they will do in the future.

Directives and advance care planning, as such, in the 1970s developed to give patients the opportunity to decline unwanted care, including life support through tube feeding or cardiopulmonary resuscitation, towards the end of life (Morrison, 2020). Advance care planning aims to help patients clearly express their values and goals, and decide how different treatments resonate with them. It is a procedure that covers a broad spectrum of variables: social and cultural assumptions; structural limitations of health and legal practices; a multi-stakeholder intervention; interactive information consultations with specialists to regulate the anxieties; and repetitive dialogue and assessment in different contexts with strategies to sort out the disagreements and implementation needs (Jimenez et al., 2018, Table 4; Luckett et al., 2015). The records that result from advance care planning are intended to empower surrogate decision-makers and medical professionals to act on behalf of a patient (that is, not to impose unwanted procedures). On the one hand, advance care directives have already proven themselves useful; as a case in point, in the US, when dealing with patients with dementia, and during the COVID-19 pandemic, especially in New York (Morrison, 2020).

On the one hand, advance care directives have already shown themselves to be beneficial; for example, during the COVID-19 pandemic, in countries with dementia patients and an elderly population (Morrison, 2020). End-of-life decisions are not straightforward or conventional; they change quickly as clinical conditions evolve. Therefore, it is challenging to achieve adequate depth in care conversations, as they consider the many permutations of real-life situations, including prognosis, disease course, comorbidities, and treatment outcomes. Individuals planning to introduce autonomous systems in multi-layered, unpredictable conflict areas should note this healthcare evidence that planning is not always easy, and that real performance does not always follow what was planned or approved (Pietrzak, 2024). There is, however, another critical difference: the distinction between autonomous systems and human beings, which can be functionally constrained. It is also possible to put forward a particular set of functions to ensure that the behaviours expected of the humans deploying them are acceptable. A patient, unlike an autonomous system, will not be able to switch its mind on an agreement or fail to act in accordance with the authority granted to it.

It might be questioned whether the medical situation provides a parallel decision-making context to a human commander controlling autonomous systems. A sceptic would say that, unlike a patient refusing treatment, human commanders do not face a physical risk when signing an advance directive, as that would change the decision calculus and even the commitment to the process. Although this is true physically, it is not true to say that the commanders are not at risk, especially regarding their moral and legal accountability for the actions of the system. Although physical damage does not occur, the possibility of moral injury is substantial for autonomy commanders when the systems under their command act in problematic ways and cause unintended damage (Phelps & Grossman, 2021). The military use of the health literature can be found instructive, even with the flaws of the analogy, since there are aspects of health choices that are vital to the future and to future command. In particular, the notions of mental ability and information delivery are applied to examine the soundness of a person's independent desire in advance directives. Human consent is not a simple yes-or-no to treatment; instead, the soundness of the decision is tied to the method used to make it. Misleading or insufficient information leading to a decision that would otherwise be rebuffed with more informed information undermines the validity of the directive. Using this justification with respect to the so-called autonomy command, commanders cannot be regarded as autonomous agents, that is, quite able to agree to a course of future action, when they are denied the entire range of relevant facts concerning a particular deployment. A commander who is given biased information to fulfil a particular agenda is not well equipped with sufficient autonomy to be morally responsible.

AUTONOMY, ETHICS, AND RESPONSIBILITY IN MILITARY COMMAND

It is necessary to remind oneself that it is not possible to build an art of war model that would serve as a scaffolding on which the commander can count at any given moment. When a leader has to live by his original equipment, he will invariably find himself outside the given models and at variance with them. Talent and genius are above the rules, and theory is incompatible with this practice, however versatile the code might be: On War (von Clausewitz, Howard, and Paret 1976). This is the character of command, as interpreted by von Clausewitz, a characteristic of war wrongly judged as uncertain, unpredictable and dynamic and therefore unable to adhere to a fixed prescription. In turn, human inventiveness, strength, and problem-solving are necessary for success. Human beings and machines working in this field should also be able to adapt to rapidly evolving situations. Moreover, it is the responsibility of human beings to oversee the people and systems they introduce. One of the core assumptions is that officers should invest in consistent and lifelong learning of the profession of arms, including military history, military ethics, military strategy, and military leadership, to excel as a tactical commander in the 21st-century society that is highly dynamic (Chambers et al., 2017; Ryan, 2022). This brings the most critical question: what actually makes moral and legal responsibility in command?

Good vs. Bad Apples

A variety of philosophers believe that only human agents, and perhaps sentient machines, have moral responsibility. In general, moral responsibility would imply three particular conditions, i.e. (1) freedom of choice and agency over the decisions (free will), (2) awareness of the situation (situational awareness), and (3) the presence of a possibility to act in the situation of events that involve ethical risk in accordance with the intent (capable action) (Crawford, 2013; Talbert, 2016). The military institutions, as Crawford investigates in her work, usually represent their staff as good apple-soldiers who will do their duty and be aware of their moral and legal obligations, giving their best performances under extreme pressure and uncertainty about information. These good apples are compared to bad apples and mad apples. Bad apples are soldiers who are aware of the expectations for their behaviour and choose to disregard the education on proper conduct, previous experience, and the demands of their military operations. On the other hand, there are mad apples, soldiers who are no longer able to understand what they should do and how they should behave. This state is commonly characterised by both character traits and environmental factors, making it less able to exercise agency, acquire knowledge, or act ethically and legally. Therefore, the responsibility of the bad apples is such that the mad apples are not. Laws also reflect the importance of mental state to culpability, with people who are not in their right frame of mind, who do not have control over their behavior, and who do not act in line with their intentions not bearing responsibility, but may be declared criminally insane, and thus subjected to limitations of their freedoms (Australasian Legal Information Institute, 2022).4. A fourth category could also be the so-called cooked apple. This refers to an otherwise sane and rational soldier being turned into a so-called moral crumple zone when operating in complex information spaces (Devitt, 2023; Elish, 2019). A situation where cooked apples blame them for making decisions, not because they are insane or lack the ability to make appropriate judgments, but because they lack a strong presence in a decision-making system.

Autonomy, AI, and Decision-Making

This thesis is impossible without free will, or the exercise of command. If commanders are not granted autonomy at the point of deploying autonomous assets, they cannot be held morally or legally accountable for the actions that result. Immanuel Kant visualised the will as the property of a will which is that it is a law to itself (notwithstanding the absence of any property of the objections of volition) (Kant, 2011). This is understood to imply that an independent person is a rational agent who follows their internal standards, rather than merely an agent acting freely. This concept of a person having his/her criterion is comparable to measures of moderation in responses to Just War doctrines and IHL, which are intertwined with it. The commanders have a legal responsibility to decide, on their own, what they perceive to be a proportionate use of force; it is an opinion, not just adherence to an objective use of force. The case law of the international community has established objectivity as a yardstick, namely that of the reasonable military commander (Hasson & Slama, 2023, p. 1; Hamurgraw, 2006, p. 100). To Kant, human reason is like a kind of compass; the way of doing good as opposed to evil according to duty, regardless of personal wants. Sovereignty is considered to be the root of human dignity and the origin of any morality. As a result, without autonomy, a commander cannot be morally or legally held responsible for the decisions taken. When a commander is pressured, influenced, or affected by automation bias, he or she has, in a sense, neglected his or her obligation as a rational and moral agent. According to Kant, a person can be autonomous only if he or she is rational and does not live in fear of punishment or any internal or external pressure. Nevertheless, it can be said by some that the requirements of autonomy put forth by Kant are overbearing on the individual.

Another explanation was offered by John Stuart Mill, who focused on the ability to make choices: He must use observation to see, reasoning and judgement to foresee, activity to gather materials to make a decision, discrimination to make a decision and when he has made a decision, firmness and self-control in maintaining his deliberate decision (Mill, 1860). Mill not only treasured individuality and spontaneity but also required decision-makers to apply all their faculties to determine which course of action is best. Although individual freedom was supreme, Mill accepted that persuasion and advice could align with autonomous choice, provided the end-user could use their own judgment in accepting the advice. This highlights the risks of rationality and autonomy that AI decision-support aids can pose. In its legally established meaning, autonomy is compromised by interference from others, except when the individual chooses to have it. The voluntariness of a decision is complicated when the decision-maker is first presented with complex and persuasive information. It is especially acute when environments are paternalistic, and the decision-maker is regarded as a subordinate to an intellectual equal (Tsai, 2014). This dynamic is depicted in the film WALL-E, in which the Captain is desperate to know his own mind amid the ship's artificial intelligence's compelling efforts (Stanton, 2008).

The advocates of nudging argue that the ethical validity of corrective manipulation is justified as long as humans can choose freely and that the relocation of information can minimise the total evils. To illustrate, in 2016, President Obama signed Executive Order 13707 (2015), which applied the principles of behavioural science by ensuring that thousands of vulnerable Americans had healthcare, rather than an opt-in/opt-out model (Lepenies & Malecka, 2018). Nonetheless, there is significant criticism of nudging, making it a highly controversial theory to follow (Kuyer & Gordijn, 2023). What matters more is choosing with ethical or legal implications in a box-ticking ritual rather than soliciting serious thought. Command is inherently connected with volition and agency in cognitive science (Haggard et al., 2002; Haggard, 2017). One has control over his or her body to the extent that he or she can command it as he or she wishes, rather than having it respond reflexively to external stimuli. Agency is an inquisitive state of mind, as in the case of a human who is planting a garden and feels accountable for the harvest six months down the road. It is necessary when the issue of autonomy is involved in high-tempo and slow-tempo functions, as this skill involves feeling agency not just in the time of acting, but over time. This normative framework on the responsibility for autonomous actions across different time scales is based on the fact that humans experience agency and responsibility for actions at different time scales.

Informed Command Decisions

Informed Consent is a cornerstone concept in medical literature. In the context of advance care planning, this principle implies that patients should be informed, to the extent possible, about the probable effects and outcomes of their decisions. In analogous terms, the commanders of military organisations should have access to professionals who could conduct serious analyses of the possible consequences of decisions on the outcomes and their likelihood. This reform shifts the usual military notion of situation awareness to be used in both fast-paced and slow-paced missions. It has to be perceived as "advanced situational awareness" - the ability to imagine and project future scenarios into enough detail to accept responsibility for them in case they transpire, even when the commander is not operating under real-time contact (Endsley, 2023). The legal weight of advance care directives in the medical profession is much stronger when patients are well informed, understand the specifics of possible treatments, and have had high-quality, long-term conversations with medical teams (Chan, 2018). On the other hand, patients who have had only ad hoc or single conversations in the past will probably receive less credibility. On the same note, during the consideration of acts of war, activities that are planned, crafted with attention to finer details, and guided by professional guidance are more likely to be considered carried out in a responsible and meaningful manner.

Commanding Autonomous Systems

The commanders are supposed to have the right mental capacity befitting their position. They need a synthesis of skills to effectively manage operations that proceed faster or slower than the speed at which humans perceive them: a high-tempo operation and a slow-tempo operation. It entails a combination of conventional military skills, including strategic, operational, and legal logic, as well as dedicated skills required to lead AI and autonomous systems. Such particular abilities are:

a) The ability to think logically under difficult hypothetical situations.
b) A profound technical knowledge of AI abilities, as well as the logic behind them, pre-written "games to play, redundancies in the operations, and controls of stopping them.
c) Efficient communication with machines that implies the power to correctly articulate command intent and interpret machine status reports and confidence measures correctly.
d) An understanding of the causal links between human orders and machine activities and responses.
e) Skills of probabilistic reasoning.

The competence of an autonomous commander must be highly evaluated in the area of training and post-training reviews of operation after-action reports. In simulations, commanders need to train to coordinate with autonomous platforms in normal time — simulations with the operational tempo set to a human cognitive speed. The practice allows them to study and plan actions, giving them the confidence to write an Advance Care Directive (ACD). It is in this process of simulation, where they have time to look back and reexamine the outcomes of hypothetical decisions, that a commander establishes the validity necessary to secure the state's ratification of their programmable agreements. The magnitude of the required training is the prerogative of every specific country. Nevertheless, when appropriate standardised tests are used, the behaviours of commanders can be compared to those of the tests. They would be able to assess specific cognitive abilities (e.g., memory, decision-making, and probabilistic reasoning) and performance in specific situations. Lastly, the "advanced control planning" is equally a matter of negation as much as action; it gets its definition from what the systems will not do. Commanders need to be responsible not only through action (commission) but also inaction (omission).

Ethical dilemmas demonstrate this: when an autonomous ship encounters an abandoned ship carrying citizens, the captain should determine whether to rescue them. Is the goodwill to lower the number of civilians more than the strategic danger of exposing the location of the vessel? The commander needs to determine whether such inaction will result in culpability.

Responsibility in Autonomous Command

To accomplish the legal meaning of command, we have to refer to the manner in which the authority is executed in contemporary military arrangements. Command is the exercise of authority in executing orders; by definition, it is a role full of responsibility for the decisions that flow from it. More importantly, either human or machine military forces need to be kept subordinate to a person in the armed forces who is lawfully a party to a war (Liivoja et al., 2022). To exercise this, commanders must process information to make decisions (van Creveld, 1985). They have to follow the orders, but at the same time they have the power to make sense of them and, where need be, to go against a plan. Such discretion is crucial not only in executing orders but also in the planning stage. The need for agile command is historically grounded. The most famous quote by Helmuth von Moltke is: Plans of war have no quarter when it comes to making first contact with the enemy (Hughes, 1993). As a result, commanders have to be agile when planning and executing. This was evident in the Falklands War battle of Goose Green, where adherence to a strict, staged, detailed plan, rather than Auftragstaktik or mission orders, almost proved disastrous for the British troops (Fitz-Gibbon, 2006). This agility need is not limited to the modern context or to a human commander, but also to autonomous systems. The commanders are free to exercise their discretion, and thus they would be responsible for the process and outcome.

In relation to the legal question of ordering autonomous systems which are physically and temporally distinct from operators, there are specific legal protocols. Since any autonomous weapons system involves a new political measure or method of warfare, the Article 36 review is triggered to determine whether its use is prohibited under international law (International Committee of the Red Cross, 2006). Nonetheless, the current international legal system does not explicitly exclude automation or autonomy in military platforms. This non-prohibition implies that it is still open to interpretation whether the boundaries of their use are met, which is the core of the command that the system must satisfy the commander's intent, establishing a straightforward legal connection between the commander and the deploying state's intent. States will resort more and more to a system where the individual in charge of the use of force is not physically and temporally near the device of action, as Liivoja et al. (2022) point out. Although it might not extend to all scenarios, even in the case of a sufficient connection that can be demonstrated with a human operator, the device will be under command because there will be a particular identifiable person who is commanding the activities of the machine and who is liable for its actions (Liivoja et al., 2022, p.32). Thus, we will assume that we are dealing with a process in which the states are evolving independent systems with this physical and temporal distance, and at the same time, attempt to maintain a satisfactory, responsible relation between the human being operating the device and the apparatus within these durations.

AI, AUTONOMY, AND COMMANDING THE FUTURE BATTLESPACE

Military command has been fundamentally changed by the emergence of AI, robotics, and autonomous systems that use AI to determine their behaviour. According to Ryan (2022), AI will reverse the force in tactical military activities (p. 42). It is possible that, when used correctly, AI can enhance lethality while simultaneously reducing the risk to deploying forces. Additionally, AI plays a paramount role in decision support, allowing synchronisation of tactical goals with broader strategic objectives and significantly speeding up the decision-making process. The use of AI simulations in planning has helped model the outcomes of many options in the shortest possible time during development, testing, evaluation, and active military operations. Ryan (2022) presents a vision of the combined efforts of the military, with hundreds of human employees sending out dozens of thousands of robotic systems in hundreds of configurations, including swarms and super swarms (systems of swarms). Such human-robot teams need to balance autonomy with the human accountability necessary for behaviours that satisfy the commander's intent. The question that has arisen is critical, as it concerns the control of battlefield robot swarms (Freedberg Jr., 2020). As a solution to this, a way is suggested to ensure that autonomous systems are properly governed, especially when they are in operation and human intervention is far-fetched, by applying an Advanced Control Directive (ACD).

ACDs regulate the interactions of autonomous systems that are not in communication with people, namely, during periods when humans cannot intervene. Examples of less-controversial directives- have lower probability of unintended harms and more certainty in the working environment are:

a) Guidelines on the robots, which were introduced into the underground tunnels to search and rescue the miners.

b) Guidelines on intelligent torpedoes that could be launched in order to immobilise a vessel.
c) Guidelines concerning autonomous counter-cybers.

The relevant question is whether the ACD framework also allows the authorisation of more controversial goals, such as picking a target and using force through an AWS. It is argued that, in a specific instance in which the use of an AWS is deemed justified, one should use an ACD to express the same and to offer proper mechanisms of scrutiny and accountability. That way will not advance a particular application of an autonomous system. Instead, there are neutral cases that show that autonomy itself is not the issue; rather, it is the particular goals, methods, or actions. The subtleties of a situation, which can be what is permitted and what is prohibited, may be identified and authorised through the formation of an ACD and Advanced Control Planning (ACP). The ACD must account for the strengths and weaknesses of autonomous systems. Therefore, the principle of autonomy command is compared to a traditional command and control (C2) paradigm created by van Creveld (1985).

Integrating AI in Command

To understand how Command and Control (C2) functions, as presented by van Creveld, are relevant to autonomous command, we need to examine the synergy and asymmetry between older frameworks and the needs of Advanced Control Planning (ACP) and the Advanced Control Directive (ACD). Advance control planning aims to ensure that commanders plan to accommodate technological limitations, maximising operational utility rather than complaining about weaknesses. The functions applied to this new paradigm, as suggested by van Creveld (1985), can be explained in terms of comparing the planning phase to the execution one:

Function 1-3 Information Collection and Processing: During the ACP stage, the military training units collect historical information on the forces, the enemy, the weather, and the terrain. This includes developing state-of-the-art physical hosting (or digital twins) to simulate and model the battlefield. In contrast, in an operation with an ACD, the autonomous system obtains real-time on-board data, which is processed to detect deviations from the baseline data transmitted in the directive. The system evaluates the scenario based on sensor data and compares it to the scenarios on the ACD.

Functions 4-5 Analysis and Decision Making: In planning (ACP), the objectives are analysed in simulated conditions by an expert team consisting of the commander, legal advisers, and technical experts, as part of the simulated scenarios (at real-time, fast-tempo, and slow-tempo speeds). They laid down parameters for the decision-making process for operations to be undertaken without human participation. The autonomous system, in its execution, utilises these predetermined means based on new situational information. It makes decisions according to the ACD but also considers internal state changes, such as the realisation that an unforeseen fuel depletion will alter the flight range.

Functions 6-8: Planning, Orders, and Monitoring. The ACP outlines the steps to execute decisions made by the system. Written orders delivered in simulated conditions are used to monitor the autonomous system's understanding of the orders. The system then runs such pre-determined orders once it has been deployed under an ACD. In contrast to human subordinates, the system monitors its implementation. It logs its actions to provide a record for the commander upon resumption of communications, rather than providing real-time feedback.

Therefore, the risks officials assign to specific commanders will exponentially alter the parameters within which autonomous systems are authorised to act. Such a planning stage requires an alternative process for issuing orders rather than issuing human commands. Although human subordinates need orders that can be interpreted and take initiative to get the job done (Storr, 2009), autonomy commands should be more analytical in nature. It would mean outlining a wide range of options and approving decision paths based on different risks in various situations. For example, in one case orders may be cautious, and in another case violent; a commander may order a high-probability, small, autonomous uncrewed aerial vehicle (UAV) to be shot down to gather intelligence, a risk they would not have exposed a human to. The resultant ACD is a legal document that operates similarly to an advance care directive in medical situations.

Commanders' Responsibility in Autonomy

What kind of military intelligence and personality is needed to operate autonomous systems? According to Clausewitz et al. (1976), intellect and temperament are important for command, but Chambers et al. (2017) summarised these findings into eight characteristics. Operationally, there is fluidity in integrating these roles. Will autonomy commanders establish an ACD and switch to real-time command in the theatre? Or will it be a bisect between old-fashioned and autonomous commanders? The capabilities and the mission probably determine this. One regulation states that top commanders must understand the ACD of each autonomous capability and assume ultimate responsibility for its deployment. This means that the commander who prepares the ACD must have close technical expertise. In contrast, senior leaders must have a summary of what is important to decide and the mission parameters. The personality of an autonomous leader differs from that of a conventional one. Although both share the same spirit of devotion to excellence and moral courage, their thinking conditions differ. The autonomy commander subsists on deliberation, giving detailed hypothetical scenarios, consultation and documentation of decision trees. The Clausewitzian commander, in contrast, can prosper in the real-time assault of battle whilst making decisions quickly under uncertainty. More importantly, it is just as important to understand when human performance will fail as it is to understand when AI is weak. Human beings might be exhausted, untrained, or overworked. ACD, which is prepared beforehand and involves predetermined situations, requires justification for autonomous deployment, including surprise, mass effects, and ISR, thereby imposing a lower cognitive burden on commanders. It is essential to avoid moral injury, which occurs when decision-makers are compelled into circumstances in which they lack agency or are forced to choose between bad options that include unforeseen harm (Enemark, 2019). When a war is, and commanders have agency and autonomy, the risk of moral injury is significantly reduced.

Nevertheless, when risk uncertainty is communicated, it may lead to decision-making paralysis (Rydmark et al., 2021). Instead of showing uncertainty as a single number, it is more meaningful to show uncertainty as a range to make decision-makers aware of the power of knowledge. It has been discovered that the practice of uncertainty information processing, free from time-related pressure, aids human beings in making quicker, real-time choices. Advance control planning should thus be equipped with probabilistic reasoning training (Stanovich et al., 2016) to expand the decision-making space and enhance the autonomy of autonomous agents. Lastly, countries must devise accountable AI regulations to which ACDs would adhere (Department of Defence Responsible AI Working Council, 2022; NATO, 2021). The ACP process is cyclical and designed to address predictable stressors, such as fatigue and complexity. Plans should be made in accordance with the objectives of the commander, risk behaviour, and personal values, as provided in International Humanitarian Law (IHL), to preserve a human-centred approach to the application of autonomous technologies in armed conflict (Pietrzak, 2023).

Autonomy Command

Technopower exercised using AI is not passively unfolding in time, but goes into the creation of time. According to Coeckelbergh (2022), AI can direct our time experience and organise our stories and lives based on data gathered and analysed. The phenomenon is taking place not only on the personal level, which requires individuals to place their personal stories into statistical shapes, but also on the societal level: our time is becoming the time of AI, and these mechanisms influence the general narrative of our societies. The military also faces the challenge of incorporating AI into the battlefield decision-making process. It aims to enhance human capabilities and offload areas of weakness to computerised operations to fasten and optimise decision-making. Nevertheless, the battlefield itself is a simulated fog of war, tricky, puzzling, and unpredictable. Machine learning (ML) and predictive analytics applied in this regard are inductive and probabilistic, which poses serious risks. For example, a sensor on an uncrewed ground vehicle (UGV) may declare with 95 per cent accuracy that an object is a truck belonging to the adversary. This confidence is not always reliable in practice but merely reflects the model's adherence to the training data.11 In the benign cases, this confidence can be a good one. Nonetheless, in a hostile setting, the same truck could be a cardboard decoy that the sensor will see as a spoof, or it could be in a cloud of mist that will confuse the ML's ability to identify the cars.

Battlespace AI should thus be backed by robust assurance measures so that commanders are aware when a system produces unreliable results. This is also quite a challenge because it is hard to know in advance that people will act unreliably- an issue compared to the seemingly unsolvable frame problem in cognitive science (Sperber & Wilson, 1995; Wilson & Sperber, 2002; Xu & Wang, 2012). This can be disastrous, as the commercial consequences of misunderstanding machine communication were evident in the Air France 447 accident (Salmon et al., 2016). Human factors and ergonomics specialists should therefore be put at the centre of designing and testing human-autonomy systems and authenticating AI across a variety of operational settings and simulations. Information integration is needed to mitigate these risks. Such methods as Bayesian inference can enable systems to triangulate reality by grading rival hypotheses. Like autonomous vehicles that rely on lidar, radars, and cameras to counteract errors, Bayesian techniques in military systems can be used to revise the likelihood of a hypothesis in the face of new information. Although this is the optimal way to construct beliefs in an ideal world, it does not mean that creative adversaries cannot use these systems to inject disinformation into multiple data streams and deceive the inference engine. In this way, military platforms need to realise that the separate AI sensors are fragile and should strategically combine the outputs. Commanders should not be oblivious to disinformation and uncertainty, and they should be regularly trained to understand evolving technological capabilities and enemy strategies.

In the future, autonomous assets will be able to be assigned tasks using AI developed based on an ordinary Command and Control (C2) language ontology (Curts & Campbell, 2005; Smith et al., 2009). This future command structure can be seen anticipating the recent inception of generative AI, including DALL-E. For example, a design lead assigned semantic keys to the AI, including descriptions of functions, concepts, actions, backgrounds, and styles, to instruct it to create a particular image of a female astronaut (Karenxcheng, 2022). Likewise, commanders in the future may request AI input based on aspects of command intention: task type (e.g., planning), objects of interest (e.g., lawful targets), objectives (e.g., secure a building), and hard constraints (e.g., 0% harm to civilians). Nevertheless, there are significant ethical risks if an AI-assisted command is based on massive transformer models. Historical human biases often exist in these models, which can be reinforced, leading to unfair and unjust decision-making (Bender et al., 2021). In turn, AI applications in military decision-making also entail personnel expertise, including ethical, safety, and legal professionals, while downplaying the significance of technical expertise. The commanders need to be experienced to review such systems critically and maintain a level of scepticism that neither overtrusts the AI output nor excessively distrusts it.

DISCUSSION

To ensure that commanders are morally and legally accountable for the decisions they make, leaders require a two-fold ability: they must have intimate knowledge of the situation at hand and a reasonable level of confidence in the accuracy of their sensors and communications. In the event of a compromise of the integrity of information obtained from these sources, the basis of command responsibility is jeopardised. The application of Fog Identification and Manipulation Methodology (FIMM) of the Advanced Framework of Simulation, Integration and Modelling (AFSIM) through the case study is a way to test this awareness. FIMM intentionally adds to the information confusion by providing inaccurate sensor information, lagging communication links, or data stream corruption (Tryhorn et al., 2021). The AFSIM environment variables can be controlled, such as sensor azimuth and elevation error signs, loss packets, and resend delays in communications. Subordinates who work based on such degenerated information are prone to taking orders that do not correspond with the will of the commander. The Fog Analysis Tool (FAT), related to it, enables analysts to introduce these fog effects across different aspects of a scenario (such as weapon sensors and fuel status) to condition agents to overcome such perturbations. Although FAT has yet to be validated, its ability to inject noise and disrupt information flow shows that risks can be modelled in simulation, thereby setting design guardrails before any autonomous implementation (Tryhorn et al., 2021).

This raises a relevant concern: can one assign moral responsibility for an autonomous fleet's future choices to a single commander? Considering the long time available during the planning stage, it could be said to have collective responsibility. The case of the legislative approach toward autonomous vehicles can be relevant for comparison. The legislation under consideration holds that, upon activation of an automated driving system, the software producer is responsible for automated errors leading to an accident (National Transport Commission, 2018). Is this the case, however? Once the human operator takes control of the vehicle, liability shifts to them. Still, there are significant distinctions between a tragic hero following the provisions of the highway regulations and a combat vehicle operating in the spirit of a warrior. This is in contrast to civilian transportation, where acts of war are, by nature, political and may involve the use of force to accomplish specific objectives with damaging human effects. The attribution of conflict actions to definite human actors will be inseparable towards the process of legitimising the actions of a belligerent (Pietrzak, 2022). It enables the global community to determine such actions based on suitability and reasonableness. Consequently, the security of the communication channel between machines and humans is an urgent requirement.

CONCLUSION

Decision-making on the battlefield is rapidly being transformed by newer technologies, demanding that evidence-based normative frameworks be created to guide the specifics of designing, developing, and adopting them. This is especially essential for the proper position of the human, rather than what they are supposed to do in the context of their deployment. The critics and the military institutions appear to agree that war should be a fundamentally human activity; thus, the technologies and systems involved must be subject to human intention and values. It is noted that humans should not lose the ability to comprehend weapons systems, control their use in the field of battle, and be legally and ethically responsible for their use. In the nexus of information, decision-making, and battle success, Keegan (2004) argues that war is about doing, not thinking. This discussion provides a plan for how the cognitive labour, which is the thinking of battle, can be implemented before the plunge, in this case, of autonomous assets. This enables the action to be carried out quickly and successfully without violating the law or incurring moral responsibility. Instead of flooding commanders with too much information within the battlespace, narrow data can be interpreted by autonomous forces that have been given specific authority to implement actions under specific circumstances, including the use of countermeasures to counter an imminent blue-force assault.

The development of the Advanced Control Plans and Directives is also used to establish the moral and legal accountability of autonomous weapons systems. The framework applies to stakeholders facing changing strategic and operational conditions, and to decision-making processes for lethality that occur in environments of congestion or in high- and low-tempo operations. In these situations, humans are unable to think out loud or even take control in real time to alter a decision made by an autonomous weapons system. Lastly, Advance Control Directives could be highly advantageous to commanders, as they would promote their autonomy and well-being and thus help avoid moral injury (Griffin et al., 2019) in the aftermath of conflict. This is done by ensuring that what is being done fits with the values and priorities of the humans who will be doing the same. Nonetheless, such instruments can only be developed within a supportive policy and military cultural setting, with informed AI and autonomy governance.

REFERENCES

Anduril. (2022). *Autonomy for every mission: Lattice is an open operating system for defense that brings autonomy to the toughest missions*. https://www.anduril.com/lattice/

Asaro, P. (2012). On banning autonomous weapon systems: Human rights, automation, and the dehumanization of lethal decision-making. *International Review of the Red Cross*, *94*(886), 687–709. DOI: 10.1017/S1816383112000768

Australasian Legal Information Institute. (2022). *Queensland consolidated acts*. Retrieved March 3, 2024, from http://www5.austlii.edu.au/au/legis/qld/consol_act/cc189994/s27.html#:~:text=(1)%20A%20person%20is%20not,to%20control%20the%20person's%20actions

Bender, E. M., Gebru, T., McMillan-Major, A., & Shmitchell, S. (2021). On the dangers of stochastic parrots: Can language models be too big? *Proceedings of the 2021 ACM Conference on Fairness, Accountability, and Transparency*, 610–623. DOI: 10.1145/3442188.3445922

Chambers, G., Smith, C., & Rees, L. (2017). *Command and control in modern warfare: The importance of talent, experience and expertise*. https://researchcentre.army.gov.au/sites/default/files/command_control_b5.pdf

Chan, H. Y. (2018). *Advance directives: Rethinking regulation, autonomy & healthcare decision-making*. Springer. DOI: 10.1007/978-3-030-00976-2

Coeckelbergh, M. (2022). *The political philosophy of AI*. Polity Press.

Crawford, N. (2013). *Accountability for killing: Moral responsibility for collateral damage in America's post-9/11 wars*. Oxford University Press. DOI: 10.1093/acprof:oso/9780199981724.001.0001

Curts, R. J., & Campbell, D. E. (2005). *Building an ontology for command & control*. 10th International Command and Control Research and Technology Symposium: The Future of C2. DOI: 10.21236/ADA464304

Dehais, F., Lafont, A., Roy, R., & Fairclough, S. (2020). A neuroergonomics approach to mental workload, engagement and human performance. *Frontiers in Neuroscience*, *14*, 268. DOI: 10.3389/fnins.2020.00268 PMID: 32317914

Department of Defence. (2023, March 6). *The first Ocius Bluebottle uncrewed surface vessels for the ADF*. https://www.minister.defence.gov.au/media-releases/2023–03–06/first-ocius-bluebottle-uncrewed-surface-vessels-adf

Department of Defense. (1978). *Managing weapon system software: Progress and problems* (Unclassified Digest of a Classified Report PSAD-78–112 B-16305e). https://www.gao.gov/products/psad-78–112

Department of Defense. (2023). *Autonomy in weapon systems* (DOD Directive 3000.09). https://www.esd.whs.mil/Portals/54/Documents/DD/issuances/dodd/300009p.PDF

Department of Defense Responsible AI Working Council. (2022). *Responsible artificial intelligence strategy and implementation pathway*. U.S. Department of Defense. https://media.defense.gov/2022/Jun/22/2003022604/-1/-1/0/Department-of-Defense-Responsible-Artificial-Intelligence-Strategy-and-Implementation-Pathway.PDF

Devitt, S. K. (2023). Bad, mad and cooked apples: Responsibility for unlawful targeting in human-AI military teams. In Schraagen, J. M. (Ed.), *Responsible use of AI in military systems*. CRC Press., https://arxiv.org/abs/2211.06326

Devitt, S. K., Gan, M., Scholz, J., & Bolia, R. S. (2021). *A method for ethical AI in defence* (DSTG-TR-3786). Defence Science and Technology Group. https://www.dst.defence.gov.au/publication/ethical-ai

Elish, M. C. (2019). Moral crumple zones: Cautionary tales in human-robot interaction. *Engaging Science, Technology, and Society*, *5*, 40–60. DOI: 10.17351/ests2019.260

Endsley, M. R. (2023). Supporting human-AI teams: Transparency, explainability, and situation awareness. *Computers in Human Behavior*, *140*, 107574. DOI: 10.1016/j.chb.2022.107574

Enemark, C. (2019). Drones, risk, and moral injury. *Critical Military Studies*, *5*(2), 150–167. DOI: 10.1080/23337486.2017.1384979

Fitz-Gibbon, S. (2006). *Not mentioned in despatches.*

Freedberg, S. J., Jr. (2020, November 9). Meet the Army's future family of robot tanks: RCV. *Breaking Defense*. https://breakingdefense.com/2020/11/meet-the-armys-future-family-of-robot-tanks-rcv/

Gebru, T., & Mitchell, M. (2022, June 17). Opinion: We warned Google that people might believe AI was sentient. Now it's happening. *The Washington Post*. https://www.washingtonpost.com/opinions/2022/06/17/google-ai-ethics-sentient-lemoine-warning/

Griffin, B. J., Purcell, N., Burkman, K., Litz, B. T., Bryan, C. J., Schmitz, M., Villierme, C., Walsh, J., & Maguen, S. (2019). Moral injury: An integrative review. *Journal of Traumatic Stress*, *32*(3), 350–362. DOI: 10.1002/jts.22362 PMID: 30688367

Gu, Y., Goez, J. C., Guajardo, M., & Wallace, S. W. (2021). Autonomous vessels: State of the art and potential opportunities in logistics. *International Transactions in Operational Research*, *28*(4), 1706–1739. DOI: 10.1111/itor.12785

Haggard, P. (2017). Sense of agency in the human brain. *Nature Reviews. Neuroscience*, *18*(4), 196–207. DOI: 10.1038/nrn.2017.14 PMID: 28251993

Haggard, P., Clark, S., & Kalogeras, J. (2002). Voluntary action and conscious awareness. *Nature Neuroscience*, *5*(4), 382–385. DOI: 10.1038/nn827 PMID: 11896397

Hancock, P. A., Vincenzi, D. A., Wise, J. A., & Mouloua, M. (2008). *Human factors in simulation and training*. CRC Press. DOI: 10.1201/9781420072846

Hasson, J., & Slama, A. (2023). *IHL's reasonable military commander standard and culture: Applying the lessons of ICL and IHRL*. Tulsa Law Review.

Hosang, J. B. (2020). *Rules of engagement and the international law of military operations*. Oxford University Press. DOI: 10.1093/oso/9780198853886.001.0001

Hughes, D. J. (Ed.). (1993). *Moltke on the art of war: Selected writings*. Random House Publishing Group.

International Committee of the Red Cross. (2006). A guide to the legal review of new weapons, means and methods of warfare: Measures to implement Article 36 of Additional Protocol I of 1977. *International Review of the Red Cross*, *88*(864). https://www.icrc.org/eng/assets/files/other/irrc_864_icrc_geneva.pdf

International Committee of the Red Cross. (2021). *ICRC position on autonomous weapons systems*. https://www.icrc.org/en/document/icrc-position-autonomous-weapon-systems

International Committee of the Red Cross. (2022). *What you need to know about autonomous weapons*. https://www.icrc.org/en/document/what-you-need-know-about-autonomous-weapons

Jacovi, A., Marasović, A., Miller, T., & Goldberg, Y. (2021). Formalizing trust in artificial intelligence: Prerequisites, causes and goals of human trust in AI. *Proceedings of the 2021 ACM Conference on Fairness, Accountability, and Transparency*, 624–635. DOI: 10.1145/3442188.3445923

Jimenez, G., Tan, W. S., Virk, A. K., Low, C. K., Car, J., & Ho, A. H. Y. (2018). Overview of systematic reviews of advance care planning: Summary of evidence and global lessons. *Journal of Pain and Symptom Management*, *56*(3), 436–459. e425. DOI: 10.1016/j.jpainsymman.2018.05.016 PMID: 29807158

Jonze, S. (Director). (2013). *Her* [Film]. Warner Bros. Pictures.

Kahneman, D. (2011). *Thinking, fast and slow*. Farrar, Straus and Giroux.

Kant, I. (2011). *Immanuel Kant: Groundwork of the metaphysics of morals*. Cambridge University Press., DOI: 10.1017/CBO9780511973741

Karenxcheng. (2022, August). *Created the first ever AI cover for Cosmopolitan magazine! #shorts* [Video]. YouTube. https://youtu.be/8fthDHDshvg

Keegan, J. (2004). *Intelligence in war: Knowledge of the enemy from Napoleon to al-Qaeda*. Random House.

Kuyer, P., & Gordijn, B. (2023). Nudge in perspective: A systematic literature review on the ethical issues with nudging. *Rationality and Society*, *35*(2), 191–230. DOI: 10.1177/10434631231155005

Lepenies, R., & Malecka, M. (2018). The ethics of behavioural public policy. In *The Routledge handbook of ethics and public policy*. Routledge. DOI: 10.4324/9781315461731-41

Lieberman, H. R., Bathalon, G. P., Falco, C. M., Morgan, C. A., Niro, P. J., & Tharion, W. J. (2005). The fog of war: Decrements in cognitive performance and mood associated with combat-like stress. *Aviation, Space, and Environmental Medicine*, *76*(7), C7–C14. PMID: 16018323

Liivoja, R., Massingham, E., & McKenzie, S. (2022). The legal requirement for command and the future of autonomous military platforms. *International Law Studies*, *99*(1), 27.

Luckett, T., Bhattarai, P., Phillips, J., Agar, M., Currow, D., Krastev, Y., & Davidson, P. M. (2015). Advance care planning in 21st century Australia: A systematic review and appraisal of online advance care directive templates against national framework criteria. *Australian Health Review : A Publication of the Australian Hospital Association*, *39*(5), 552–560. DOI: 10.1071/AH14187 PMID: 26027723

Maan, A. K., & Cobaugh, P. (2018). *Introduction to narrative warfare: A primer and study guide*. Narrative Strategies LLC.

McKenzie, S. (2020). When is a ship a ship? Use by state armed forces of uncrewed maritime vehicles and the United Nations Convention on the Law of the Sea. *Melbourne Journal of International Law*, *21*(2), 373–402.

McKenzie, S. (2021). Autonomous technology and dynamic obligations: Uncrewed maritime vehicles and the regulation of maritime military surveillance in the exclusive economic zone. *Asian Journal of International Law*, *11*(1), 146–175. DOI: 10.1017/S2044251321000011

Mettraux, G. (2009). *The law of command responsibility*. Oxford University Press. DOI: 10.1093/acprof:oso/9780199559329.001.0001

Mill, J. S. (1860). *On liberty* (2nd ed.). John W. Parker & Son.

Morrison, S. R. (2020). Advance directives/care planning: Clear, simple, and wrong. *Journal of Palliative Medicine, 23*(7), 878–879. DOI: 10.1089/jpm.2020.0272 PMID: 32453620

National Transport Commission. (2018). *Changing driving laws to support automated vehicles*. https://www.ntc.gov.au/transport-reform/ntc-projects/changing-driving-laws-support-AVs

NATO. (2021, October 22). *NATO releases first-ever strategy for artificial intelligence*. https://www.nato.int/cps/en/natohq/news_187934.htm

Phelps, W., & Grossman, D. (2021). *On killing remotely: The psychology of killing with drones*. Little, Brown and Company.

Pietrzak, P. (2022). *Introducing the idea of ontology in statu nascendi to the broader international relations theory. International Conference Proceeding Series: International Conference on Economics and Social Sciences*. ECLSS., https://www.eclss.org/publicationsfordoi/abst11act8boo8kIE%26SS2022_antalya.pdf

Pietrzak, P. (2023). Approaching regional conflicts through the prism of ontology in statu nascendi: The new compartmentalization of the IR theory. *In Statu Nascendi – Journal of Political Philosophy and International Relations, 6*(1).

Pietrzak, P. (2024). The sixth great debate in international relations theory revolves around clarity: Ontology in statu nascendi as a cutting-edge tool for conflict management. In Pietrzak, P. (Ed.), *Dealing with regional conflicts of global importance* (pp. 1–27). IGI Global., DOI: 10.4018/978-1-6684-9467-7.ch001

Ricoeur, P. (1980). Narrative time. *Critical Inquiry, 7*(1), 169–190. DOI: 10.1086/448093

Ryan, M. (2022). *War transformed: The future of twenty-first-century great power competition and conflict*. Naval Institute Press.

Rydmark, J., Kuylenstierna, J., & Tehler, H. (2021). Communicating uncertainty in risk descriptions: The consequences of presenting imprecise probabilities in time critical decision-making situations. *Journal of Risk Research, 24*(5), 629–644. DOI: 10.1080/13669877.2020.1801807

Salmon, P. M., Walker, G. H., & Stanton, N. A. (2016). Pilot error versus sociotechnical systems failure: A distributed situation awareness analysis of Air France 447. *Theoretical Issues in Ergonomics Science*, *17*(1), 64–79. DOI: 10.1080/1463922X.2015.1106618

Schneider, S. (2019). *Artificial you*. Princeton University Press.

Sharkey, N. (2012). Killing made easy: From joysticks to politics. In *Robot ethics: The ethical and social implications of robotics* (pp. 111–128). MIT Press.

Smith, B., Miettinen, K., & Mandrick, W. (2009). The ontology of command and control (C2). *Proceedings of the 14th International Command and Control Research and Technology Symposium (ICCRTS)*.

Smith, R. P. (1981). Boredom: A review. *Human Factors*, *23*(3), 329–340.1. DOI: 10.1177/001872088102300308 PMID: 7016721

Sperber, D., & Wilson, D. (1995). *Relevance: Communication and cognition* (2nd ed.). Wiley-Blackwell.

Stanovich, K. E., & West, R. F. (2000). [3] Individual differences in reasoning: Implications for the rationality debate? *The Behavioral and Brain Sciences*, *23*(5), 645–726. DOI: 10.1017/S0140525X00003435 PMID: 11301544

Stanovich, K. E., West, R. F., & Toplak, M. E. (2016). *The rationality quotient: Toward a test of rational thinking*. MIT Press. DOI: 10.7551/mitpress/9780262034845.001.0001

Stanton, A. (Director). (2008). *Wall-E* [Film]. Walt Disney Pictures; Pixar Animation Studios.

Storr, J. (2009). *The human face of war*. A&C Black.

Talbert, M. (2016). *Moral responsibility: An introduction*. John Wiley & Sons.

Tryhorn, D., Dill, R., Hodson, D. D., Grimaila, M. R., & Myers, C. W. (2021). Modeling fog of war effects in AFSIM. *The Journal of Defense Modeling and Simulation*, *20*(2), 131–146. DOI: 10.1177/15485129211041963

Tsai, G. (2014). Rational persuasion as paternalism. *Philosophy & Public Affairs*, *42*(1), 78–112. DOI: 10.1111/papa.12026

van Creveld, M. (1985). *Command in war*. Harvard University Press.

von Clausewitz, C., Howard, M., & Paret, P. (1976). *On war*. Princeton University Press. DOI: 10.1515/9781400837403

Wilson, D., & Sperber, D. (2002). Relevance theory. In *Handbook of pragmatics*.

Xu, Y., & Wang, P. (2012). The frame problem, the relevance problem, and a package solution to both. *Synthese*, *187*(1), 43–72. DOI: 10.1007/s11229-012-0117-8

ADDITIONAL READINGS

Boulanin, V., & Verbruggen, M. (2017). *Mapping the development of autonomy in weapon systems*. Stockholm International Peace Research Institute (SIPRI). https://www.sipri.org/publications/2017/other-publications/mapping-development-autonomy-weapon-systems

Cummings, M. L. (2017). *Artificial intelligence and the future of warfare*. Chatham House for the Royal Institute of International Affairs. https://www.chathamhouse.org/2017/01/artificial-intelligence-and-future-warfare

Horowitz, M. C. (2019). When speed kills: Lethal autonomous weapon systems, deterrence and stability. *The Journal of Strategic Studies*, *42*(6), 764–788. DOI: 10.1080/01402390.2019.1621174

Payne, K. (2018). Artificial intelligence: A revolution in strategic affairs? *Survival*, *60*(5), 7–32. DOI: 10.1080/00396338.2018.1518374

Roff, H. M., & Moyes, R. (2016). *Meaningful human control, artificial intelligence and autonomous weapons*. Article 36. http://www.article36.org/wp-content/uploads/2016/04/MHC-AI-and-AWS-FINAL.pdf

Scharre, P. (2018). *Army of none: Autonomous weapons and the future of war*. W. W. Norton & Company.

Sparrow, R. (2007). Killer robots. *Journal of Applied Philosophy*, *24*(1), 62–77. DOI: 10.1111/j.1468-5930.2007.00346.x

KEY TERMS AND DEFINITIONS

Advance Control Directive (ACD): A system of governance in which human beings will issue directives on the actions of autonomous systems in the future, which respect the legal and ethical accountability boundaries by prescribing the autonomous systems to act ethically and legally before they can even start their operations.

Type 1 vs. Type 2 Reasoning: The difference between fast, intuitive, unconscious thinking (Type 1) and slow, deliberate conscious reasoning (Type 2), the former arguably necessary to make moral decisions, but not quickening up.

Moral Crumple Zone: An idea that was used to state a scenario in which a human operator suffers the moral and legal burden of the failures of a complex autonomous system that he effectively had limited control over.

Autonomy Command: An advanced command paradigm in which the leaders would be expected to have specialised abilities in probabilistic reasoning, system logic, and risk assessment to guide independent assets under real-time stress.

Article 36 Review: A legal review necessary under the 1977 Additional Protocol I to the Geneva Conventions to enquire whether a new weapon, means, or method of warfare is illegal under international law.

Predictive Situation Awareness: The ability of the commanders to foresee and plot future events at a level of detail during the planning stage, so that they will accept responsibility even without real-time contact in the eventuality of such events.

Fog Identification and Manipulation Methodology (FIMM): A simulation method of carefully placing data errors, sensor noise, and communication delays ("fog") and seeing how human-machine teams cope with uncertainty and poor information.

Chapter 7
Moral Code:
Ethical Software Engineering for AI-Driven Battlefield Healthcare

Hanh Le
http://orcid.org/0009-0007-8172-0809
University of the Cumberlands, USA

ABSTRACT

As warfare becomes increasingly defined by algorithms and autonomous systems, the moral weight of software development extends far beyond lines of code. Nowhere is this more apparent than in AI-driven battlefield healthcare, where diagnostic models, triage algorithms, and telemedicine platforms can determine who lives or dies. This chapter explores the ethical, technical, and human dimensions of software engineering for military medical AI. It argues that secure, explainable, and accountable software systems are indispensable for preserving trust, upholding international humanitarian law, and ensuring the moral integrity of code deployed in combat zones. By examining software verification, human-in-the-loop design, and ethical auditing frameworks, this chapter offers a blueprint for moral resilience in the age of software-defined warfare.

INTRODUCTION

In modern operational environments, software has become the decisive layer that enables coordination, data-driven decision-making, and medical response in complex, high-pressure situations. Artificial intelligence (AI) is now an integral component of modern healthcare delivery in conflict and disaster zones, powering systems that perform automated triage, diagnostic support, patient monitoring, and telemedicine coordination. These AI-driven systems are designed to operate

DOI: 10.4018/979-8-3373-6264-9.ch007

in unstable and resource-constrained settings, where reliability, latency, and fault tolerance are as critical as clinical accuracy.

From an engineering perspective, the growing reliance on AI in battlefield healthcare introduces both technical opportunities and ethical constraints. Machine learning models can rapidly analyze functional data, prioritize medical interventions, and optimize logistics in real time. However, these same systems also carry significant risks when deployed in unpredictable conditions. A misclassified injury, a communication delay, or a corrupted sensor input can lead to life-critical errors. In such cases, software quality, model robustness, and ethical design become inseparable from moral responsibility.

This chapter approaches the problem from a software engineering and AI systems perspective, arguing that ethical software development is essential for ensuring safety, trustworthiness, and compliance in AI-driven medical applications. The notion of moral code is treated as both a technical and ethical construct: it represents the disciplined integration of ethical reasoning into architecture, design, coding standards, and verification processes. In practice, this involves embedding moral and operational constraints directly into algorithms, ensuring traceability through version control and documentation, and validating that AI outputs align with established medical ethics and humanitarian principles.

AI-driven battlefield healthcare systems must therefore be engineered with explainability, transparency, and human oversight as core design parameters, not as afterthoughts. Software verification, model validation, and fail-safe mechanisms are not only technical safeguards but also moral imperatives in systems that influence life-and-death outcomes.

This chapter applies three conceptual frameworks: Deskilling of Jobs Theory, Job Characteristics Theory, and Technology Threat Avoidance Theory to examine how automation affects professional competence, human–machine interaction, and trust in AI systems. By synthesizing these perspectives, it proposes a structured approach to ethical software engineering in defense healthcare contexts, where algorithmic precision must coexist with accountability, auditability, and human-centered design.

Ultimately, the goal is to establish a Moral Code Architecture. This framework unites technical reliability with ethical accountability in AI systems designed to operate on the battlefield and in the emergency medical field. Through this lens, software engineering becomes not only a technical discipline but also a moral practice, shaping how humanity delegates decision-making power to machines in the most critical environments imaginable.

This chapter is organized as follows: Section 2 presents the research questions, purpose, and scope of the study. Section 3 establishes the conceptual foundations through the background and theoretical frameworks that guide the analysis. Section 4 provides a comprehensive literature review on software development, the phenomenon of deskilling, and the integration of generative AI in programming practices. Section 5 describes the qualitative research methodology used to explore these themes. Section 6 details the key findings of the study, while Section 7 offers an in-depth discussion of their implications. Section 8 presents practical solutions and recommendations derived from the findings. Finally, Section 9 concludes the chapter by summarizing the insights gained and identifying directions for future research.

RESEARCH QUESTIONS

To investigate how ethical software engineering practices influence the design, implementation, and deployment of AI-driven battlefield healthcare systems, this study is structured around three central research questions. Each question examines a core dimension of what may be termed the ethical architecture of software-defined warfare: reliability, responsibility, and transparency. Together, these dimensions represent the moral and technical foundation upon which trustworthy AI healthcare systems must be built.

The first dimension, reliability, concerns the technical robustness and operational resilience of AI systems under extreme, unpredictable conditions. It explores how software engineers verify, validate, and stress-test algorithms to ensure dependable performance even when sensor inputs degrade, communication links fail, or contextual data shifts unexpectedly.

The second dimension, responsibility, examines how moral and professional accountability are distributed across the AI development lifecycle, from data collection and model training to field deployment and post-decision auditing. It considers how engineers, medical officers, and command structures share responsibility for the outcomes produced by autonomous or semi-autonomous systems.

The third dimension, transparency, addresses the necessity of explainability, documentation, and auditability in AI-driven medical decision-making. It investigates how developers can design systems that are interpretable to human operators, compliant with ethical review standards, and traceable through secure logging and version control mechanisms.

Collectively, these three dimensions guide the study's central inquiry: how can software engineers integrate ethical reasoning into the core design principles of AI-driven battlefield healthcare systems to ensure both operational effectiveness and moral integrity? Through this lens, ethical software engineering becomes not only a technical requirement but also a framework for sustaining human trust and accountability in the era of intelligent automation.

RQ1: How can secure, ethical software development practices be implemented in AI-driven battlefield healthcare systems?

RQ2: How can human oversight and fail-safe mechanisms be effectively integrated into AI-driven triage, diagnostic, and telemedicine platforms?

RQ3: What mechanisms can ensure transparency, accountability, and moral responsibility in software-defined warfare environments?

PURPOSE AND SCOPE

Together, these research questions guide an inquiry into the moral code of software-defined warfare, as well as the principles and practices that ensure AI-driven battlefield healthcare technologies uphold humanitarian values. The study aims to construct a conceptual framework linking ethical software engineering, AI governance, and moral responsibility, offering both theoretical insights and practical guidelines for engineers, policymakers, and defense ethicists.

The focus of this chapter is to examine how ethical software engineering serves as the moral foundation for developing AI-driven battlefield healthcare systems, including triage applications, diagnostic algorithms, and telemedicine platforms designed for combat environments. As warfare becomes increasingly algorithmic and data-driven, the chapter argues that moral integrity must be engineered into the software itself, not treated as an afterthought or external policy concern.

Specifically, this chapter explores how the principles of security, transparency, explainability, and human accountability can be systematically embedded into military medical AI. The discussion emphasizes that software engineers working in defense contexts are not merely technical practitioners but moral agents, whose design decisions directly influence the preservation of human life and adherence to humanitarian ethics in warfare.

Through the integration of three theoretical lenses: Deskilling of Jobs Theory, Job Characteristics Theory, and Technology Threat Avoidance Theory, the chapter analyzes how automation affects human skill, professional motivation, and moral responsibility. It contends that ethical AI design must aim to preserve human expertise, sustain meaningful engagement, and reinforce user trust through transparent and accountable software architectures.

Ultimately, the chapter's central focus is the development of a Moral Code Architecture, a conceptual and practical model for guiding ethical software engineering in AI-driven battlefield healthcare. This architecture seeks to align technical excellence with moral purpose, ensuring that every line of code written for combat medicine contributes not only to operational effectiveness but also to the ethical treatment of human life under the laws of armed conflict.

BACKGROUND

AI has become a transformative element in modern military operations, particularly in battlefield healthcare. Systems powered by AI, such as autonomous triage algorithms, diagnostic assistants, and telemedicine platforms, are now being deployed to assist medics and healthcare teams in making rapid, high-stakes decisions under extreme conditions. These systems process vast amounts of real-time data to recommend treatment priorities, assess vital signs, and allocate limited medical resources with remarkable efficiency, in environments where every second counts. AI technologies promise to enhance decision-making accuracy and expand human capabilities beyond traditional limitations.

Yet, as AI takes on increasingly complex and consequential roles, its integration into battlefield healthcare raises critical ethical and technical questions. How can software engineers ensure that these systems function reliably and transparently in unpredictable environments? How can developers encode humanitarian principles into algorithms that may influence life-or-death outcomes? The shift toward software-defined decision-making means that ethical accountability now extends beyond the battlefield and into the software development lifecycle. A single line of code, an unverified data model, or a bias embedded in an algorithm can determine the fate of those in need of urgent care. Thus, software engineering in this context is not merely a technical pursuit; it is an act of moral agency.

Inadequate or poorly governed AI systems pose substantial risks, including data breaches, decision bias, and failures in diagnostic accuracy. These challenges make it imperative to integrate ethical safeguards directly into the design and deployment of AI-driven medical systems. Secure coding, continuous validation, and transparent documentation must be treated as essential components of ethical engineering, not optional features. The goal is to design systems that are not only technically sound but also aligned with the humanitarian principles of safety, fairness, and accountability.

This chapter examines how ethical software engineering principles can be systematically embedded throughout the lifecycle of AI-driven battlefield healthcare systems, from conceptual design and model training to deployment and post-decision auditing. It emphasizes the importance of explainability, robustness, and moral traceability in software that influences medical outcomes in operational environments. By drawing upon the Deskilling of Jobs Theory, Job Characteristics Theory, and Technology Threat Avoidance Theory, the discussion establishes a conceptual framework called the Moral Code Architecture. This framework integrates moral responsibility with technical reliability, offering a model for the development of trustworthy AI systems that uphold human dignity and professional integrity in military medical practice.

THEORETICAL FRAMEWORK

Three theoretical foundations guide this study: Deskilling of Jobs Theory, Job Characteristics Theory, and Technology Threat Avoidance Theory. Together, they provide a multidimensional analytical lens for examining how AI transforms human agency, ethical engagement, and system trust within the domain of battlefield healthcare software engineering. Each theory contributes a distinct ethical and technical perspective, forming the conceptual pillars of the proposed Moral Code Architecture.

Deskilling of Jobs Theory

The Deskilling of Jobs Theory (Braverman, 1974) asserts that automation can diminish human expertise and reduce workers' autonomy over their labor processes. Within the context of AI-driven battlefield healthcare, this theory illuminates the risk of moral and cognitive deskilling among both software engineers and medical personnel who increasingly depend on automated recommendations. As AI systems assume greater decision-making authority, developers may become detached from ethical reflection, adopting the implicit belief that "the code knows best." This detachment can lead to the erosion of moral sensitivity and critical judgment. However, the theory also allows for the possibility of reskilling, wherein developers strengthen their ethical literacy, socio-technical awareness, and moral reasoning to retain meaningful control over automation. In this sense, the design of ethical AI systems must not only mitigate technical errors but also preserve the moral agency of their human collaborators.

Job Characteristics Theory

Job Characteristics Theory (Hackman & Oldham, 1976) posits that motivation, satisfaction, and responsibility arise from five core job dimensions: autonomy, task significance, skill variety, feedback, and task identity. Applied to ethical software engineering, this framework emphasizes that moral accountability flourishes when engineers have autonomy over design decisions related to transparency, bias mitigation, and safety validation. Conversely, overly rigid command structures or top-down ethical compliance protocols can suppress moral initiative and foster ethical passivity. Designing software development environments that encourage moral autonomy, iterative feedback, and ethical creativity enhances both professional motivation and system integrity. Such environments align individual ethical reflection with institutional mission, reinforcing the moral reliability of AI technologies deployed in life-critical settings.

Technology Threat Avoidance Theory

The Technology Threat Avoidance Theory (Liang & Xue, 2009) provides insight into how individuals respond to perceived technological risks through protective or preventive actions. In the sphere of AI healthcare, both engineers and clinicians often recognize potential threats, system malfunctions, algorithmic bias, or diagnostic error. These perceptions influence their levels of ethical vigilance and trust. While concern about system failure may motivate stronger safeguards, excessive reliance on technical defenses can lead to moral complacency, where ethical reasoning is replaced by procedural compliance. A balanced approach is therefore essential: effective ethical resilience arises from coupling robust technical defenses with reflective moral judgment, ensuring that the pursuit of safety does not obscure the necessity of moral accountability.

Integration: The Moral Code Architecture

Synthesizing these three theoretical lenses yields the Moral Code Architecture, a holistic ethical engineering framework that integrates skill preservation, motivational integrity, and technological assurance. The architecture conceptualizes ethical software engineering as a three-layer system, each anchored by a corresponding theoretical foundation:

Table 1. The three-layer model of ethical software engineering

Layer	Primary Focus	Theoretical Foundation	Ethical Engineering Outcome
Human-Centric Layer	Skill preservation and moral agency	Deskilling of Jobs Theory	Prevents automation-induced moral deskilling through human-in-the-loop design
Organizational Layer	Work design and professional motivation	Job Characteristics Theory	Fosters engagement, autonomy, and ethical accountability in DevSecOps teams
Technological Layer	Threat mitigation and system trust	Technology Threat Avoidance Theory	Enhances transparency, explainability, and moral assurance through secure design

Table 1 outlines the three-layer ethical engineering framework that guides the study. Each layer represents a distinct but interconnected dimension of ethical software engineering practice. The Human-Centric Layer focuses on skill preservation and moral agency, drawing on the Deskilling of Jobs Theory to prevent automation-induced moral deskilling through human-in-the-loop design. The Organizational Layer emphasizes work design and professional motivation, grounded in the Job

Characteristics Theory, to promote engagement, autonomy, and ethical accountability within DevSecOps teams. Finally, the Technological Layer centers on threat mitigation and system trust, guided by the Technology Threat Avoidance Theory, enhancing transparency, explainability, and moral assurance through secure and responsible design (see Table 1).

Framework Narrative

In this architecture, human skills (derived from Deskilling Theory) anchor moral agency; job meaning (from Job Characteristics Theory) sustains motivation and ethical engagement; and technological assurance (from the Technology Threat Avoidance Theory) reinforces user trust and responsible system interaction. The intersection of these domains: skill, meaning, and assurance, defines the ethical nucleus of AI-driven battlefield healthcare systems.

At this intersection emerges the principle of moral code engineering: the deliberate embedding of ethical reasoning, transparency mechanisms, and fail-safe safeguards within software that directly affects human life. The framework posits that the more software development practices support continuous learning, motivational autonomy, and security transparency, the more morally resilient and trustworthy the resulting AI systems become.

The conceptual model can be visually represented as a three-circle Venn diagram illustrating the convergence of the three theories:

- **Deskilling of Jobs Theory**: *Skill and Agency*
- **Job Characteristics Theory**: *Meaning and Motivation*
- **Technology Threat Avoidance Theory**: *Trust and Assurance*

At their intersection lies the core outcome: Ethical Software Engineering for AI-Driven Battlefield Healthcare. Surrounding the circles are dynamic feedback loops labeled Transparency, Human Oversight, and Moral Accountability, symbolizing the continuous interaction between human, organizational, and technological systems. The model emphasizes that moral responsibility in AI warfare emerges not from a single component, neither algorithm nor operator, but from the synergistic interaction of these ethical subsystems.

Theoretical Propositions

From the integrated model, three theoretical propositions emerge to guide the study. The first proposition asserts that ethical software engineering practices that prioritize skill preservation can mitigate the deskilling effects of AI automation in battlefield healthcare. By maintaining human expertise and embedding continuous learning within development processes, organizations can prevent overreliance on automated systems and sustain critical decision-making capabilities in high-pressure environments.

The second proposition suggests that job designs emphasizing autonomy, constructive feedback, and moral significance enhance both professional motivation and ethical awareness among DevSecOps teams. When developers perceive their work as ethically meaningful and technically impactful, they are more likely to internalize moral responsibility within the software lifecycle.

Finally, the third proposition contends that transparent, secure, and explainable AI systems reduce perceived technological threats, fostering moral confidence and user trust in high-stakes medical contexts. Through explainability and accountability, such systems can strengthen the human–AI partnership essential for responsible innovation in military healthcare.

Figure 1. The moral code architecture framework

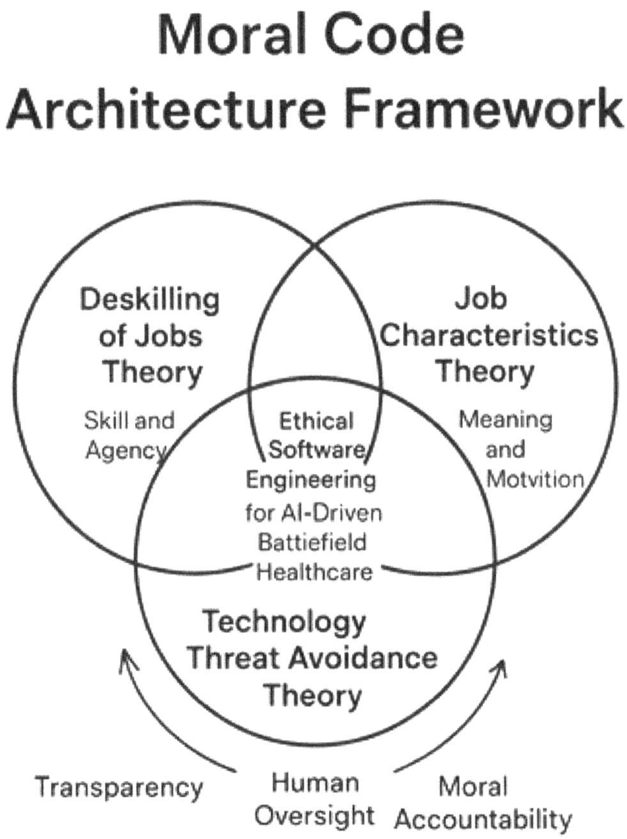

LITERATURE REVIEW

The integration of AI into healthcare introduces unprecedented ethical and technical challenges. While AI promises faster triage, improved diagnostics, and autonomous support for combat medics, it also presents risks associated with algorithmic bias, system failure, and moral accountability. As warfare becomes increasingly software-defined, the ethics of code itself, how it is written, tested, and deployed, become a determinant of life and death (Cummings, 2021; Decker et al., 2023).

AI and Battlefield Healthcare Systems

AI-driven applications are rapidly transforming battlefield medicine through capabilities such as automated triage, diagnostic decision support, predictive casualty evacuation, and telemedicine under austere conditions. Studies show that AI-assisted triage systems can analyze physiological data, imaging, and environmental cues to prioritize care and allocate scarce medical resources in real time (Kuttan et al., 2025; Nichols et al., 2023). Similarly, autonomous and semi-autonomous diagnostic systems promise faster detection of hemorrhage, trauma severity, or infectious exposure, thereby augmenting the decision-making of combat medics (Cole et al., 2025).

Despite these benefits, the operational reliability of AI systems in battlefield contexts remains a persistent concern. Consequently, the reliability, explainability, and interoperability of these systems are central to ensuring that decision support tools behave predictably under uncertain and high-stakes conditions. Explainability, in particular, is vital to trust and accountability: clinicians and medics must understand the reasoning behind algorithmic recommendations to validate outputs and intervene when systems err (Ribeiro et al., 2016).

Ethical deployment of military medical AI requires fail-safe design, robust data integrity, and human-in-the-loop verification, all of which are emphasized in the Department of Defense's (2021) Responsible AI principles. These principles: responsibility, equity, traceability, reliability, and governability, require that AI tools be not only technically sound but also ethically defensible throughout their lifecycle. Scholars of military ethics and human-centered AI argue that oversight mechanisms, audit trails, and algorithmic transparency must be integrated from the design phase to deployment to prevent catastrophic decision errors (Stahl, 2021). When AI systems misclassify injuries, delay triage, or provide erroneous treatment guidance, the consequences can be fatal, undermining both mission effectiveness and humanitarian law obligations.

Therefore, ethical software design in battlefield medicine is not solely a technical mandate but a humanitarian imperative, anchored in the moral responsibility to safeguard life in war. Integrating ethical engineering into AI system design ensures that autonomy never supersedes accountability and that the deployment of intelligent systems aligns with the core principles of medical ethics, including beneficence, nonmaleficence, and respect for human dignity (Stahl, 2021).

Software Verification and Validation in Military AI

Verification and validation (V&V) for artificial intelligence (AI) enabled healthcare and military systems require a fundamental rethinking of traditional assurance practices. Unlike conventional software that follows deterministic logic, AI models learn and evolve from data, introducing uncertainty, opacity, and contextual dependence. As Sommerville (2020) observed, classical testing and verification methods are insufficient for data-driven models that continuously adapt to new inputs. Consequently, V&V in this domain must evolve to address issues of reliability, explainability, and accountability across dynamic operational environments.

Recent work reframes the V&V paradigm under the broader concept of Test, Evaluation, Verification, and Validation (TEVV), which integrates software testing with data auditing, robustness assessment, and lifecycle monitoring. The National Institute of Standards and Technology (2025) emphasizes that TEVV must be context-sensitive and generate auditable evidence about model performance, safety, and limitations across the system's life cycle. In healthcare and combat medicine, these requirements translate to ensuring that AI systems maintain consistent diagnostic or triage accuracy under stress, environmental noise, and shifting data distributions (Cole et al., 2025).

Military medicine presents additional V&V challenges due to its high-risk and adversarial contexts. Scholars have argued that continuous validation in such settings must include adversarial testing, red teaming, and explainable model auditing to expose vulnerabilities prior to deployment (Majumdar et al., 2025). Red-team exercises and threat simulations are increasingly recognized as indispensable for uncovering weaknesses that standard benchmark tests overlook. These practices are consistent with the Department of Defense's (DoD, 2021) mandate for ethical AI development, which identifies five core principles: responsibility, equity, traceability, reliability, and governability, as essential attributes of trustworthy AI.

To operationalize these practices, V&V for military AI must be continuous rather than episodic. Constant monitoring, periodic re-validation, and automated alerting for model drift are necessary to ensure sustained performance and compliance with ethical standards (National Institute of Standards and Technology, 2025). This ongoing approach reflects the integration of V&V within DevSecOps pipelines, enabling rapid feedback between testing, deployment, and field performance (Department of Defense, 2021). Such practices represent a form of ethical engineering discipline designed not only to mitigate harm but also to preserve human dignity in high-stakes, software-defined warfare.

Human-in-the-Loop and Moral Responsibility

Human oversight in AI-assisted decision-making remains central to preserving ethical and operational accountability, particularly in high-stakes battlefield healthcare. Autonomous medical systems, while capable of rapid triage and clinical support, risk moral disengagement if human judgment is bypassed (Cummings, 2021; Floridi & Sanders, 2024). The integration of a human-in-the-loop (HITL) framework ensures that life-or-death decisions, such as casualty prioritization or emergent interventions, remain morally anchored within human oversight, aligning technological efficiency with ethical responsibility (Cole et al., 2025; Stahl, 2021).

However, introducing human oversight in high-pressure military environments presents its own challenges. Cognitive workload, situational awareness demands, and trust calibration can affect decision quality, particularly when AI outputs are complex or opaque (Hackman & Oldham, 1976; Liang & Xue, 2009). Ribeiro et al. (2016) emphasize the importance of explainable AI (XAI) as a tool for maintaining operator trust and interpretability, ensuring that human supervisors can understand, interrogate, and validate algorithmic recommendations.

Furthermore, ethical software engineering in military healthcare requires embedding HITL principles into design and deployment processes. This process includes adversarial testing and red-teaming exercises to identify failure points and mitigate risks prior to deployment (Majumdar et al., 2025; Mulchandani & Shanahan, 2022). It also necessitates alignment with institutional ethical frameworks and governance, as outlined in the Department of Defense (2021) Responsible AI guidance, emphasizing responsibility, equity, traceability, reliability, and governability. By combining these principles with human oversight, military healthcare systems can leverage AI to augment care without abdicating ethical accountability (Braverman, 1974; Cole et al., 2025; Kuttan et al., 2025).

Ultimately, effective HITL implementation balances automation efficiency with human cognitive capacity, transparency, and interpretability. Ethical resilience in AI-enabled battlefield medicine is achieved not solely through technical sophistication, but through continuous, context-sensitive human engagement, ensuring that AI serves as a tool to enhance, rather than replace, moral and operational judgment (Decker et al., 2023; Sommerville, 2020; Stahl, 2021).

Transparency and Ethical Auditing in AI-Enabled Military Healthcare

Transparency in AI systems is critical for fostering ethical accountability, operational trust, and human oversight, particularly in high-stakes battlefield healthcare applications. Clear documentation of data provenance, model decision logic, and system limitations allows both internal and external stakeholders to evaluate not only the technical performance of AI but also its alignment with ethical and humanitarian principles (Cole et al., 2025; Stahl, 2021). Ethical auditing in this context involves systematic assessments of algorithmic accuracy, fairness, and compliance with the principles of medical ethics, including beneficence, nonmaleficence, and respect for human dignity (Department of Defense, 2021; Floridi & Sanders, 2024).

Traceability of AI workflows, from training data to deployment artifacts, enables organizations to hold both developers and operators accountable in the event of system failure or misclassification. Code repositories, version histories, and operational logs can serve as moral and legal evidence, providing an auditable chain that supports responsible decision-making (Cummings, 2021; Majumdar et al., 2025; National Institute of Standards and Technology, 2025). Such transparency is particularly vital in military medicine, where automated triage and clinical decision-support systems directly influence life-or-death outcomes (Cole et al., 2025; Kuttan et al., 2025).

Moreover, embedding transparency into software engineering processes aligns with the broader objectives of ethical AI governance. By making system behavior interpretable and auditable, developers and commanders can ensure that AI augments rather than overrides human moral judgment, providing a robust mechanism for continuous ethical oversight (Braverman, 1974; Decker et al., 2023; Sommerville, 2020). This combination of technical traceability and ethical auditing reinforces trust, mitigates risk, and upholds accountability in AI-enabled battlefield healthcare systems.

Ethical Dimensions of Software-Defined Warfare

Software-defined warfare increasingly depends on autonomous and semi-autonomous systems that operate in extreme, high-stakes environments with minimal direct human oversight. These systems, including AI-enabled battlefield healthcare platforms, require designers and operators to consider both operational performance and ethical accountability (Cole et al., 2025; Mulchandani & Shanahan, 2022). Scholars argue that responsibility in such contexts extends beyond battlefield commanders to encompass software engineers, data scientists, and system architects whose design choices directly shape AI behavior and its consequences (Cummings, 2021; Stahl, 2021).

This perspective reframes moral accountability as a distributed process, embedded in the architecture of algorithms, data pipelines, validation procedures, and operational protocols (Department of Defense, 2021; Floridi & Sanders, 2024). Ethical software engineering in software-defined warfare thus becomes a form of moral code-writing, requiring developers to anticipate potential harm, bias, and unintended consequences before deployment (Majumdar et al., 2025; Sommerville, 2020). By embedding ethical considerations into system design, verification, and validation, AI-enabled platforms can align operational efficiency with humanitarian principles and legal obligations, ensuring that technological sophistication does not come at the cost of moral accountability (Braverman, 1974; Kuttan et al., 2025).

In this context, ethical oversight is not a post-hoc process but a continuous, proactive dimension of the software lifecycle, integrating human judgment, transparency, and rigorous auditing throughout development, deployment, and operation (Decker et al., 2023; National Institute of Standards and Technology, 2025). Such an approach ensures that AI systems in software-defined warfare remain both effective and ethically defensible.

Synthesis

The literature reviewed underscores an apparent convergence between software ethics, military AI governance, and healthcare accountability. Across studies, scholars consistently argue that ethical considerations must be embedded throughout the technical lifecycle, from algorithm design and data pipeline development to testing, deployment, and continuous monitoring (Cummings, 2021; Department of Defense, 2021; Floridi & Sanders, 2024). This moral infrastructure ensures that AI systems do not merely perform tasks efficiently but do so in alignment with humanitarian principles, legal norms, and clinical best practices (Cole et al., 2025; Kuttan et al., 2025; Stahl, 2021).

In the context of battlefield healthcare, the ethical stakes are particularly high. AI-enabled triage, diagnostic support, and telemedicine platforms operate in environments where misclassification, delay, or opaque reasoning can have life-or-death consequences. Consequently, human-in-the-loop frameworks, transparency mechanisms, and ethical auditing procedures are not optional; they are integral to system reliability and moral accountability (Decker et al., 2023; Majumdar et al., 2025; Ribeiro et al., 2016). Ethical software engineering in this context requires developers to anticipate bias, system failure, and unintended consequences at every stage, transforming each line of code, each test case, and each deployment decision into an expression of moral responsibility (Braverman, 1974; Sommerville, 2020).

As military healthcare becomes increasingly automated, the critical question shifts from *whether* AI should assist human medics to *how* AI can be designed, governed, and audited to ensure that its foundations uphold operational reliability, legal compliance, and humanitarian ethics. Integrating these principles creates AI systems that not only enhance mission effectiveness but also reinforce ethical accountability as a core feature of software-defined warfare (Cole et al., 2025; Mulchandani & Shanahan, 2022).

Gaps in the Literature

Despite the growing interdisciplinary interest in artificial intelligence (AI), military ethics, and digital healthcare, significant gaps persist in the existing body of literature. These deficiencies stem from structural, methodological, and conceptual limitations that constrain scholarly efforts to address the moral, operational, and technical complexities of AI-enabled battlefield medicine. Collectively, these gaps highlight the need for a more unified and empirically grounded understanding of how ethical principles can be embedded into the design, deployment, and governance of military medical AI systems.

One significant gap involves the limited integration of ethics into software engineering frameworks. While scholars have examined AI ethics in healthcare (Stahl, 2021) and explored governance principles in defense contexts (Department of Defense, 2021; Mulchandani & Shanahan, 2022), few studies explicitly bridge these domains with the practical realities of software verification, validation, and DevSecOps lifecycles. This disconnect results in a fragmented understanding of how moral reasoning can be operationalized through coding standards, testing protocols, and model auditing (Floridi & Sanders, 2024; Sommerville, 2020). Without an applied ethical architecture, technical compliance often overshadows moral accountability, leaving software engineers without clear frameworks for embedding humanitarian values into algorithmic design.

A second gap lies in the scarcity of empirical research conducted in military and emergency healthcare environments. Much of the current literature relies on theoretical or policy-based analyses rather than field-based or experimental studies. As a result, little is known about how AI systems perform ethically and operationally under the uncertainty, stress, and chaos characteristic of combat or crisis conditions (Cole et al., 2025; Kuttan et al., 2025). Few investigations examine how human operators interact with AI-driven diagnostic or triage tools in real or simulated battlefield scenarios, limiting insight into critical factors such as trust calibration, situational awareness, and shared accountability in human–machine decision-making (Cummings, 2021).

Another gap concerns the absence of standardized ethical auditing and accountability mechanisms. Although organizations such as the National Institute of Standards and Technology (2025) have proposed general guidelines for testing and verification, no universal framework currently exists for ethical auditing of AI systems in military healthcare. Transparency and explainability are often discussed as abstract design ideals rather than measurable or enforceable standards. Consequently, there is still no consensus on what constitutes valid ethical verification of AI software in contexts where decisions carry life-or-death consequences (Majumdar et al., 2025; Ribeiro et al., 2016).

Interdisciplinary collaboration also remains underdeveloped. Much of the scholarship isolates technical research from ethical and humanistic inquiry, resulting in fragmented analyses that fail to capture the full socio-technical complexity of military AI deployment. Few studies integrate perspectives from philosophy, software engineering, and military medicine into a coherent analytical model. This disciplinary separation inhibits the development of holistic frameworks capable of addressing the interconnected issues of algorithmic bias, operational reliability, and humanitarian accountability (Braverman, 1974; Floridi & Sanders, 2024).

Finally, the emerging nature of software-defined warfare and AI governance contributes to the literature gap. The rapid evolution of AI technologies in defense and healthcare contexts has outpaced academic assessment, leaving scholarship often speculative and reactive rather than empirically grounded. Concepts such as software-defined warfare and autonomous medical decision support remain fluid, with limited data available from actual deployments (Cole et al., 2025; Mulchandani & Shanahan, 2022). Consequently, most research explores ethical implications in theoretical terms instead of analyzing longitudinal or mixed-method evidence from real-world use.

Together, these gaps underscore the urgent need for interdisciplinary frameworks and empirical validation strategies that integrate ethical reasoning directly into the technical foundations of AI systems supporting battlefield and emergency medicine.

METHODOLOGY

This study followed established qualitative research practices, employing a hermeneutic phenomenological approach to explore how software developers and IT managers experience the integration of generative AI in their professional workflows. Hermeneutic phenomenology, as defined by Peoples (2021), seeks to uncover the meaning behind lived experiences by interpreting how individuals make sense of emerging phenomena, in this case, the transformation of roles, skills, and responsibilities in AI-augmented development environments.

Semi-structured, open-ended interview questions were used to guide the inquiry, allowing participants the freedom to share rich, personal reflections while ensuring thematic consistency across responses. Triangulation was employed by engaging participants from varied industries, seniority levels, and functional roles, enhancing the credibility and depth of the findings through multiple perspectives on how AI is reshaping software development and IT governance.

To ensure the trustworthiness of the data, member checking was conducted with selected participants, who were invited to review and validate preliminary interpretations of their narratives. This process supported accurate representation of participants' views and strengthened the integrity of the thematic analysis (Riazi et al., 2023). Throughout the study, the researcher engaged in reflexive journaling to document analytic decisions, reflect on positionality, and minimize potential bias during interpretation.

Ethical standards were rigorously upheld. Participant confidentiality was protected by de-identifying transcripts and securely storing data on encrypted, password-protected devices. All participants provided informed consent and were assured of their voluntary participation and right to withdraw at any stage. Transparency, respect for participant autonomy, and adherence to IRB-approved protocols were central to the ethical framework guiding this research.

Research Design

This study employed a qualitative research design centered on in-depth, semi-structured interviews to investigate how software developers and IT managers experience the integration of generative AI tools in their professional environments. A qualitative approach was selected to capture rich, contextualized narratives and to explore how individuals interpret and adapt to the evolving roles, responsibilities, and ethical challenges introduced by AI. Rather than seeking to generalize findings to a broader population, the study aimed to uncover nuanced patterns, subjective perceptions, and meaning-making processes that emerge as professionals engage with AI technologies in real-world software development and IT management settings.

Participant Selection

Participants were recruited using purposive sampling, targeting individuals with direct experience in software development, IT management, and the use of AI tools in professional settings. The sample included twelve participants, comprising mid-level and senior software developers, IT managers, and technical leads from both public and private sector organizations. This range of roles and perspectives enabled a deeper exploration of how generative AI technologies, such as ChatGPT,

GitHub Copilot, and other LLM-based tools, are being integrated, evaluated, and governed across varied development environments.

Recruitment was conducted through professional networks, LinkedIn outreach, and referrals from industry colleagues. All participants had hands-on experience using AI-assisted tools for code generation, system design, or IT oversight. Prior to each interview, participants received an informed consent form detailing the study's purpose, their rights, and assurances of confidentiality. Participation was entirely voluntary, with the option to withdraw at any point, and no financial compensation was provided.

Data Collection

Semi-structured, in-depth data were collected through interviews conducted via Zoom, each lasting between 45 and 75 minutes. The interview protocol was designed to explore participants' lived experiences with AI integration in software development and IT management. Key areas of inquiry included interactions with AI-assisted tools like GitHub Copilot and ChatGPT, perceptions of changing roles and responsibilities, the impact of AI on skill development and autonomy, risk management strategies, and expectations for future human–AI collaboration.

The interviews followed a flexible format, allowing participants to elaborate on topics they found most relevant, while still ensuring consistency across interviews. This structure enabled the researcher to probe deeper into emerging themes and clarify complex insights related to workflow adaptation, ethical concerns, and governance practices. All interviews were recorded with participant consent and transcribed verbatim for systematic analysis.

Analytical Approach

Data analysis followed a thematic coding approach that integrated both inductive and deductive strategies. Transcripts were initially reviewed in their entirety to identify recurring ideas, language use, and areas of emphasis across participant narratives. Preliminary codes were generated from participant responses and iteratively refined through close reading and memo-writing. The coding process was informed by the theoretical frameworks guiding this study: Deskilling Theory, Job Characteristics Theory, and Technology Threat Avoidance Theory, which provided a structured lens for interpreting themes related to skill transformation, role adaptation, motivation, and risk perception. Emergent patterns were then organized into higher-order themes to capture how developers and IT managers make meaning of AI integration in their day-to-day work.

Coding and organization of thematic categories were conducted using qualitative data analysis software (MaxQDA), which facilitated data visualization and ensured consistency in identifying themes across interviews.

Trustworthiness

This study established best practices in qualitative research to ensure the credibility and integrity of the findings. Triangulation was used by comparing data across participants with diverse roles, including developers, IT managers, and technical leads, to capture a range of perspectives on AI integration. Member checking was conducted with select participants, who were invited to review and validate initial interpretations of their interview responses, ensuring that their perspectives were accurately represented. The researcher also engaged in reflexive journaling throughout the study to document analytical decisions, acknowledge positionality, and minimize interpretive bias during data analysis.

In addition, the study upheld strict ethical research standards, including confidentiality, transparency, and participant autonomy. All data were stored on password-protected, encrypted devices, and identifying information was removed from transcripts prior to coding and analysis. These measures supported the trustworthiness of the research and ensured that participants' contributions were handled with care and respect.

Ethical Considerations

The study received ethical approval from the appropriate Institutional Review Board (IRB) from the University of the Cumberlands prior to data collection. All participants provided informed consent after being briefed on the study's purpose, procedures, and their rights. Confidentiality was maintained using pseudonyms, and identifying information was removed from all transcripts. Data were securely stored on encrypted, password-protected devices. Participants were informed of their right to withdraw at any time without penalty, and all interactions were conducted with transparency and respect for participant autonomy.

FINDINGS

RQ1: Secure and Ethical Software Development Practices

The study found that ethical software development for battlefield healthcare requires a fusion of technical precision and moral foresight. Developers emphasized that secure coding practices, when aligned with ethical auditing and explainable AI, produced more trustworthy systems for critical decision-making in combat zones. Teams adopting DevSecOps frameworks with integrated ethical validation achieved higher confidence in software reliability and moral integrity. Moreover, participants noted a growing recognition that ethical competence, not just technical expertise, is now a key skill among military software engineers.

However, many developers reported challenges balancing speed and ethical rigor under military pressure. Agile sprints often prioritize functionality and delivery timelines, leading to a form of moral deskilling, where ethical reflection is deprioritized. While secure-by-design methods and automated code reviews improved software resilience, developers still lacked clear ethical guidelines to navigate dilemmas involving AI autonomy and human safety.

RQ2: Human Oversight and Fail-Safe Design

Findings revealed that human-in-the-loop (HITL) mechanisms remain the cornerstone of safe and ethical battlefield healthcare systems. Respondents agreed that AI triage and diagnostic tools must never operate in complete autonomy during life-critical missions. The most effective systems combined adaptive human oversight with real-time explainability features, allowing medics to understand, contest, or override AI recommendations.

Despite advances in user interface design, many systems failed to communicate risk levels or uncertainty effectively, increasing operator dependence on algorithmic output, a potential pathway toward deskilling of clinical judgment. Developers also highlighted the importance of cognitive ergonomics: systems designed for usability under duress significantly improved moral and operational decision quality.

RQ3: Transparency, Accountability, and Moral Responsibility

Transparency and accountability emerged as the most critical yet underdeveloped dimensions of ethical battlefield AI. While technical traceability tools, such as version control, audit logs, and model documentation, were widely used, few systems implemented ethical traceability, which records the rationale for moral choices embedded in software design. Participants stressed that accountability in warfare must extend beyond system failure to include moral accountability for algorithmic harm.

Teams that practiced collaborative moral documentation, recording decisions about bias mitigation, fail-safes, and interpretability, demonstrated more substantial alignment with ethical warfare principles. However, the lack of standardized accountability frameworks and the absence of shared moral responsibility across hierarchical military structures limited transparency. Developers expressed a need for interdisciplinary oversight boards to review not only software defects but also moral risk exposure.

DISCUSSION

The findings underscore the moral and organizational complexity of developing AI-driven healthcare systems in militarized contexts. Three theoretical lenses: Deskilling of Jobs Theory, Job Characteristics Theory, and Technology Threat Avoidance Theory, help explain how ethical and technical imperatives interact in this environment.

1. Deskilling and Reskilling in Ethical Engineering

According to the Deskilling of Jobs Theory, automation can erode human expertise by shifting critical decision-making from humans to machines. In this study, developers and medics described subtle deskilling effects: overreliance on automated triage and diagnostic AI reduced opportunities for reflective ethical judgment. Yet, paradoxically, the same technology also reskilled engineers by demanding higher ethical literacy and cross-disciplinary competence. AI-driven warfare thus redefines technical skills to include moral reasoning, algorithmic transparency, and socio-technical awareness.

2. Redesigning Work for Moral Engagement

The Job Characteristics Theory highlights that autonomy, task significance, and feedback foster intrinsic motivation. Applied here, ethical software engineering benefits from designs that enhance moral autonomy and professional significance. Developers who had greater decision authority in integrating explainability and oversight reported more substantial ethical commitment. Conversely, rigid command structures and siloed development workflows diminished ownership of moral outcomes. Designing for job enrichment, through ethical reflection checkpoints and participatory design reviews, supports both technical excellence and moral engagement.

3. Technology Threat Avoidance and Ethical Resilience

Drawing on Technology Threat Avoidance Theory, the study shows that fear of system misuse or malfunction motivates both ethical caution and defensive design. Developers who perceived higher threat levels (e.g., potential AI error in combat triage) adopted more proactive fail-safe mechanisms and robust validation procedures. However, excessive reliance on technical safeguards can lead to moral complacency, where developers assume ethical issues are solved through code rather than judgment. Balancing threat perception with moral resilience, the capacity to confront uncertainty and ethical ambiguity, is essential for sustaining trustworthy AI systems.

4. Integrating the Moral Code Architecture

The synthesis of these frameworks leads to what this study calls a Moral Code Architecture, a conceptual foundation that aligns software design, human oversight, and moral accountability. This architecture envisions AI-driven battlefield healthcare as a human-centered, ethically auditable, and failure-tolerant ecosystem, where each line of code reflects humanitarian values. Transparent documentation, participatory ethics boards, and continuous learning cycles are the pillars of this architecture.

5. Implications for Policy and Practice

The discussion suggests a paradigm shift from compliance-based ethics toward embedded moral engineering. Military and healthcare organizations must institutionalize ethical review in software pipelines, ensure explainability standards, and foster developer moral agency. Ethical leadership, not just regulation, will determine whether AI on the battlefield preserves life or compromises it.

SOLUTIONS AND RECOMMENDATIONS

Building upon the synthesis of literature and identified gaps, this section outlines practical and theoretical solutions for designing, developing, and governing ethical AI systems in battlefield and emergency healthcare contexts. These recommendations are organized around three guiding research questions (RQ1–RQ3), each addressing a critical dimension of responsible AI deployment, secure and ethical system design, strengthened human oversight, and enhanced transparency and moral accountability.

Together, they propose an integrated roadmap for aligning technological innovation with humanitarian and ethical imperatives.

RQ1: Building Secure and Ethical AI Healthcare Systems

The first research question focuses on strategies for constructing secure and ethically aligned AI healthcare systems. This strategy requires embedding moral reasoning and ethical accountability throughout the software development process. Within the DevSecOps framework, ethical risk assessments and impact reviews should be incorporated into every sprint cycle to ensure that system design anticipates moral dilemmas rather than reacts to them post-deployment. Battlefield ethics committees can play a vital oversight role by evaluating algorithmic fairness, bias mitigation, and compliance with humanitarian principles before AI tools are approved for use. Moreover, establishing explainability standards is essential; all AI-driven triage and diagnostic applications should offer interpretable outputs that allow clinicians to understand the system's reasoning and limitations. Cross-disciplinary collaboration among ethicists, software engineers, and medical professionals is equally critical, ensuring that ethical principles are not isolated from technical design or clinical application but are instead integrated within continuous code review and testing cycles.

RQ2: Strengthening Human Oversight and Fail-Safe Mechanisms

The second research question addresses how to preserve human oversight and ensure robust fail-safe mechanisms in AI-assisted battlefield medicine. Adaptive human-in-the-loop (HITL) protocols are necessary to maintain human authority in final decision-making, especially in high-stakes scenarios where algorithmic recommendations influence patient survival outcomes. Redundant safeguards, such as manual override options and fallback systems, should be embedded into AI architectures to prevent single points of failure.

Real-time explainability dashboards can enhance operational transparency by displaying model confidence scores, uncertainty levels, and reasoning pathways to the user. Complementing these technical measures, cognitive training programs must prepare military medics and clinicians to identify algorithmic bias, interpret model limitations, and intervene when ethical or operational risks arise. Together, these measures ensure that AI systems augment rather than replace human judgment, preserving moral accountability in critical care decisions.

RQ3: Ensuring Transparency and Moral Responsibility

The third research question explores how transparency and moral responsibility can be institutionalized in AI-driven medical systems. Ethical documentation frameworks should be developed to systematically record decision rationales, data provenance, and model evolution throughout the software lifecycle. This documentation enables both traceability and ethical auditing, providing verifiable evidence when system performance or moral integrity is questioned. Regular moral audits, conducted internally and by external reviewers, can evaluate AI systems for fairness, safety, and potential bias, ensuring compliance with international medical and humanitarian standards.

Furthermore, transparent accountability chains must be established across both technical and command levels, defining who bears responsibility for decisions and outcomes. Finally, moral simulation training can serve as a proactive educational tool, exposing developers, medics, and commanders to ethically challenging scenarios in controlled environments. Through these interventions, AI healthcare systems in military contexts can evolve toward greater transparency, moral robustness, and operational trustworthiness.

Integrative Recommendation

The chapter recommends institutionalizing the Moral Code Architecture, embedding ethical foresight directly into system design, testing, and deployment. This architecture ensures that battlefield healthcare AI remains secure, human-centered, and morally defensible.

CONCLUSION AND FUTURE RESEARCH

The increasing reliance on artificial intelligence in battlefield healthcare presents profound ethical and operational challenges. This chapter has examined how secure, ethical software engineering can safeguard both human life and moral accountability within the high-stakes domain of software-defined warfare. Findings from this study reveal that ethical competence, transparency, and human oversight are as vital to the integrity of battlefield healthcare systems as technical performance or algorithmic accuracy.

The integration of Deskilling of Jobs Theory, Job Characteristics Theory, and Technology Threat Avoidance Theory provided a multidimensional understanding of how technology reshapes professional identity, moral agency, and risk perception among software developers and military healthcare personnel. The evidence suggests that while automation can erode certain aspects of human judgment, it simultaneously introduces opportunities for moral reskilling, in which ethical reasoning becomes a new professional competency. Ethical software engineering thus emerges not merely as a technical discipline but as a moral craft, one that encodes humanitarian principles directly into the architecture of AI systems.

The proposed Moral Code Architecture represents a unifying conceptual model that bridges the technical and ethical dimensions of AI-driven battlefield healthcare. This architecture emphasizes three interdependent pillars:

1. **Transparency**, through explainable algorithms and moral documentation.
2. **Human Oversight**, through adaptive human-in-the-loop protocols; and
3. **Accountability**, through auditable decision-making frameworks and moral traceability.

When integrated into DevSecOps pipelines and institutional policy, this framework ensures that AI technologies not only function securely but also align with the moral imperatives of medical ethics and international humanitarian law.

Future Research Directions

Future research should build upon these findings by employing both qualitative and quantitative methods to explore how moral and technical values can effectively coexist within fast-paced, mission-critical development environments. A critical direction for future inquiry involves strengthening ethical competency and training among developers, engineers, and medical personnel. Longitudinal studies could evaluate how ongoing moral education, scenario-based ethics workshops, and simulation-based training influence ethical awareness, decision-making, and accountability over time. Such research would help determine whether ethical reflexivity can be sustained amid the operational pressures of military healthcare innovation.

Another important area of exploration is cross-theater validation. Comparative studies conducted across various military, humanitarian, and emergency response contexts could reveal how differing cultural norms, logistical constraints, and regulatory standards shape the implementation of moral software design practices. These studies would enhance the generalizability of ethical AI frameworks and help align international standards of responsible AI deployment in the battlefield and crisis medicine.

Finally, future research should advance the study of human–AI ethical interaction models. This model includes developing frameworks that formalize shared accountability between human operators and AI systems, emphasizing the calibration of cognitive trust, the distribution of moral responsibility, and adaptive learning in dynamic and high-risk medical environments. Such models could provide empirical grounding for balancing automation efficiency with human judgment, ensuring that the integration of AI into battlefield healthcare remains both operationally effective and ethically sound.

Ultimately, the future of AI in warfare medicine depends on how societies and institutions translate ethical intent into technical reality. Building software for life-preserving missions demands more than precision and efficiency; it requires conscience, foresight, and an unwavering commitment to human dignity. Through the Moral Code Architecture, ethical software engineering can become both a safeguard and a guidepost for the evolving moral terrain of software-defined warfare.

REFERENCES

Braverman, H. (1974). *Labor and monopoly capital*. Monthly Review Press. DOI: 10.14452/MR-026-03-1974-07_1

Cole, R., Simmons, S., Duncan, J., Cole, J., & Peacock, J. G. (2025). Readying military medicine for AI-enabled warfare. *Military Medicine*, usaf460. Advance online publication. DOI: 10.1093/milmed/usaf460 PMID: 41003652

Cummings, M. L. (2021). Rethinking the maturity of artificial intelligence in safety-critical settings. *AI Magazine, 42*(1). Obtenido de https://ojs.aaai.org/aimagazine/index.php/aimagazine/article/view/7394

Decker, S. E., Matthieu, M. M., Smith, B. N., & Landes, S. J. (2023). Barriers and Facilitators to Dialectical Behavior Therapy Skills Groups in the Veterans Health Administration. *Military Medicine*, *189*(5-6), 1055–1063. DOI: 10.1093/milmed/usad123 PMID: 37104810

Department of Defense. (2021). *Implementing responsible artificial intelligence in the Department of Defense*. Obtenido de Office of the Secretary of Defense: https://media.defense.gov/2021/May/27/2002730593/-1/-1/0/IMPLEMENTING-RESPONSIBLE-ARTIFICIAL-INTELLIGENCE-IN-THE-DEPARTMENT-OF-DEFENSE.PDF?utm_source=chatgpt.com

Floridi, L., & Sanders, J. (2024). Minds and Machines. *On the morality of artificial agents, 14*(3), 349–379. DOI: 10.1023/B:MIND.0000035461.63578.9d

Hackman, J. R., & Oldham, G. R. (1976). Motivation through the design of work: Test of a theory. *Organizational Behavior and Human Performance*, *16*(2), 250–279. DOI: 10.1016/0030-5073(76)90016-7

Kuttan, N., Pundkar, A., Gadkari, C., Patel, A., & Kumar, A. (2025). Transforming emergency medicine with artificial intelligence: From triage to clinical decision support. *Multidisciplinary Reviews*, *8*(10), 2025285. Advance online publication. DOI: 10.31893/multirev.2025285

Liang, H., & Xue, Y. (2009). Avoidance of information technology threats: A theoretical perspective. *Management Information Systems Quarterly*, *33*(1), 71–90. DOI: 10.2307/20650279

Majumdar, S., Pendleton, B., & Gupta, A. (2025). Red Teaming AI Red Teaming. https://doi.org//arXiv.2507.05538DOI: 10.48550

Mulchandani, N., & Shanahan, J. (2022). *Software-Defined Warfare: Architecting the DOD's Transition to the Digital Age.* https://www.csis.org/analysis/software-defined-warfare-architecting-dods-transition-digital-age

National Institute of Standards and Technology. (2025). *Outline: Proposed zero draft for a standard on AI TEVV (Test, Evaluation, Validation, and Verification).* Obtenido de https://www.nist.gov/system/files/documents/2025/07/15/Outline_%20Proposed%20Zero%20Draft%20for%20a%20Standard%20on%20AI%20TEVV-for-web.pdf

Nichols, R., Horstman, J., & Nitz, I. (2023). Army Health Systems Doctrine and Training in Relation to Antibiotics: A Systematic Review. *Military Medicine, 188*(7-8), 1649–1655. DOI: 10.1093/milmed/usac210 PMID: 35830415

Peoples, K. (2021). *How to write a phenomenological dissertation: A step-by-step guide* (Vol. 56). SAGE.

Riazi, M., Rezvani, R., & Ghanbar, H. (2023). Trustworthiness in L2 writing research: A review and analysis of qualitative articles in the Journal of Second Language Writing. *Research Methods in Applied Linguistics, 2*(3), 100065. Advance online publication. DOI: 10.1016/j.rmal.2023.100065

Ribeiro, M. T., Singh, S., & Guestrin, C. (2016). "Why should I trust you?": Explaining the predictions of any classifier. *Proceedings of the 22nd ACM SIGKDD International Conference on Knowledge Discovery and Data Mining.* DOI: 10.1145/2939672.2939778

Sommerville, I. (2020). *Software Engineering* (10th ed.). Pearson.

Stahl, B. C. (2021). *Artificial Intelligence for a Better Future.* Springer. DOI: 10.1007/978-3-030-69978-9

KEY TERMS AND DEFINITIONS

Software-Defined Warfare (SDW): Modern military operations that rely on software, data, and AI systems to make decisions and control missions.

AI-Driven Battlefield Healthcare: Using artificial intelligence to support medical care in combat zones, such as triage apps, remote diagnosis, and decision-making tools that help save lives.

Moral Code Architecture (MCA): A design approach that builds ethical values, like honesty, safety, and accountability, into software from the start.

Deskilling Jobs Theory: The idea that when technology takes over complex tasks, workers may lose some of their original skills or independence.

Job Characteristics Theory (JCT): A theory that explains how the design of a job affects motivation and satisfaction, based on how meaningful, independent, and feedback-driven the work is.

Technology Threat Avoidance Theory: A theory that looks at how people respond when they feel technology might be unsafe, risky, or harmful.

DevSecOps: A method of building software that combines development, security, and operations to make systems safer and more reliable.

Human-in-the-Loop (HITL): A system design where humans stay involved in decisions made by AI, especially in critical areas like medicine or defense.

Ethical Explainability: The ability to clearly understand and explain how an AI system made its decision.

Algorithmic Accountability: The responsibility of developers and organizations to make sure AI systems are fair, safe, and used ethically.

Chapter 8
Humanitarian Dilemmas of AI's Role in Warfare and Medicine:
Insights From India and the United States

Naresh Prajapati
http://orcid.org/0000-0002-0752-3651
Bennett University, Greater Noida, India

Anwesha Ghosh
http://orcid.org/0000-0003-4167-2121
Amity Law School, Amity University, Mumbai, India

Nathan Darren Manuk
http://orcid.org/0009-0000-9716-2505
Amity Law School, Amity University, Kolkata, India

ABSTRACT

This chapter examines the humanitarian dilemmas of artificial intelligence (AI) in warfare and medicine, with a focus on India and the United States. It explores how AI technologies enhance diagnostics, treatment, and defence operations while raising profound ethical concerns around accountability, equity, and human oversight. By comparing AI adoption in two contrasting national contexts, the chapter highlights systemic risks such as data bias, misdiagnosis, and autonomous weapon use. It argues for responsible innovation rooted in ethical foresight, inclusive policy, and sustained human agency, ensuring that AI serves, rather than displaces, human values.

DOI: 10.4018/979-8-3373-6264-9.ch008

INTRODUCTION

In recent years, intelligent systems have moved from the margins of research labs and industrial innovation into the fabric of everyday life. From personalised recommendations on digital platforms to voice assistants managing household routines, the presence of these systems, often referred to as artificial intelligence, has become both subtle and significant. They are now embedded in essential sectors such as healthcare and defence, where their influence extends far beyond convenience (Hirani et al., 2024). These technologies are not just tools of efficiency; they are decision-making systems that increasingly affect how societies deliver care, respond to crises, and defend themselves (Rashid & Kausik, 2024). This growing influence raises urgent humanitarian questions: What are the consequences of using intelligent systems to make decisions about human life? Who is accountable when things go wrong? Moreover, most importantly, how do we balance technological progress with the human values that sustain ethical societies?

Understanding this dilemma requires situating the discussion within specific national contexts. This chapter focuses on the deployment of AI-driven technologies in two critical domains, medicine and warfare, in both India and the United States (Robinson et al., 2021). These countries represent contrasting realities. India, a developing nation with a vast population, continues to grapple with uneven access to healthcare, resource limitations, and growing digital infrastructure. Intelligent technologies here are often positioned as solutions to longstanding gaps in care delivery and public health administration (Gu et al., 2021). In contrast, the United States, with its advanced technological infrastructure and policy apparatus, approaches AI in healthcare and defence with greater institutional capacity and regulation, but not without ethical complexity. Together, these contexts provide a meaningful comparative framework to evaluate both the potential and the perils of integrating such systems into life-critical decisions (Bitko, 2025).

The importance of this study lies in examining these technologies not just as innovations but as actors in humanitarian environments (Zarei et al., 2024). In medicine, intelligent systems are now used to detect cancerous cells, interpret radiological images, manage digital health records, and predict disease outbreaks. In warfare, their roles are more contentious, guiding drones, analysing surveillance data, and sometimes, making decisions that may lead to loss of life. In both cases, there is a growing concern about the absence of empathy, the dilution of accountability, and the challenge of regulating systems that are often opaque in design and operation. These are not just technological challenges; they are moral and institutional dilemmas that directly impact human dignity and rights (Althobiti et al., 2025).

Despite this significance, there remains a concerning gap in the existing body of research. Much of the academic and policy literature on AI tends to generalise its effects or focus heavily on its technical capabilities. Ethical discussions, when they occur, are often abstract, removed from the lived realities of different national systems and cultural contexts (Riley, 2012). For instance, the regulatory and infrastructural needs of Indian hospitals deploying AI for rural diagnostics are vastly different from those of American institutions working with genomics and personalised medicine. Likewise, the governance challenges faced by defence sectors in countries with limited oversight capacity differ markedly from those with established ethical frameworks. There is, therefore, a need for research that is grounded, comparative, and attentive to the socio-political conditions under which these technologies are adopted (Roy et al., 2023).

The background for this inquiry is both timely and urgent. In India, the launch of initiatives such as the Ayushman Bharat Digital Mission reflects a governmental push towards digitalising healthcare to increase access and efficiency. Tools powered by intelligent systems are increasingly used for tuberculosis detection, cancer diagnostics, ICU monitoring, and even predicting seasonal outbreaks of diseases like malaria and dengue (Sharma et al., 2023). In the United States, similar technologies have been implemented with a focus on personalised care, public health surveillance, and emergency preparedness, especially highlighted during the COVID-19 pandemic. Simultaneously, both countries are integrating autonomous systems into defence planning, raising questions about the ethics of delegation, human oversight, and the potential for civilian harm (Donelle et al., 2023).

This chapter aims to investigate these humanitarian dilemmas by exploring how intelligent systems are deployed in medicine and warfare in India and the United States. It seeks to unpack the ethical, cultural, and institutional implications of these deployments. Specifically, this study examines whether these systems enhance human welfare or inadvertently compromise it by replacing judgment with calculation, empathy with efficiency (McGowan et al., 2020). The objective is not to evaluate AI as a technology in isolation but to understand its role within social systems that vary in structure, equity, and readiness. By drawing on evidence from policy documents, healthcare programs, public health data, and defence strategies, this chapter offers a critical and context-sensitive exploration of how intelligent systems shape human outcomes in profoundly sensitive domains (Radanliev, 2025).

In doing so, this chapter contributes to a broader conversation about responsible innovation. It argues that the development and deployment of intelligent systems must be guided by more than functionality; they must be rooted in ethical foresight, cultural relevance, and institutional responsibility. This is especially necessary in humanitarian sectors, where decisions affect not only efficiency but the quality and sanctity of human life (Buonocore et al., 2024).

UNDERSTANDING ARTIFICIAL INTELLIGENCE IN HUMAN TERMS

In conversations about contemporary technology, few concepts have attracted as much attention and confusion as artificial intelligence. Despite its increasing influence in sectors as varied as education, entertainment, governance, warfare, and medicine, artificial intelligence (AI) is often discussed in abstract, futuristic, or overly technical terms. This section aims to clarify what AI truly means in practical, human-centred terms, drawing attention to its growing social relevance and the ways it interacts with daily life (Tai, 2020).

At its core, AI refers to systems designed to perform tasks that would ordinarily require human intelligence. These tasks include recognising speech, interpreting images, analysing data patterns, making decisions, and even engaging in conversation. Unlike traditional machines or programs that operate strictly on user input, AI systems are built to "learn" from data and improve their performance over time. Importantly, these systems are not conscious or sentient; instead, they are trained to simulate specific human-like capabilities using algorithms and vast datasets. In doing so, they offer both speed and scale that human labour alone cannot match (K, 2024).

The social relevance of AI is immediately visible in everyday contexts. Virtual assistants like Siri, Alexa, or Google Assistant rely on speech recognition and natural language processing to carry out commands, from setting reminders to answering factual queries. Recommendation algorithms on platforms like Netflix, YouTube, and Spotify track viewing or listening habits to suggest content that aligns with user preferences. Similarly, AI-powered tools help sort spam from legitimate emails, optimise traffic flows in smart cities, assist in customer service via chatbots, and even enable facial recognition in security systems. These applications, while individually modest, collectively reflect a growing integration of AI into the intimate routines of daily life (Krishna, 2023).

Importantly, AI does not operate as a monolithic concept. Within the broader field, there is a significant distinction between two foundational categories of artificial intelligence: narrow (or limited-purpose) AI and general (or adaptive, human-like) AI. Understanding the difference is essential for evaluating not only what AI can currently do but also the implications of where it might go in the future (Gignac & Szodorai, 2024).

Narrow AI, also referred to as weak AI, is designed to carry out specific tasks. These systems are goal-oriented and operate within a predefined set of rules and data structures. For example, an AI system that can detect tumours in X-rays is trained on thousands of annotated medical images. It learns to associate specific visual patterns with malignancy and flags similar instances in new images. However, this system cannot apply its knowledge to unrelated tasks, such as diagnosing neurological disorders or interpreting handwritten notes. Narrow AI systems function within a tightly scoped environment; they do not possess contextual awareness or broader cognitive flexibility. Their "intelligence" is limited to pattern recognition and decision rules optimised for a particular use case (Kalani & Anjankar, 2024).

Despite their limitations, narrow AI systems have proven immensely effective in high-stakes sectors. In public health, they are used to monitor outbreaks, predict disease spread, and analyse medical imagery. In transportation, narrow AI powers navigation apps that factor in real-time traffic and road closures. Financial institutions use them to detect fraudulent transactions based on anomalous spending behaviour. The precision and speed with which these systems process data make them invaluable in domains where human decision-making is time-sensitive but prone to error (Nancholas, 2023).

In contrast, general AI, or what scholars' term artificial general intelligence (AGI), refers to systems that aim to replicate broader aspects of human cognition. Unlike narrow AI, which is bound to a singular domain, AGI aspires to be adaptable. It can apply knowledge learned in one area to a different context, much like a human being might draw on general problem-solving skills to navigate unfamiliar situations. AGI would not just recognise patterns but understand them in context; it would be capable of reasoning, abstract thinking, and even moral judgment (Joshi, 2024).

While the idea of AGI has been a subject of speculation and research, it remains largely aspirational. No existing system has yet demonstrated accurate general intelligence. Even the most sophisticated language models and autonomous systems today rely on massive data inputs and structured training rather than independent understanding. Moreover, the development of AGI brings with it not only technical challenges but deep philosophical and ethical concerns. Can a machine truly "understand" empathy or intention? Should machines be trusted to make moral decisions? These questions are no longer hypothetical, particularly as AI systems are introduced into sectors involving life-critical decisions, such as battlefield simulations or intensive care triage (Raman et al., 2025).

To illustrate this distinction in a more grounded way, consider two scenarios: In the first, a radiology tool is used to detect signs of tuberculosis from chest X-rays in rural Indian hospitals. It performs this task with high accuracy, reducing delays in diagnosis where human radiologists are unavailable. This is an example of narrow AI: it excels in a specialised task but is incapable of doing anything outside of that role. In the second scenario, imagine a healthcare assistant who not only interprets a patient's medical records but also considers socio-economic background, emotional state, and personal values before recommending a treatment plan. Such an agent would require AGI-level capabilities, contextual awareness, ethical reasoning, and emotional intelligence, none of which exist in current systems (Cordova-Pozo & Rouwette, 2023).

For now, most AI systems deployed across India and the United States remain within the domain of narrow AI. However, the ambitions of policymakers, technologists, and corporations often exceed the limitations of current technology. This disjuncture creates a societal tension: people begin to trust machines with tasks they are not fully capable of handling, often due to a lack of transparency or oversimplification of what AI is. The danger lies not in the technology itself but in misunderstanding its scope, misplacing trust, or failing to build adequate ethical safeguards (*The 2025 AI Index Report | Stanford HAI*, 2025).

Thus, a humanised understanding of artificial intelligence begins with clarity. AI is not magic, nor is it inherently intelligent in the way humans are. It is a set of tools, powerful, precise, and expanding, designed to simulate specific aspects of human reasoning. As such, they require careful design, ethical constraints, and ongoing oversight. As society integrates these tools into healthcare, defence, and governance, the responsibility lies not in the hands of the machines, but in those who build, deploy, and regulate them (Bai, 2025).

THE RELEVANCE OF AI IN MODERN SOCIETY

Artificial intelligence (AI) has become a defining force in contemporary society, subtly restructuring the way individuals live, work, communicate, and make decisions. Far from a futuristic concept confined to science fiction, AI is now embedded in infrastructures and services that define modernity, ranging from healthcare diagnostics and urban mobility systems to banking algorithms and virtual education tools (Rawas, 2024). The practical relevance of AI lies not in its novelty but in its operational ubiquity and its capacity to process, synthesise, and act upon large datasets with speed and precision that far exceeds human capabilities. As societies become more interconnected and data-driven, the role of AI as a facilitator of convenience and efficiency is increasingly critical (Colther & Doussoulin, 2024).

One of AI's most immediate and visible contributions is its enhancement of everyday efficiency. In communication, AI technologies power predictive text, real-time translation, spam filtering, and intelligent personal assistants like Siri and Alexa. These systems interpret human language and respond in ways that mimic conversational norms, creating seamless digital interactions. In professional environments, AI-driven email categorisation, task automation, and meeting transcription tools reduce cognitive load, enabling users to focus on decision-making rather than administrative processes. While these tasks may seem minor, their aggregate impact on productivity is significant (Coombs et al., 2020).

AI has also made profound contributions in the domain of public safety and security. In law enforcement and urban management, AI is used for facial recognition, predictive policing, and surveillance analytics. Smart CCTV systems, for example, can identify suspicious behaviour and alert authorities in real time. Though controversial, these tools are promoted in some jurisdictions as solutions to rising urban crime and overloaded human surveillance teams. Similarly, in cybersecurity, machine learning algorithms detect and neutralise digital threats by recognising patterns that signal data breaches or fraud attempts, often before human analysts can respond (Fontes et al., 2022).

In the health sector, AI's relevance is arguably most pronounced and ethically charged. From medical imaging interpretation to drug discovery, intelligent systems play a growing role in expediting diagnosis and improving treatment accuracy. AI-based tools have been especially impactful in low-resource environments, such as rural India, where human specialists are scarce. Diagnostic platforms trained on radiology or pathology datasets can evaluate medical images and flag anomalies, such as potential malignancies, with high consistency. AI has also been crucial in pandemic management. During the COVID-19 crisis, AI helped governments monitor infection trends, identify outbreak clusters, and model containment strategies (Bouderhem, 2024).

Another important area of application is urban mobility and transportation. AI powers real-time traffic management systems, autonomous vehicle navigation, and route optimisation for logistics. Applications like Google Maps rely on AI to process geospatial data and user behaviour to deliver efficient routing. In India's congested metros and the US interstate systems alike, AI has helped reduce commute times, lower fuel consumption, and enhance road safety. Some cities have begun to pilot AI-enabled traffic signals that adapt to traffic flow, adjusting in real-time based on congestion levels (Saki & Soori, 2025).

Despite these contributions, the growing reliance on AI invites critical reflection, particularly regarding dependency, employment disruption, and data governance. One of the most urgent challenges is the displacement of human labour. As machines become capable of performing tasks previously assigned to skilled professionals, ranging from customer service to legal document review, concerns about widespread job loss have intensified. In sectors like banking, healthcare, and logistics, companies increasingly prefer AI systems due to their cost-efficiency and scalability. This shift necessitates urgent policy interventions around retraining, upskilling, and inclusive economic transition strategies (Jose et al., 2025).

Dependency is another point of concern. As societies delegate more cognitive and operational tasks to machines, the risk of de-skilling or reduced human judgment grows. Overreliance on GPS navigation, for instance, has been shown to diminish spatial memory in younger users. In healthcare, blind trust in diagnostic algorithms may lead to reduced clinical reasoning among physicians. This phenomenon, sometimes referred to as "automation complacency," can result in critical errors when machines fail, and humans are unprepared to intervene (Spatola, 2024).

The most pressing issue in the public discourse around AI is data privacy. AI systems are trained and operate based on access to vast amounts of personal data, including health records, location history, voice recordings, and biometric information. The ethical problem arises not only from how this data is collected but also how it is stored, shared, and monetised, often without informed consent. In countries like India, where digital literacy and regulatory safeguards are still developing, individuals are particularly vulnerable to data misuse. In the United States, despite more robust legal structures like HIPAA and GDPR influences, the scale of data harvesting by private tech firms continues to outpace regulation (Martin & Zimmermann, 2024).

In light of these challenges, it becomes imperative to frame AI not as a replacement for human intelligence but as a tool for augmentation and support. While machines excel at processing high volumes of information and identifying patterns, they lack emotional intelligence, ethical reasoning, and contextual understanding. These qualities are critical in fields like education, caregiving, and diplomacy, sectors that demand human empathy and judgment. For example, while AI might suggest optimal chemotherapy based on tumour genetics, the conversation about treatment options and patient anxieties must remain a human responsibility (Dégallier-Rochat et al., 2022).

Therefore, the design and deployment of AI systems must be guided by principles of human-centred computing, where technology enhances rather than displaces the human role. This approach insists on transparency in how algorithms function, accountability mechanisms for AI decisions, and participatory governance in technology policy. Scholars and policymakers increasingly advocate for algorithmic impact assessments, akin to environmental assessments, which evaluate the social and ethical consequences of deploying AI in sensitive domains (Sigfrids et al., 2023).

To conclude, the relevance of AI in modern society is both evident and multifaceted. It facilitates convenience, enhances efficiency, and holds transformative potential across sectors. At the same time, it raises profound ethical and practical concerns that cannot be ignored. A nuanced understanding requires not only celebration of technological progress but also critical engagement with its risks and responsibilities. In this light, AI must be treated not as a substitute for human agency, but as a companion to it, valuable, powerful, and in need of continual human oversight (Brandao, 2025).

AI IN INDIAN HEALTHCARE: CURRENT LANDSCAPE AND PROGRESS

India's healthcare system has long faced structural challenges rooted in its vast geography, socio-economic disparities, and rapidly growing population. With over 1.4 billion citizens, the strain on public health infrastructure has become increasingly visible in both urban and rural contexts. The uneven distribution of healthcare professionals, particularly in remote and underserved areas, contributes to delayed diagnosis, limited access to specialised care, and a disproportionate disease burden across the population. These systemic limitations have created space for technological innovations, particularly artificial intelligence (AI), to emerge as potential enablers of healthcare equity and efficiency (Kumar, 2023).

In recent years, AI has played a transformative role in addressing critical gaps in diagnostics, public health surveillance, pandemic management, and personalised treatment planning. India's adoption of AI in healthcare, although still nascent in many respects, reflects an intersection between technological innovation and the urgent need for scalable, cost-effective health solutions (Faiyazuddin et al., 2025).

Addressing Diagnostic Inequity: AI in Radiology and Pathology

One of the most impactful uses of AI in Indian healthcare has been in diagnostic imaging, particularly in radiology and pathology. India has fewer than one radiologist per 100,000 people, a ratio that falls far short of meeting national demand, especially in tier 2 and tier 3 cities. To bridge this gap, AI-enabled tools have been developed to interpret chest X-rays and CT scans for diseases such as tuberculosis (TB), which continues to be a significant public health concern (Kapoor et al., 2024).

These AI models, trained on large, annotated datasets, can detect radiographic features of TB with accuracy comparable to junior radiologists. By rapidly screening images and prioritising high-risk cases, such systems help expedite diagnosis and treatment initiation. This is particularly valuable in rural clinics where radiologists are not always present, and where delays in diagnosis can lead to disease progression and community transmission (Pinto-Coelho, 2023).

Similar applications have emerged in histopathology, where AI algorithms assist pathologists in interpreting complex tissue images to identify cancers. Algorithms now support pattern recognition in biopsy samples, flagging abnormalities with high consistency and reducing the time burden on human experts. These systems not only increase throughput in overwhelmed pathology labs but also reduce the variability in diagnostic accuracy that can occur due to human fatigue or inexperience (Ruane et al., 2025).

AI's Role During the COVID-19 Pandemic

The COVID-19 pandemic served as a critical inflexion point for India's digital health strategy, bringing AI into the centre of public health planning. During the height of the crisis, AI was used for infection tracking, contact tracing, and predictive modelling to guide policy interventions. AI-supported dashboards integrated real-time data from testing centres, hospitals, and government databases to help identify hotspots and forecast future surges, enabling more targeted lockdowns and resource deployment (Narayan et al., 2024).

Public communication was another key area where AI systems proved indispensable. The Indian Ministry of Health deployed chatbot systems, such as the MyGov Corona Helpdesk on WhatsApp, to answer citizen queries, reduce misinformation, and disseminate verified health advice in multiple regional languages. These AI-based services helped reduce the burden on human health workers and ensured that information reached populations with limited access to formal healthcare.

Additionally, AI models helped manage clinical triage by identifying high-risk patients through electronic health records, guiding hospitals to allocate limited ventilators and ICU beds more effectively. Such capabilities demonstrated that intelligent systems could not only support logistical coordination but also play a role in clinical decision support under emergency conditions (Gore & Olawade, 2024).

Predictive Tools for Outbreak Surveillance

India's battle with vector-borne diseases like malaria and dengue has persisted for decades, exacerbated by climate variability and urban density. AI is now being used to enhance predictive surveillance systems that anticipate outbreaks before they escalate (Kumar et al., 2024).

Using historical data on rainfall, temperature, population movement, and previous case distribution, AI models have been trained to predict when and where outbreaks are likely to occur. These systems enable health authorities to pre-position mosquito nets, insecticides, and medical supplies in at-risk areas. For instance, pilot studies in Maharashtra and Odisha have demonstrated that predictive models can anticipate dengue spikes weeks in advance, allowing local governments to act preemptively rather than reactively (Latif et al., 2023).

This predictive power is critical for a country like India, where underreporting and data fragmentation often hamper timely response. AI helps centralise and analyse disparate datasets, creating a unified intelligence layer that improves the country's ability to anticipate and control outbreaks (Meadows et al., 2022).

Personalised Oncology: From Standardisation to Precision

One of the most promising developments in Indian healthcare has been the integration of AI into oncology, particularly in tailoring treatments based on individual patient profiles. In public and private cancer centres, AI systems are now being used to analyse tumour genetics, clinical history, and imaging data to guide personalised therapy plans (Das et al., 2024).

This transition from "one-size-fits-all" protocols to precision medicine is significant in India, where cancer incidence is rising but access to oncologists and genomic counsellors is limited. AI tools help in identifying the most effective combination of chemotherapy, radiation, or immunotherapy for specific cancer subtypes. Moreover, they assist in monitoring patient response, predicting side effects, and adjusting dosages accordingly (Parvizpour et al., 2023).

Some Indian hospitals have also deployed AI in ICU settings, where real-time monitoring systems analyse patient vitals to predict potential deterioration hours before it becomes clinically apparent. These alerts allow clinicians to intervene earlier, potentially reducing mortality in critical care units (Chaudhari et al., 2025).

Barriers and the Path Ahead

Despite these advancements, AI adoption in Indian healthcare remains uneven and is hindered by several factors. Digital infrastructure gaps, especially in rural areas, limit the implementation of cloud-based AI tools. Moreover, data quality and availability remain significant challenges. AI systems require large volumes of clean, well-annotated health data, which is often lacking in Indian public hospitals due to inconsistent record-keeping and the absence of standardised health information systems (Chettri et al., 2025).

The Ayushman Bharat Digital Mission (ABDM) represents a significant step toward solving this problem by aiming to create interoperable, digital health IDs for every citizen. However, ethical concerns about data privacy, algorithmic bias, and the transparency of decision-making systems must be addressed through robust governance frameworks. There is also a pressing need for capacity-building initiatives to train clinicians, technicians, and administrators in working collaboratively with AI tools (Mishra et al., 2024).

AI's integration into India's healthcare ecosystem is no longer an aspirational project but a practical imperative. By enhancing diagnostic capabilities, enabling real-time pandemic responses, predicting disease outbreaks, and personalising treatment pathways, AI systems are actively reshaping the way healthcare is delivered across the country. While challenges remain, particularly around infrastructure, data governance, and ethical deployment, the direction is clear. AI is not replacing human expertise; it is augmenting it. In doing so, it holds the potential to democratise access to high-quality care, reduce systemic inefficiencies, and ultimately, improve health outcomes across India's diverse population (Bhavana & Mithra, 2025).

NATIONAL INITIATIVES: AYUSHMAN BHARAT DIGITAL MISSION

India's healthcare system has historically been characterised by fragmentation, both in service delivery and record-keeping. Patients often move between public and private hospitals without any continuity of care. Medical records are frequently lost or duplicated, and there is limited coordination between diagnostic labs, pharmacies, and clinics. Recognising the urgent need for digital transformation, the Government of India launched the Ayushman Bharat Digital Mission (ABDM) in 2021. This initiative aims to create a seamless, unified, and patient-centric digital health ecosystem across the country, designed to improve both accessibility and accountability in healthcare delivery (Kumar, 2023).

Background and Goals of ABDM

The ABDM is an extension of the broader Ayushman Bharat program, which seeks to achieve Universal Health Coverage (UHC) by providing equitable and affordable health services. The digital mission specifically addresses the technological infrastructure needed to support this ambition. The primary goal of ABDM is to build a robust digital health infrastructure that connects patients, healthcare providers, insurers, diagnostic services, and government health schemes into a single, interoperable system (Sharma et al., 2023b).

At the core of this initiative is the vision to ensure that health data travels with the patient, enabling timely and accurate diagnosis, treatment, and follow-up care, regardless of where or by whom care is provided. According to Bhatt and Bhatt (2025), this digital backbone is essential for improving decision-making in health services and ensuring continuity of care, especially in a diverse and mobile population like India's (Sarkar, 2022).

ABHA ID and Digital Health Records

One of the most significant features of ABDM is the introduction of the Ayushman Bharat Health Account (ABHA) ID. This is a unique 14-digit identification number assigned to everyone, enabling them to link and store their health records across different healthcare facilities and digital platforms. The ABHA ID allows users to consolidate prescriptions, diagnostic reports, discharge summaries, and treatment plans into a single, lifetime-accessible repository (Paliwal et al., 2023).

This system transforms health records from scattered paper files into longitudinal digital health histories, accessible to authorised providers with patient consent. Such continuity can be life-saving, particularly in emergencies where access to prior medical history is essential. ABHA offers a scalable and standardised platform that ensures both accessibility and security, bridging the information divide between patients and providers (Broyles et al., 2016).

By enabling interoperability, the ABHA system minimises redundant diagnostic tests, enhances diagnostic accuracy, and promotes evidence-based care. It also empowers individuals to become active participants in their healthcare journeys, an important cultural shift in a system where patients have traditionally been passive recipients of care (Li et al., 2022).

Centralised Healthcare Database: Benefits and Strategic Value

The centralised database envisioned under ABDM has far-reaching implications for public health management. When anonymised and aggregated, this data can provide macro-level insights into disease patterns, treatment efficacy, health behaviours, and system performance. Policymakers can use such data to forecast disease burdens, allocate resources more efficiently, and evaluate program outcomes. For instance, early trends in antibiotic resistance or seasonal outbreaks can be identified more quickly, allowing proactive interventions rather than reactive crisis management (Raj et al., 2023).

At the micro level, a digital trail of individual health interactions can improve clinical outcomes. Physicians can avoid adverse drug interactions, track chronic conditions more effectively, and customise interventions based on past treatment responses. As noted, these capabilities are especially relevant for managing non-communicable diseases like diabetes and hypertension, which require continuous monitoring and lifestyle adjustments (Amri & Abed, 2023).

Integration With Telemedicine and e-Pharmacy

ABDM is designed to work synergistically with emerging digital healthcare services, particularly Telemedicine and e-pharmacy platforms. The pandemic catalysed the adoption of teleconsultation services in India, and ABDM provides the necessary backend to make these services more secure, efficient, and integrated. Using the ABHA ID, a patient in a rural village can receive a video consultation with a specialist in a metropolitan hospital, share diagnostic reports in real time, and receive e-prescriptions, all within the same digital ecosystem (Chandrakar, 2024).

This integration ensures that the telemedicine interaction is not a one-off transaction but part of the patient's continuous care history. Moreover, pharmacies can access e-prescriptions directly and verify authenticity, reducing the risk of misuse and over-the-counter sales without supervision. These linkages help build a trust-based digital health architecture, which is essential for scaling access to care while maintaining clinical rigour (Chen et al., 2023).

Limitations and Ethical Concerns

While the ABDM is a promising step toward health system modernisation, it raises critical questions regarding data privacy, digital access, and ethical oversight. Centralised health records, if not adequately protected, are vulnerable to breaches that could expose sensitive personal information. Civil society groups have raised concerns regarding the potential misuse of health data for commercial or discriminatory purposes, particularly in the absence of a robust data protection law in India (Mishra et al., 2024b).

There are also issues of digital exclusion. A significant proportion of India's population lacks digital literacy or access to smartphones and the internet. Without parallel efforts to bridge these digital divides, ABDM risks entrenching existing inequities rather than eliminating them. Vulnerable populations, such as the elderly, the illiterate, or residents of remote areas, may be unable to benefit fully from digital services unless these are accompanied by physical outreach and support mechanisms (Tewathia et al., 2020).

Moreover, consent mechanisms, though technically present, can be challenging to understand, especially for users unfamiliar with digital systems. There is a need to develop context-sensitive consent frameworks that account for local languages, health literacy levels, and cultural norms around trust and authority. Ethical implementation must go beyond regulatory compliance and engage with the lived realities of the people it intends to serve (Lee et al., 2023).

The Ayushman Bharat Digital Mission represents a landmark shift in India's approach to health governance, moving from siloed, manual systems to an integrated, data-driven ecosystem. By centralising health information, enabling interoperability, and promoting patient agency, ABDM has the potential to significantly improve the quality, efficiency, and equity of care. Its integration with Telemedicine and digital pharmacy services further strengthens its utility in a resource-constrained but digitally ambitious nation (Inamdar et al., 2025).

However, realising this potential will require continued investment not only in technology but also in trust-building, digital literacy, ethical oversight, and inclusive design. As India builds the digital infrastructure of its future healthcare system, it must ensure that human dignity, consent, and equity remain at its core (Lohr, 2025).

AI-DRIVEN HEALTHCARE IN THE UNITED STATES

The United States has emerged as a global frontrunner in integrating artificial intelligence (AI) into healthcare. This leadership stems from a unique confluence of institutional readiness, vast biomedical datasets, academic-industry partnerships, and a regulatory ecosystem that, while still evolving, supports technological experimentation. From diagnostics and drug development to pandemic surveillance and personalised therapies, AI is no longer a speculative asset in the American healthcare system; it is becoming central to its operation (Thacharodi et al., 2024).

National AI Governance and Institutional Support

Proactive governance frameworks have facilitated the advancement of AI in the U.S. healthcare sector. Beginning with the Obama-era Big Data Research and Development Initiative and continuing with executive orders under subsequent administrations, the U.S. government has promoted responsible innovation in AI through policies that balance advancement with ethical oversight (Hassan et al., 2024). In 2020, the National Artificial Intelligence Initiative Act was enacted to support interagency coordination and establish research centres focused on trustworthy AI development (Jensen, 2021).

In healthcare specifically, agencies like the National Institutes of Health (NIH) and Food and Drug Administration (FDA) have incorporated AI into regulatory and funding strategies. The NIH's Bridge2AI Program and All of Us Research Initiative are two examples of data-intensive programs designed to train AI models on large, diverse health datasets. The FDA, on the other hand, has begun piloting flexible approval frameworks for adaptive algorithms used in diagnostics and clinical decision support (Ferguson & Langer, 2021).

AI in Diagnostics, Drug Development, and Genomic Research

The most visible impact of AI in U.S. healthcare has been in diagnostics. AI-driven tools are used to interpret radiological images, such as X-rays, MRIs, and CT scans, with accuracy approaching or exceeding that of human radiologists. For instance, convolutional neural networks trained on thousands of annotated images are now deployed in hospital imaging suites to detect early signs of lung nodules, breast tumours, and cardiovascular anomalies. These tools act as second readers, reducing diagnostic errors and expediting clinical workflows (Olawade et al., 2024).

AI has also catalysed transformation in drug discovery, where traditional pipelines often require 10–15 years and billions of dollars. Machine learning models now help identify potential drug compounds by analysing massive datasets of molecular interactions and clinical outcomes. During the COVID-19 pandemic, AI accelerated the repurposing of existing medications and facilitated the modelling of viral protein structures to support vaccine development (Bilotta et al., 2025).

Equally significant is AI's contribution to genomic research. Precision medicine initiatives across the U.S., such as those led by academic medical centres and institutions like the Broad Institute, have used AI to interpret genetic data for personalised therapy selection. AI algorithms can identify mutations linked to rare genetic diseases, inform cancer risk profiles, and predict how patients may respond to specific treatments. This integration of AI and genomics marks a shift from population-level interventions to personalised, data-informed care (Álvarez-Machancoses et al., 2020).

Public Health Surveillance and Emergency Response

AI's utility extends beyond individual care into population-level surveillance and emergency response. The U.S. Centres for Disease Control and Prevention (CDC) and the Department of Health and Human Services (HHS) have implemented AI systems to monitor disease outbreaks, track vaccination rates, and forecast epidemiological trends. For example, AI-powered dashboards were pivotal during the COVID-19 pandemic in identifying community transmission hotspots, allocating ventilators, and projecting healthcare capacity (Parums, 2023).

More recently, AI has been integrated into wastewater surveillance programs, where it analyses viral concentrations to predict regional outbreaks of infectious diseases, including emerging COVID variants and influenza. These predictive capabilities allow governments to act before healthcare systems are overwhelmed. AI is also instrumental in mapping health inequities by analysing social determinants of health, insurance claims, and environmental exposure data to identify communities at higher risk (Krishnan & Devarajan, 2025).

Personalised Treatment Protocols and Clinical Decision Support

One of the most promising frontiers of AI in the United States is the development of personalised treatment protocols. By synthesising patient-specific data, ranging from electronic health records (EHRs) and lab results to wearable device data, AI models generate individualised treatment plans. For example, oncology departments at major U.S. hospitals now employ AI systems that assess tumour genetics, treatment history, and predicted drug response to tailor chemotherapy regimens. These tools reduce trial-and-error approaches and minimise adverse effects (Olawade, Wada, et al., 2024).

In cardiology, AI models predict heart attack risk by analysing electrocardiogram (ECG) patterns, lifestyle data, and cholesterol profiles. In diabetes management, predictive algorithms suggest insulin dosages and dietary modifications in real-time, often through mobile applications connected to glucose monitors. Importantly, AI does not replace the physician in these scenarios; it augments clinical judgment, helping practitioners make more informed decisions under time constraints (Khurshid et al., 2025).

Clinical decision support (CDS) tools are being embedded into EHR systems across U.S. hospitals to alert physicians about potential drug interactions, identify gaps in care (e.g., overdue screenings), and suggest evidence-based interventions. As these tools mature, the vision is a healthcare environment where machine intelligence enhances human expertise without compromising empathy or patient autonomy (Solomon et al., 2023).

Private-Sector Innovation and Global Technological Leadership

The United States owes much of its AI momentum to its private sector, which includes both established tech giants and dynamic health-tech startups. Companies like Google (via DeepMind), Microsoft, IBM Watson Health, and Amazon Web Services (AWS) have invested heavily in AI solutions tailored to healthcare applications. DeepMind's AI system AlphaFold, which accurately predicts protein folding, has revolutionised molecular biology and drug discovery worldwide (Smith, 2025).

Startups such as Tempus, PathAI, and Aidoc have gained traction for developing AI platforms that support real-time diagnostics, remote triage, and digital pathology. These innovations are frequently incubated within the venture capital ecosystem and tested through collaborations with U.S. hospitals and academic medical centres. The entrepreneurial culture, combined with a strong IP protection framework, enables rapid experimentation and scaling (Madhiraju, 2025).

Moreover, the United States remains a global thought leader in AI ethics and governance, contributing to multilateral guidelines on algorithmic fairness, explainability, and bias reduction. Through its participation in the OECD AI Policy Observatory and the Global Partnership on AI, the U.S. sets normative standards for AI use in medicine, influencing international frameworks and export models (*OECD AI Policy Observatory Portal*, 2025).

AI has become a cornerstone of healthcare transformation in the United States, integrating deeply into diagnostics, genomics, drug development, personalised care, and public health infrastructure. The strength of its innovation ecosystem, bolstered by federal initiatives and a vibrant private sector, has placed the U.S. at the forefront of medical AI globally. However, challenges remain. Issues such as algorithmic bias, digital health disparities, and regulatory complexity continue to demand attention (Puri, 2024).

The American model demonstrates that while AI can drastically enhance healthcare efficiency and precision, its implementation must be carefully governed to ensure equity, transparency, and human-centred care. Ultimately, the goal is not to automate medicine but to empower clinicians and patients alike with intelligent tools that elevate the standard of care (Varnosfaderani & Forouzanfar, 2024).

ETHICAL IMPLICATIONS IN WARFARE AND MEDICINE

As artificial intelligence becomes increasingly embedded in critical decision-making systems, its applications in warfare and medicine prompt profound ethical scrutiny. Both domains involve life-or-death outcomes. However, while the settings differ, one shaped by conflict, the other by care, the deployment of intelligent systems in each raises fundamental concerns around accountability, human dignity, and the moral boundaries of delegation. The central question is not merely what AI can do, but what it should do, and whether humans can ethically permit machines to participate in decisions traditionally governed by human judgment (Rashid et al., 2023).

The Moral Burden of Intelligent Machines

At the heart of the ethical debate is the moral burden of delegating decisions to machines that involve human lives. While AI systems offer remarkable capabilities, speed, precision, and data processing power, they do not possess empathy, moral reasoning, or contextual sensitivity. In high-stakes environments like surgery or battlefield targeting, these human traits are not peripheral; they are central. As highlighted, AI systems operate on statistical patterns, not moral frameworks, and this divergence creates tension when machines are asked to make or support consequential decisions (Bigman & Gray, 2018).

The moral unease stems from the fear that, in outsourcing decisions to AI, society may abdicate ethical responsibility. Who is to be held accountable when an autonomous drone misidentifies a civilian structure as a military target? Or when a cancer diagnosis is missed by an AI-based radiology tool, resulting in delayed treatment? These scenarios reflect not just technological failure but moral and institutional gaps (Hassen, 2025).

AI in Warfare: The Dangers of Automation Without Accountability

The use of AI in warfare represents one of the most contested ethical frontiers. Intelligent military systems, ranging from autonomous drones to algorithmic targeting systems, have been developed and deployed by several nations, including the United States, China, Israel, and Russia. These systems are often praised for reducing human casualties among soldiers, enabling precision strikes, and minimising the psychological burden on combatants (Johnson, 2022).

However, such claims are counterbalanced by serious concerns about autonomy, bias, and legal responsibility. Autonomous weapons, often referred to as "killer robots," operate without real-time human control, raising the question: Can a machine be ethically entrusted with the authority to kill? (De Ágreda, 2020).

According to Asaro (2020), the lack of transparency in AI targeting decisions can lead to untraceable errors. Machine learning algorithms used in drone warfare are often trained on incomplete or biased data, potentially misclassifying civilians or infrastructure as threats. Moreover, such systems may not adequately interpret contextual variables—like the presence of non-combatants, local customs, or cultural symbols—leading to morally unjustifiable outcomes (Asaro, 2020).

International humanitarian law demands distinction and proportionality in military operations, principles complex to embed into computational logic. While the Geneva Conventions emphasise the necessity of human judgment in warfare, AI systems may act without ethical nuance, especially when deployed at scale or in contested territories with weak oversight mechanisms. This has led organisations such as the International Committee for Robot Arms Control to advocate for global bans on lethal autonomous weapons, insisting on the necessity of "meaningful human control (Dutt, 2018)."

AI in Medicine: Errors, Empathy, and Ethical Risks

While the application of AI in medicine is driven by noble goals, efficiency, accuracy, and expanded access, it is not without ethical risks. AI diagnostic tools, robotic surgical systems, and clinical decision-support software now play active roles in patient care. However, these systems are inherently probabilistic. They learn from past data but cannot interpret the present context with human intuition (Poalelungi et al., 2023).

One critical concern is the potential for misdiagnosis. For instance, an AI trained primarily on datasets from Western populations may misclassify symptoms in Indian or African patients due to demographic discrepancies in the training data. A false-negative in detecting cancer or a missed cardiac abnormality in an ECG can result in fatal outcomes. When these tools are seen as infallible, clinicians may defer too heavily to machine output, sidelining their own judgment (Y. Li et al., 2025).

Moreover, AI systems lack the capacity for empathy and narrative understanding. Medicine is not merely a technical pursuit; it is relational. When AI chatbots or diagnostic apps replace human interaction, patients may feel depersonalised or neglected. As Bhatt and Bhatt (2025) observe, this shift may erode trust in healthcare systems, particularly among older or more vulnerable populations (Montemayor et al., 2021).

The question of informed consent also becomes complex in AI-mediated care. Patients often do not understand how these systems function, how their data is used, or the limitations of AI-generated recommendations. Ethical practice demands that patients are not only informed about procedures but also about the nature of the tools being used in their diagnosis or treatment (Hurley et al., 2024).

The Need for Human Oversight

Across both domains, warfare and medicine, the necessity of human oversight emerges as a non-negotiable ethical safeguard. Human oversight ensures that decisions involving human life are not reduced to computational outputs but are situated within moral and social frameworks (Dhillon et al., 2023).

In military contexts, this means that autonomous systems must be subject to real-time human authorisation. Military operators must retain the ability to intervene, halt, or redirect AI actions based on changing information or ethical considerations. In healthcare, clinicians must be trained not only to use AI but to evaluate its recommendations critically. They must balance machine-suggested pathways with their own clinical expertise, patient narratives, and ethical judgment (Konert & Balcerzak, 2021).

Scholars like Fuster et al. (2025) advocate for a hybrid model of decision-making, where AI acts as an assistant, not a replacement. This model demands the co-design of AI systems with ethicists, clinicians, and affected communities to ensure alignment with societal values (Fuster et al., 2025).

Voices From the Frontlines: Ethical Reflections

Real-life cases underscore the complexity of AI ethics. In the U.S., a widely cited case involved an AI-driven cancer diagnostic tool that overestimated patient survival rates due to flawed training data, leading to aggressive treatments with limited benefit. Clinicians later testified that they were pressured to rely on the system due to hospital mandates, despite personal doubts (Farhud & Zokaei, 2021).

In Gaza and Yemen, reports have emerged of drone strikes based on AI-assisted targeting systems that hit non-military buildings, including schools and hospitals.

Investigations revealed that the AI had been trained on outdated satellite data and lacked contextual understanding of humanitarian zones. Survivors and aid workers interviewed by Human Rights Watch (2024) described the strikes as "faceless violence", an act without accountability or remorse (Birch, 2024).

These voices, of clinicians, patients, soldiers, and civilians, serve as moral reminders that AI must never operate in ethical vacuums. Its deployment must be grounded in a commitment to dignity, transparency, and accountability (Vaidya & Bobdey, 2021).

The ethical implications of AI in warfare and medicine demand continuous vigilance. While intelligent systems offer powerful enhancements, their use in life-and-death decisions requires deep ethical reflection and strong human oversight. Whether deployed in a hospital or a conflict zone, AI must remain a tool under human command, not a moral actor. As societies increasingly rely on these systems, the challenge is to embed them not only with technical competence but with ethical conscience, something only humans can provide (Helkala et al., 2023).

POLICY, PREPAREDNESS, AND THE PATH AHEAD

As artificial intelligence (AI) becomes embedded in the structural functions of healthcare, defence, and governance, it is no longer sufficient to evaluate its impact solely in terms of efficiency or innovation. What is increasingly urgent is the development of comprehensive frameworks, legal, ethical, institutional, and cultural, that can guide its deployment in socially responsible ways. While AI has opened unprecedented possibilities for solving complex problems, its power must be regulated through principled, participatory governance that foregrounds human dignity and public good (Papagiannidis et al., 2025).

Legal and Ethical Frameworks: More than Regulation

Across democracies like India and the United States, the conversation around AI has shifted from technological capability to governance readiness. However, the legal frameworks required to manage AI's societal impact are still in early stages. In India, for example, the Digital Personal Data Protection Act (2023) marks a significant step forward in regulating personal data use, but specific provisions addressing algorithmic bias, medical liability in AI-driven diagnostics, or the chain of accountability in autonomous systems remain vague. The United States, while more mature in its regulatory discourse, also lacks a unified AI-specific law; governance remains scattered across federal agencies and executive orders (Zaidan & Ibrahim, 2024).

Ethical frameworks must do more than prevent harm; they must also promote fairness, inclusion, and trust. Ethical AI governance requires a shift from reactive compliance models to proactive ethical design, where principles like transparency, non-discrimination, and explainability are embedded from the outset. This is particularly relevant in healthcare, where decisions made by AI systems can influence access to treatment, insurance eligibility, and diagnosis accuracy. Similarly, in military contexts, ethical frameworks must ensure that the deployment of intelligent weapons adheres to international humanitarian law and preserves human oversight (Narayan et al., 2024b).

The absence of such clear, enforceable frameworks risks creating an ecosystem where innovation outpaces accountability, a gap that could deepen public distrust and expose vulnerable populations to disproportionate risk.

Cross-Sector Collaboration: Building Shared Responsibility

AI governance cannot be siloed within individual sectors. The complexities involved demand collaboration between governments, private industries, research institutions, civil society, and end users. In India, initiatives like the National AI Portal and the Responsible AI for Social Empowerment (RAISE) Summit reflect early efforts to build multi-stakeholder engagement. These platforms are crucial for aligning technological innovation with public policy, citizen rights, and equitable development goals (Ghosh et al., 2025).

Similarly, in the U.S., interagency collaborations between the National Institutes of Health (NIH), Department of Defence (DoD), and academic consortia aim to create standardised AI evaluation protocols and ethical guidelines (Thomsen & Knudsen, 2025). Notably, such frameworks must not only include technical experts but also bioethicists, patient representatives, disability advocates, and minority groups, who can identify blind spots and raise concerns that may not be visible to developers or policymakers (Thomsen & Knudsen, 2025).

Shared responsibility also means jointly investing in risk assessments, data governance standards, and public communication strategies. Without this collective infrastructure, AI systems, especially those in medicine and defence, will remain vulnerable to misuse, unintended harm, or public backlash.

Inclusion and Access: Bridging the Digital Divide

As AI becomes more entrenched in public services, there is a growing risk that its benefits may bypass marginalised populations. The promise of AI to democratise access to healthcare or education can quickly unravel if the underlying systems reflect structural inequities—such as biased data, linguistic exclusion, or limited internet access.

In India, a significant portion of the population lacks digital literacy or even basic access to smartphones and internet connectivity. AI-based telemedicine platforms or health record portals can alienate older adults, non-literate patients, or residents of rural regions unless consciously designed for inclusion. In the U.S., studies have shown that AI systems trained on predominantly white, urban patient data underperform in diagnosing illnesses among Black and Latino communities (Dixon et al., 2024).

Ethical policy must address these disparities through inclusive data curation, linguistic accessibility, public outreach, and digital infrastructure investment. Furthermore, algorithmic audits, periodic evaluations of how AI systems perform across demographics, should become standard in both public and private sector deployments.

Ultimately, the test of any AI policy is not how efficient the system is but how equitably it serves.

Investing in Awareness, Education, and Human Responsibility

The future of AI policy also hinges on cultural readiness. The rapid integration of AI into healthcare and public infrastructure has outpaced public understanding of how these systems function or how they impact individual rights. There is an urgent need for AI literacy campaigns that can empower citizens to ask informed questions, give meaningful consent, and hold institutions accountable. Healthcare professionals, for example, require training not just in using AI tools, but in understanding their limitations, ethical implications, and legal liabilities. In defence, personnel must be equipped to intervene in autonomous system operations, balancing efficiency with ethical judgment. Education systems, from school curricula to professional development programs, must reorient to foster critical thinking, ethical reasoning, and digital citizenship. As Fuster et al. (2025) emphasise, AI literacy is no longer a technical skill; it is a civic competence (Fuster et al., 2025).

Moreover, the human role in AI deployment must be framed not as supervisory, but as morally central. AI is not a replacement for human agency, empathy, or moral accountability. Policies must ensure that ultimate responsibility remains with people, not programs.

Framing a Sustainable and Humanitarian AI Future

The path ahead must prioritise sustainability, not only environmental, but also social and ethical. AI should serve long-term public welfare, not short-term institutional gains. This calls for embedding humanitarian values into algorithmic design, procurement policies, and deployment strategies.

In both India and the U.S., public policy must evolve toward AI that is socially embedded, sensitive to context, reflective of collective values, and responsive to community needs. This involves decentralising innovation, supporting local tech ecosystems, and promoting open-source platforms that communities can adapt and govern themselves (Rosário & Boechat, 2025).

We must reject deterministic narratives that frame AI's evolution as inevitable or beyond human control. Policy is not about catching up to machines; it is about shaping their role in the service of human dignity, equity, and solidarity. The deployment of AI in healthcare, warfare, and public infrastructure offers tremendous possibilities, but also profound responsibilities. Policy and preparedness must move beyond compliance checklists to a deeper ethical vision: one that centres human values, protects the vulnerable, and builds systems that earn public trust. Through cross-sector collaboration, inclusive design, ethical regulation, and civic education, we can frame an AI future that is not only intelligent but humane (Dahlin, 2023).

CONCLUSION

The accelerating integration of artificial intelligence across sectors as distinct as medicine and warfare reveals both the promise and the peril of intelligent systems. Through this chapter, we have explored the humanitarian dilemmas posed by AI, particularly within the contexts of healthcare and defence, and underscored the tension between technological advancement and ethical responsibility.

In medicine, AI has already demonstrated its potential to improve diagnostic accuracy, support personalised treatment plans, and enhance public health surveillance. In India, AI-enabled radiology, oncology, and pandemic response systems have helped bridge the structural gaps caused by population pressure and resource limitations. In the United States, AI supports cutting-edge genomics research, clinical decision-making, and pharmaceutical innovation. However, these benefits are not immune to misuse or inequity. Misdiagnoses, data biases, lack of transparency, and patient exclusion remain serious concerns, particularly in underserved populations.

In warfare, the deployment of autonomous weapons and algorithmic targeting tools raises more profound moral questions: Can machines be allowed to make life-or-death decisions? Who bears responsibility when errors occur in combat zones? These questions defy technical solutions and demand a robust framework of ethical reflection, legal accountability, and international regulation. AI may enhance national security, but without meaningful human oversight, it risks eroding the very humanitarian principles that guide just conflict.

Across these contexts, a central insight emerges: AI is not an autonomous moral actor; it is a human construct. The tools we design, the data we train on, and the systems we authorise all reflect human decisions, biases, and intentions. Therefore, the responsibility for ethical AI does not lie with the technology itself, but with the societies that deploy it.

Moving forward, the path to a more equitable and humane AI future requires deliberate choices. We must ground innovation in ethical frameworks that prioritise inclusion, transparency, and justice. We must invest in digital literacy and ensure that public engagement, not just technical expertise, shapes the development and deployment of AI systems. Above all, we must ensure that AI serves human welfare, not replaces it.

As we stand at the crossroads of innovation and responsibility, this is not a call for technological restraint, but for ethical imagination. We must reframe AI not as a substitute for human intelligence, but as a companion to human values. Only then can it become a tool not of disruption or dominance, but of care, dignity, and shared progress.

REFERENCES

Althobiti, M., Nhung, T. T. T., Verma, S., Albugami, R. R., & Kumar, R. (2025). Artificial intelligence and biosensors: Transforming cancer diagnostics. *Medicine in Novel Technology and Devices*, *27*, 100378. DOI: 10.1016/j.medntd.2025.100378

Álvarez-Machancoses, Ó., Galiana, E. J. D., Cernea, A., De La Viña, J. F. S., & Fernández-Martínez, J. L. (2020). <p>On the Role of Artificial Intelligence in Genomics to Enhance Precision Medicine</p>. *Pharmacogenomics and Personalized Medicine*, *13*, 105–119. DOI: 10.2147/PGPM.S205082 PMID: 32256101

Amri, M. M., & Abed, S. A. (2023). The Data-Driven Future of Healthcare: A review. *Mesopotamian Journal of Big Data*, *2023*, 68–74. DOI: 10.58496/MJBD/2023/010

Asaro, P. (2020). Autonomous weapons and the ethics of artificial intelligence. In *Ethics of Artificial Intelligence* (pp. 212–236). online edn, Oxford Academic. https://doi.org/DOI: 10.1093/oso/9780190905033.003.0008

Bai, Y. (2025). Moral judgment in the age of artificial intelligence: How humans perceive ethical decisions made by machines. *Humanistic Management Journal*, *10*(3), 361–385. Advance online publication. DOI: 10.1007/s41463-025-00215-0

Bhavana, S., & Mithra, S. G. S. (2025). Revolutionizing healthcare with an in-depth analysis of AI's transformative impact in India. *Social Sciences & Humanities Open*, *11*, 101467. DOI: 10.1016/j.ssaho.2025.101467

Bigman, Y. E., & Gray, K. (2018). People are averse to machines making moral decisions. *Cognition*, *181*, 21–34. DOI: 10.1016/j.cognition.2018.08.003 PMID: 30107256

Bilotta, M., Rocca, R., & Alcaro, S. (2025). Next-Generation Drug Discovery: The AI revolution in pharmaceutical research. *Artificial Intelligence in the Life Sciences*, *100149*, 100149. Advance online publication. DOI: 10.1016/j.ailsci.2025.100149

Birch, M. (2024). Who did that? AI assisted targeting and the lowering of thresholds in Gaza. *Medicine, Conflict, and Survival*, *40*(2), 97–100. DOI: 10.1080/13623699.2024.2364937 PMID: 38915179

Bitko, G. (2025, May 15). Powering Tomorrow: How Strategic AI infrastructure Policy Can Secure America's technological edge. *Forbes*. https://www.forbes.com/sites/gordonbitko/2025/05/15/powering-tomorrow-how-strategic-ai-infrastructure-policy-can-secure-americas-technological-edge/

Bouderhem, R. (2024). Shaping the future of AI in healthcare through ethics and governance. *Humanities & Social Sciences Communications*, *11*(1), 416. Advance online publication. DOI: 10.1057/s41599-024-02894-w

Brandao, P. R. (2025). The impact of artificial intelligence on modern society. *AI*, *6*(8), 190. DOI: 10.3390/ai6080190

Broyles, D., Crichton, R., Jolliffe, B., Sæbø, J. I., & Dixon, B. E. (2016). Shared longitudinal health records for clinical and population health. In *Elsevier eBooks* (pp. 149–162). https://doi.org/DOI: 10.1016/B978-0-12-803135-3.00010-4

Buonocore, F., Annosi, M. C., De Gennaro, D., & Riemma, F. (2024). Digital transformation and social change: Leadership strategies for responsible innovation. *Journal of Engineering and Technology Management*, *74*, 101843. DOI: 10.1016/j.jengtecman.2024.101843

Chandrakar, M. (2024). Telehealth and digital tools enhancing healthcare access in rural systems. *Discover Public Health*, *21*(1), 144. Advance online publication. DOI: 10.1186/s12982-024-00271-1

Chaudhari, N., Bhade, M., Havelikar, U., & Patel, V. P. (2025). Artificial intelligence and robotics in intensive care units (ICUs): A review of critical care innovations. *Intelligent Hospital*, *1*(2), 100010. DOI: 10.1016/j.inhs.2025.100010

Chen, E., Prakash, S., Reddi, V. J., Kim, D., & Rajpurkar, P. (2023). A framework for integrating artificial intelligence for clinical care with continuous therapeutic monitoring. *Nature Biomedical Engineering*, *9*(4), 445–454. DOI: 10.1038/s41551-023-01115-0 PMID: 37932379

Chettri, S. K., Deka, R. K., & Saikia, M. J. (2025). Bridging the gap in the adoption of trustworthy AI in Indian healthcare: Challenges and opportunities. *AI*, *6*(1), 10. DOI: 10.3390/ai6010010

Colther, C., & Doussoulin, J. P. (2024). Artificial intelligence: Driving force in the evolution of human knowledge. *Journal of Innovation & Knowledge*, *9*(4), 100625. DOI: 10.1016/j.jik.2024.100625

Coombs, C., Hislop, D., Taneva, S. K., & Barnard, S. (2020). The strategic impacts of Intelligent Automation for knowledge and service work: An interdisciplinary review. *The Journal of Strategic Information Systems*, *29*(4), 101600. DOI: 10.1016/j.jsis.2020.101600

Cordova-Pozo, K., & Rouwette, E. A. (2023). Types of scenario planning and their effectiveness: A review of reviews. *Futures*, *149*, 103153. DOI: 10.1016/j.futures.2023.103153

Dahlin, E. (2023). Think Differently We must! An AI Manifesto for the Future. *AI & Society*, *39*(3), 1423–1426. DOI: 10.1007/s00146-022-01620-x

Das, S. K., Dasgupta, R. K., Roy, S. D., & Shil, D. (2024). AI in Indian healthcare: From roadmap to reality. *Intelligent Pharmacy*, *2*(3), 329–334. DOI: 10.1016/j.ipha.2024.02.005

De Ágreda, Á. G. (2020). Ethics of autonomous weapons systems and its applicability to any AI systems. *Telecommunications Policy*, *44*(6), 101953. DOI: 10.1016/j.telpol.2020.101953

Dégallier-Rochat, S., Kurpicz-Briki, M., Endrissat, N., & Yatsenko, O. (2022). Human augmentation, not replacement: A research agenda for AI and robotics in the industry. *Frontiers in Robotics and AI*, *9*, 997386. DOI: 10.3389/frobt.2022.997386 PMID: 36267424

Dhillon, N., Jeon, N., Gurkan, U., Gupta, A. S., Bonomo, R., Drummy, L., Zhang, M., & Chance, M. (2023). Military medicine and medical research as a source of inspiration and innovation to solve national security and health challenges in the 21st century. *Pathogens & Immunity*, *8*(1), 51–63. DOI: 10.20411/pai.v8i1.596 PMID: 37799210

Dixon, D., Sattar, H., Moros, N., Kesireddy, S. R., Ahsan, H., Lakkimsetti, M., Fatima, M., Doshi, D., Sadhu, K., & Hassan, M. J. (2024). Unveiling the Influence of AI Predictive analytics on patient Outcomes: A Comprehensive Narrative review. *Cureus*, *16*(5), e59954. DOI: 10.7759/cureus.59954 PMID: 38854327

Donelle, L., Comer, L., Hiebert, B., Hall, J., Shelley, J. J., Smith, M. J., Kothari, A., Burkell, J., Stranges, S., Cooke, T., Shelley, J. M., Gilliland, J., Ngole, M., & Facca, D. (2023). Use of digital technologies for public health surveillance during the COVID-19 pandemic: A scoping review. *Digital Health*, *9*, 20552076231173220. DOI: 10.1177/20552076231173220 PMID: 37214658

Dutt, K. (2018, October 11). *World leaders in legal education and research - Legal Wires*. Legal Wires - World Leaders in Legal Education and Research. https://legal-wires.com/columns/necessity-and-proportionality-in-international-humanitarian-law-legal-boundaries-in-armed-conflict/

Faiyazuddin, M., Rahman, S. J. Q., Anand, G., Siddiqui, R. K., Mehta, R., Khatib, M. N., Gaidhane, S., Zahiruddin, Q. S., Hussain, A., & Sah, R. (2025). The Impact of Artificial Intelligence on Healthcare: A comprehensive review of advancements in diagnostics, treatment, and operational efficiency. *Health Science Reports*, *8*(1), e70312. DOI: 10.1002/hsr2.70312 PMID: 39763580

Farhud, D. D., & Zokaei, S. (2021). Ethical issues of artificial intelligence in medicine and healthcare. *Iranian Journal of Public Health*, *50*(11), i–v. DOI: 10.18502/ijph.v50i11.7600 PMID: 35223619

Ferguson, S. M., & Langer, L. J. (2021). The U.S. National Institutes of Health – founding a national biomedical "Innovation Ecosystem.". *Journal of Commercial Biotechnology*, *26*(1), 72–82. DOI: 10.5912/jcb972 PMID: 33859535

Fontes, C., Hohma, E., Corrigan, C. C., & Lütge, C. (2022). AI-powered public surveillance systems: Why we (might) need them and how we want them. *Technology in Society*, *71*, 102137. DOI: 10.1016/j.techsoc.2022.102137

Fuster, V., Swirski, F. K., & Nadkarni, G. N. (2025). Conventional and precision medicine: Opposites or complementary ends? *Frontiers in Science*, *3*, 1701495. Advance online publication. DOI: 10.3389/fsci.2025.1701495

Ghosh, A., Saini, A., & Barad, H. (2025). Artificial intelligence in governance: Recent trends, risks, challenges, innovative frameworks and future directions. *AI & Society*, *40*(7), 5685–5707. DOI: 10.1007/s00146-025-02312-y

Gignac, G. E., & Szodorai, E. T. (2024). Defining intelligence: Bridging the gap between human and artificial perspectives. *Intelligence*, *104*, 101832. DOI: 10.1016/j.intell.2024.101832

Gore, M. N., & Olawade, D. B. (2024). Harnessing AI for public health: India's roadmap. *Frontiers in Public Health*, *12*, 1417568. DOI: 10.3389/fpubh.2024.1417568 PMID: 39399702

Gu, D., Andreev, K., & Dupre, M. E. (2021). Major trends in population growth around the world. *China CDC Weekly*, *3*(28), 604–613. DOI: 10.46234/ccdcw2021.160 PMID: 34594946

Hassan, M., Borycki, E. M., & Kushniruk, A. W. (2024). Artificial intelligence governance framework for healthcare. *Healthcare Management Forum*, *38*(2), 125–130. DOI: 10.1177/08404704241291226 PMID: 39470044

Hassen, M. Z. (2025). The risk of moral outsourcing: Why artificial intelligence cannot and should not make our ethical decisions. *EthAIca*, *4*, 428. DOI: 10.56294/ai2025428

Helkala, K. M., Lucas, G., Barrett, E., & Syse, H. (2023). Editorial: Ethical challenges in AI-enhanced military operations. *Frontiers in Big Data*, *6*, 1229252. Advance online publication. DOI: 10.3389/fdata.2023.1229252 PMID: 37405063

Hirani, R., Noruzi, K., Khuram, H., Hussaini, A. S., Aifuwa, E. I., Ely, K. E., Lewis, J. M., Gabr, A. E., Smiley, A., Tiwari, R. K., & Etienne, M. (2024). Artificial Intelligence and Healthcare: A Journey through History, Present Innovations, and Future Possibilities. *Life (Basel, Switzerland)*, *14*(5), 557. DOI: 10.3390/life14050557 PMID: 38792579

Hurley, M. E., Lang, B. H., Kostick-Quenet, K. M., Smith, J. N., & Blumenthal-Barby, J. (2024). Patient consent and the right to notice and explanation of AI systems used in health care. *The American Journal of Bioethics : AJOB*, *25*(3), 102–114. DOI: 10.1080/15265161.2024.2399828 PMID: 39288291

Inamdar, A. V., Kale, M., Deogaonkar, A., Dariya, S. S., & Dey, A. K. (2025). A unified health ecosystem in India: Exploring the Ayushman Bharat digital mission. *International Journal of Community Medicine and Public Health*, *12*(8), 3813–3819. DOI: 10.18203/2394-6040.ijcmph20252497

Jensen, P. A. (2021, January 1). *Intro*. blog.biocomm.ai. https://blog.biocomm.ai/2021/01/01/national-artificial-intelligence-initiative-overseeing-and-implementing-the-united-states-national-ai-strategy/t

Johnson, J. (2022). The AI Commander Problem: Ethical, political, and psychological Dilemmas of Human-Machine Interactions in AI-enabled Warfare. *Journal of Military Ethics*, *21*(3–4), 246–271. DOI: 10.1080/15027570.2023.2175887

Jose, B., Cherian, J., Verghis, A. M., Varghise, S. M., S, M., & Joseph, S. (2025). The cognitive paradox of AI in education: Between enhancement and erosion. *Frontiers in Psychology*, *16*, 1550621. DOI: 10.3389/fpsyg.2025.1550621 PMID: 40297599

Joshi, K. (2024). Artificial General Intelligence (AGI): A Comprehensive review. *Journal of the Epidemiology Foundation of India*, *2*(3), 93–96. DOI: 10.56450/JEFI.2024.v2i03.004

K, S. (2024, October 3). *AI: Simulating Human Intelligence for Real-World Tasks | InfoSecured.ai*. InfoSecured.ai. https://www.infosecured.ai/i/ai-definitions/artificial-intelligence-simulating-human-intelligence/

Kalani, M., & Anjankar, A. (2024). Revolutionizing Neurology: The role of Artificial intelligence in advancing diagnosis and treatment. *Cureus*, *16*(6), e61706. DOI: 10.7759/cureus.61706 PMID: 38975469

Kapoor, N., Sanjana, S. N., Davalagi, S. B., Balu, P. S., & Sethia, S. (2024). AI Horizons in Indian Healthcare: A vision for transformation and equity. *Indian Journal of Community Medicine : Official Publication of Indian Association of Preventive & Social Medicine*, *49*(Suppl 2), S210–S216. DOI: 10.4103/ijcm.ijcm_806_24 PMID: 40124859

Khurshid, S., Friedman, S. F., Al-Alusi, M. A., Kany, S., Sommers, T., Anderson, C. D., Ho, J. E., McManus, D. D., Borowsky, L. H., Ashburner, J. M., Lubitz, S. A., Atlas, S. J., Maddah, M., Singer, D. E., Ellinor, P. T., Khurshid, S., Friedman, S. F., Al-Alusi, M. A., Kany, S., & Ellinor, P. T. (2025). Artificial intelligence-enabled analysis of handheld single-lead electrocardiograms to predict incident atrial fibrillation: An analysis of the VITAL-AF randomized trial. *NPJ Digital Medicine*, *8*(1), 776. Advance online publication. DOI: 10.1038/s41746-025-02164-2 PMID: 41299008

Konert, A., & Balcerzak, T. (2021). Military autonomous drones (UAVs) - from fantasy to reality. Legal and Ethical implications. *Transportation Research Procedia*, *59*, 292–299. DOI: 10.1016/j.trpro.2021.11.121

Krishna, V. V. (2023). A I and contemporary challenges: The good, bad and the scary. *Journal of Open Innovation*, *10*(1), 100178. DOI: 10.1016/j.joitmc.2023.100178

Krishnan, A., & Devarajan, Y. (2025). Sustainable wastewater treatment with AI: A path to enhanced hygiene and environmental protection. *Process Safety and Environmental Protection*, *201*, 107487. DOI: 10.1016/j.psep.2025.107487

Kumar, A. (2023). The transformation of the Indian healthcare system. *Cureus*, *15*(5), e39079. DOI: 10.7759/cureus.39079 PMID: 37378105

Kumar, G., Baharia, R., Singh, K., Gupta, S. K., Joy, S., Sharma, A., & Rahi, M. (2024). Addressing challenges in vector control: A review of current strategies and the imperative for novel tools in India's combat against vector-borne diseases. *BMJ Public Health*, *2*(1), e000342. DOI: 10.1136/bmjph-2023-000342 PMID: 40018116

Latif, S. D., Hazrin, N. B., Koo, C. H., Ng, J. L., Chaplot, B., Huang, Y. F., El-Shafie, A., & Ahmed, A. N. (2023). Assessing rainfall prediction models: Exploring the advantages of machine learning and remote sensing approaches. *Alexandria Engineering Journal*, *82*, 16–25. DOI: 10.1016/j.aej.2023.09.060

Lee, A. R., Koo, D., Kim, I. K., Lee, E., Kim, H. H., Yoo, S., Kim, J., Choi, E. K., & Lee, H. (2023). Identifying facilitators of and barriers to the adoption of dynamic consent in digital health ecosystems: A scoping review. *BMC Medical Ethics*, *24*(1), 107. DOI: 10.1186/s12910-023-00988-9 PMID: 38041034

Li, E., Clarke, J., Ashrafian, H., Darzi, A., & Neves, A. L. (2022). The Impact of Electronic Health record interoperability on safety and Quality of Care in High-Income Countries: Systematic review. *Journal of Medical Internet Research*, *24*(9), e38144. DOI: 10.2196/38144 PMID: 36107486

Li, Y., Yi, X., Fu, J., Yang, Y., Duan, C., & Wang, J. (2025). Reducing misdiagnosis in AI-driven medical diagnostics: A multidimensional framework for technical, ethical, and policy solutions. *Frontiers in Medicine*, *12*, 1594450. DOI: 10.3389/fmed.2025.1594450 PMID: 41244777

Lohr, K. D. (2025). Digital Literacy and Access: Equity from a global and local perspective. *New Directions for Adult and Continuing Education*, *2025*(185), 39–43. DOI: 10.1002/ace.20559

Madhiraju, P. (2025, June 9). *How AI is reshaping the role of pathologists and radiologists in oncology*. Business-News-Today.com. https://business-news-today.com/how-ai-is-reshaping-the-role-of-pathologists-and-radiologists-in-oncology/

Martin, K. D., & Zimmermann, J. (2024). Artificial intelligence and its implications for data privacy. *Current Opinion in Psychology*, *58*, 101829. DOI: 10.1016/j.copsyc.2024.101829 PMID: 38954851

McGowan, C. R., Baxter, L., DuBois, M., Sheather, J., Khondaker, R., Cummings, R., & Watkins, K. (2020). Preparing humanitarians to address ethical problems. *Conflict and Health*, *14*(1), 72. DOI: 10.1186/s13031-020-00319-4 PMID: 33292441

Meadows, A. J., Oppenheim, B., Guerrero, J., Ash, B., Badker, R., Lam, C. K., Pardee, C., Ngoon, C., Savage, P. T., Sridharan, V., Madhav, N. K., & Stephenson, N. (2022). Infectious disease underreporting is predicted by Country-Level preparedness, politics, and pathogen severity. *Health Security*, *20*(4), 331–338. DOI: 10.1089/hs.2021.0197 PMID: 35925788

Mishra, U. S., Yadav, S., & Joe, W. (2024). The Ayushman Bharat Digital Mission of India: An assessment. *Health Systems and Reform*, *10*(2), 2392290. DOI: 10.1080/23288604.2024.2392290 PMID: 39437234

Montemayor, C., Halpern, J., & Fairweather, A. (2021). In principle obstacles for empathic AI: Why we can't replace human empathy in healthcare. *AI & Society*, *37*(4), 1353–1359. DOI: 10.1007/s00146-021-01230-z PMID: 34054228

Nancholas, B. (2023, October 13). *Narrow artificial intelligence: advantages, disadvantages, and the future of AI - University of Wolverhampton*. University of Wolverhampton. https://online.wlv.ac.uk/narrow-artificial-intelligence-advantages-disadvantages-and-the-future-of-ai/

Narayan, A., Bhushan, I., & Schulman, K. (2024a). India's evolving digital health strategy. *NPJ Digital Medicine*, *7*(1), 284. DOI: 10.1038/s41746-024-01279-2 PMID: 39414924

Narayan, A., Bhushan, I., & Schulman, K. (2024b). India's evolving digital health strategy. *NPJ Digital Medicine*, *7*(1), 284. DOI: 10.1038/s41746-024-01279-2 PMID: 39414924

OECD AI Policy Observatory Portal. (2025). https://oecd.ai/en/

Olawade, D. B., David-Olawade, A. C., Wada, O. Z., Asaolu, A. J., Adereni, T., & Ling, J. (2024). Artificial intelligence in healthcare delivery: Prospects and pitfalls. *Journal of Medicine, Surgery, and Public Health*, *3*, 100108. DOI: 10.1016/j.glmedi.2024.100108

Olawade, D. B., Wada, O. Z., Odetayo, A., David-Olawade, A. C., Asaolu, F., & Eberhardt, J. (2024). Enhancing mental health with Artificial Intelligence: Current trends and future prospects. *Journal of Medicine, Surgery, and Public Health*, *3*, 100099. DOI: 10.1016/j.glmedi.2024.100099

Paliwal, S., Parveen, S., Singh, O., Alam, M. A., & Ahmed, J. (2023). The role of Ayushman Bharat Health Account (ABHA) in Telehealth: a new frontier of smart healthcare delivery in India. In *Lecture notes in networks and systems* (pp. 388–406). https://doi.org/DOI: 10.1007/978-3-031-47451-4_28

Papagiannidis, E., Mikalef, P., & Conboy, K. (2025). Responsible artificial intelligence governance: A review and research framework. *The Journal of Strategic Information Systems*, *34*(2), 101885. DOI: 10.1016/j.jsis.2024.101885

Parums, D. V. (2023). Editorial: Infectious Disease Surveillance Using Artificial Intelligence (AI) and its Role in Epidemic and Pandemic Preparedness. *Medical Science Monitor*, *29*, e941209. DOI: 10.12659/MSM.941209 PMID: 37259578

Parvizpour, S., Beyrampour-Basmenj, H., Razmara, J., Farhadi, F., & Shamsir, M. S. (2023). Cancer treatment comes to age: From one-size-fits-all to next-generation sequencing (NGS) technologies. *BioImpacts*, *14*(4), 29957. DOI: 10.34172/bi.2023.29957 PMID: 39104623

Pinto-Coelho, L. (2023). How Artificial intelligence is shaping medical Imaging Technology: A survey of Innovations and applications. *Bioengineering (Basel, Switzerland)*, *10*(12), 1435. DOI: 10.3390/bioengineering10121435 PMID: 38136026

Poalelungi, D. G., Musat, C. L., Fulga, A., Neagu, M., Neagu, A. I., Piraianu, A. I., & Fulga, I. (2023). Advancing Patient Care: How Artificial intelligence is transforming healthcare. *Journal of Personalized Medicine*, *13*(8), 1214. DOI: 10.3390/jpm13081214 PMID: 37623465

Puri, G. (2024, September 20). How AI is ushering in a new era of value in US healthcare - healthcare business today. *Healthcare Business Today*. https://www.healthcarebusinesstoday.com/ai-new-era-value-us-healthcare/

Radanliev, P. (2025). AI Ethics: Integrating transparency, fairness, and privacy in AI development. *Applied Artificial Intelligence*, *39*(1), 2463722. Advance online publication. DOI: 10.1080/08839514.2025.2463722

Raj, G. M., Dananjayan, S., & Agarwal, N. (2023). Inception of the Indian Digital Health Mission: Connecting. . .the. . .Dots. *Health Care Science*, *2*(5), 345–351. DOI: 10.1002/hcs2.67 PMID: 38938588

Raman, R., Kowalski, R., Achuthan, K., Iyer, A., & Nedungadi, P. (2025). Navigating artificial general intelligence development: Societal, technological, ethical, and brain-inspired pathways. *Scientific Reports*, *15*(1), 8443. DOI: 10.1038/s41598-025-92190-7 PMID: 40069265

Rashid, A. B., Kausik, A. K., Sunny, A. a. H., & Bappy, M. H. (2023). Artificial intelligence in the military: An overview of the capabilities, applications, and challenges. *International Journal of Intelligent Systems, 2023*(1). https://doi.org/DOI: 10.1155/2023/8676366

Rashid, A. B., & Kausik, M. K. (2024). AI revolutionizing industries worldwide: A comprehensive overview of its diverse applications. *Hybrid Advances*, *7*, 100277. DOI: 10.1016/j.hybadv.2024.100277

Rawas, S. (2024). AI: The future of humanity. *Discover Artificial Intelligence*, *4*(1), 25. Advance online publication. DOI: 10.1007/s44163-024-00118-3

Riley, W. J. (2012). *Health disparities: Gaps in access, quality and affordability of medical care*. https://pmc.ncbi.nlm.nih.gov/articles/PMC3540621/

Robinson, K., McKenna, B., & Rooney, D. (2021). The Relationship of Risk to Rules, Values, Virtues, and Moral Complexity: What We can Learn from the Moral Struggles of Military Leaders. *Journal of Business Ethics*, *179*(3), 749–766. DOI: 10.1007/s10551-021-04874-5 PMID: 34177014

Rosário, A. T., & Boechat, A. C. (2025). How sustainable leadership can leverage sustainable development. *Sustainability (Basel)*, *17*(8), 3499. DOI: 10.3390/su17083499

Roy, A. D., Das, D., & Mondal, H. (2023). The Tribal health system in India: Challenges in healthcare delivery in comparison to the global healthcare systems. *Cureus, 15*(6), e39867. DOI: 10.7759/cureus.39867 PMID: 37404413

Ruane, S., Sirinukunwattana, K., Kotanska, A., & Aberdeen, A. (2025). Developing AI-powered tools for histopathology: Opportunities, challenges and new friends along the way. *Diagnostic Histopathology, 31*(5), 277–283. DOI: 10.1016/j.mpdhp.2025.03.003

Saki, S., & Soori, M. (2025). Artificial intelligence, machine learning and deep learning in advanced transportation systems, a review. *Multimodal Transportation, 5*(1), 100242. DOI: 10.1016/j.multra.2025.100242

Sarkar, I. N. (2022). Transforming health data to Actionable Information: Recent progress and future opportunities in health information exchange. *Yearbook of Medical Informatics, 31*(01), 203–214. DOI: 10.1055/s-0042-1742519 PMID: 36463879

Sharma, R. S., Rohatgi, A., Jain, S., & Singh, D. (2023). The Ayushman Bharat Digital Mission (ABDM): Making of India's Digital Health Story. *CSI Transactions on ICT, 11*(1), 3–9. DOI: 10.1007/s40012-023-00375-0

Sigfrids, A., Leikas, J., Salo-Pöntinen, H., & Koskimies, E. (2023). Human-centricity in AI governance: A systemic approach. *Frontiers in Artificial Intelligence, 6*, 976887. DOI: 10.3389/frai.2023.976887 PMID: 36872934

Smith, B. (2025, September 18). *Winning the AI race: Strengthening U.S. capabilities in computing and innovation*. Microsoft on the Issues. https://blogs.microsoft.com/on-the-issues/2025/05/08/winning-the-ai-race/

Solomon, J., Dauber-Decker, K., Richardson, S., Levy, S., Khan, S., Coleman, B., Persaud, R., Chelico, J., King, D., Spyropoulos, A., & McGinn, T. (2023). Integrating clinical decision support into electronic health record systems using a novel platform (EvidencePoint): Developmental study. *JMIR Formative Research, 7*, e44065. DOI: 10.2196/44065 PMID: 37856193

Spatola, N. (2024). The efficiency-accountability tradeoff in AI integration: Effects on human performance and over-reliance. *Computers in Human Behavior Artificial Humans, 2*(2), 100099. DOI: 10.1016/j.chbah.2024.100099

Tai, M. C. (2020). The impact of artificial intelligence on human society and bioethics. *Tzu-Chi Medical Journal, 32*(4), 339. DOI: 10.4103/tcmj.tcmj_71_20 PMID: 33163378

Tewathia, N., Kamath, A., & Ilavarasan, P. V. (2020). Social inequalities, fundamental inequities, and recurring of the digital divide: Insights from India. *Technology in Society*, *61*, 101251. DOI: 10.1016/j.techsoc.2020.101251

Thacharodi, A., Singh, P., Meenatchi, R., Ahmed, Z. H. T., & Kumar, R. R. S. (2024). Revolutionizing healthcare and medicine: The impact of modern technologies for a healthier future—A comprehensive review. *Health Care Science*, *3*(5), 329–349. DOI: 10.1002/hcs2.115 PMID: 39479277

The 2025 AI Index Report | Stanford HAI. (2025). https://hai.stanford.edu/ai-index/2025-ai-index-report

Thomsen, M. D. T., & Knudsen, L. E. (2025). Editorial: Reviews in: regulatory science 2024-2025. *Frontiers in Medicine*, *12*, 1705977. DOI: 10.3389/fmed.2025.1705977 PMID: 41189882

Vaidya, R., & Bobdey, S. (2021). Medical ethics during armed conflicts: Dilemmas of a physician soldier. *Medical Journal, Armed Forces India*, *77*(4), 377–381. DOI: 10.1016/j.mjafi.2021.08.013 PMID: 34594063

Varnosfaderani, S. M., & Forouzanfar, M. (2024). The role of AI in Hospitals and Clinics: Transforming Healthcare in the 21st century. *Bioengineering (Basel, Switzerland)*, *11*(4), 337. DOI: 10.3390/bioengineering11040337 PMID: 38671759

Zaidan, E., & Ibrahim, I. A. (2024). AI governance in a complex and rapidly changing regulatory landscape: A Global perspective. *Humanities & Social Sciences Communications*, *11*(1), 1121. Advance online publication. DOI: 10.1057/s41599-024-03560-x

Zarei, H., Baharmand, H., Bashiri, M., & Madanian, S. (2024). Technological advancements in humanitarian aid. *International Journal of Disaster Risk Reduction*, *109*, 104527. DOI: 10.1016/j.ijdrr.2024.104527

KEY TERMS AND DEFINITIONS

Artificial Intelligence (AI): A computational system designed to perform functions typically requiring human intelligence, including pattern recognition, problem solving, and data-driven decision-making through algorithms that learn from experience.

Narrow Artificial Intelligence (Narrow AI): An AI system developed to execute a single, specialised task, such as medical image interpretation, without the capacity to generalise knowledge beyond its limited domain.

Artificial General Intelligence (AGI): A theoretical model of AI that possesses human-like cognitive abilities, including autonomous reasoning, contextual understanding, and moral judgment; not yet realised in contemporary technology.

Clinical Decision Support Systems (CDSS): AI-enabled digital tools integrated into clinical workflows to assist healthcare professionals by providing diagnostic suggestions, identifying drug interactions, and recommending evidence-based treatment options.

Algorithmic Bias: A systematic distortion in AI outputs caused by skewed, incomplete, or unrepresentative training data, often leading to discriminatory or inaccurate outcomes in healthcare or security settings.

Autonomous Weapons Systems (AWS): AI-driven military technologies capable of selecting and engaging targets with limited or no human oversight, raising significant concerns under international humanitarian and human rights law.

Meaningful Human Control: A governance principle requiring that human operators retain oversight, interpretive authority, and final decision-making power over AI systems, particularly in ethically sensitive environments such as medicine and warfare.

Electronic Health Records (EHRs): Digitally stored and interoperable patient medical information that supports continuity of care, reduces redundancy, and enhances accuracy across healthcare providers and institutions.

Ayushman Bharat Digital Mission (ABDM): A digital health initiative in India that seeks to establish a unified, interoperable ecosystem through electronic health records, provider registries, and patient-linked digital health identities.

Ayushman Bharat Health Account (ABHA ID): A unique, patient-specific digital identifier enabling individuals to securely link, store, and share their medical records across diverse healthcare facilities and digital platforms with informed consent.

Predictive Analytics: The use of AI-based modelling techniques to anticipate future outcomes, such as disease outbreaks or patient deterioration, by analysing historical patterns and real-time datasets.

Public Health Surveillance Systems: AI-supported platforms used to monitor population-level health indicators, disease trends, and epidemiological patterns to facilitate timely interventions and informed policy planning.

Precision Medicine: A data-informed approach to healthcare that uses AI to tailor treatment strategies to individual genetic, clinical, and lifestyle profiles, moving beyond generalised therapeutic protocols.

Ethical AI Governance: A structured framework of norms, policies, and institutional mechanisms designed to ensure that AI technologies operate transparently, fairly, and responsibly, safeguarding human dignity and rights.

Automation Complacency: A behavioural tendency in which human users become overly reliant on AI systems, resulting in reduced vigilance and critical judgment, particularly in clinical or security-critical environments.

Algorithmic Transparency: The extent to which the processes, decision logic, and data pathways of AI systems can be clearly understood, interpreted, and scrutinised by users, regulators, and stakeholders.

Chapter 9
From Nothingness to il y a:
Recontextualizing "Just War" in Heidegger and Levinas

Stavros Panagiotou
http://orcid.org/0000-0002-5584-0272
University of Athens, Greece

ABSTRACT

Martin Heidegger and Emmanuel Levinas engage in a profound exploration of the tension between il y a (there is) and nothingness in their philosophy of war. This chapter examines the philosophical roots of war, tracing its derivation from nothingness to il y a through the perspectives of Heidegger and Levinas. Human existence, distinct from other beings, is marked by self-awareness and intentionality, shaping its ontological and ethical framework. Heidegger, emphasizing fundamental ontology, asserts that existence precedes essence. For Dasein is inseparable from meaning, constituting the foundation of reality through immanence. Il y a, in this view, signifies a persistent presence, a 'will to power' grounded in the categories of Dasein, leaving no space for an existence beyond being. Levinas, by contrast, reconfigures il y a through ethical transcendence, shifting from an ontological to an ethical framework. He exteriorizes the subject by situating being within an eschatological and messianic horizon that extends towards infinity.

DOI: 10.4018/979-8-3373-6264-9.ch009

1. INTRODUCTION

«Everyone is the other, and no one is himself». (Heidegger, 1962)
«The alterity of the other does not result from its identity, but constitutes it: the other is the Other». Levinas (1969, p. 251)

This chapter aims at recontextualizing the intersection of transcendence, immanence and ethical responsibility within the developing discourse of phenomenology with regard to the meaning and interpretation of the peculiar concept of *il y a du quotidien* and its relation to nothingness (Levinas, 2003, p. 9). The ontological-transcendental debate, which has its roots in Greek philosophy (add some references of the relation between Levinas and the ancient Greek philosophy), underwent a profound development in phenomenological research, particularly through Edmund Husserl and Martin Heidegger.[1] Husserl's transcendental phenomenology introduced the concept of to supply the transcendental ego as a fundamental investigation of being. However, this expression is too strong and, *in extremis*, contradictory: The question remains: was not the very idea of the Kantian transcendental born in Enlightenment epoch and would it not be taken up by Husserl? Moreover, from the first section of the *Krisis* - cited by the author in the footnote as justification for this position - no 'anti-Enlightenment' attitude emerges, rather an allergy to the positivist restriction of the knowable and of meaning. Hence, the traditional line of thought concerning the term transcendental ego needs to be reconsidered.[2] Heidegger, who moved on to existential phenomenology, emphasized the ontological movement of being and developed his analysis of existence in *Being and Time* (Heidegger, 1962, p. 31).

In the first part, a critique of Heidegger's fundamental ontology is attempted through the lens of Emmanuel Levinas' ethics as first philosophy, with an emphasis on Levinas' engagement with Husserl's transcendental phenomenology rather than an outright rejection of metaphysics (Levinas, 1998b, pp. 1-11, 123-132). This argument is justified and achieved through the reorientation of the *il y a*, from the ontological dimension of Heidegger's *Dasein* towards an ethical language derived from Husserl's intentionality (Bergo, 2019, pp. 80-82).[3] Levinas questions Heidegger's reduction of Being to being-in-the-world and his neglect of transcendental intentionality. He argues that Heidegger marginalizes the *a priori* dimensions of selfhood and instead overemphasizes its in-time and in-the-world (*Dasein*) dimensions (Fagenblat 2019, pp. 103-110; Bergo 2005, pp. 141-177). Thus, Levinas proposes an ethical-transcendental framework in which ethics and relationality take precedence over fundamental ontology, positioning responsibility towards the other at the center of human existence, prior to freedom or free will.[4] Levinas' interpretation of *il y a* criticizes Heidegger's view of Cartesian *cogito-trust* and the lack of a transcendental dialectic.[5]

2. METHODOLOGY AND LITERATURE REVIEW

The ontological-transcendental debate, which has been thoroughly explored in phenomenology, has its roots in Greek philosophy. Thinkers such as Parmenides (Levinas, 1969; 1987b),[6] Heraclitus (Levinas, 1987b),[7] Socrates (Petropoulos, 2021, pp. 95-97; Landmann, 1941, pp. 15-42; Murray, 1988, pp. 501-518; Moran, 2021, pp. 98-117), Plato and Aristotle (Russon, 2021, pp. 155-168), as well as Plotinus,[8] Augustine (Lo, 2018, pp. 50-61; Hannan, 2012, pp. 75-91), John Scotus (Widomski, 1986, pp. 15–28; McGrath, 2003, pp. 339-358), and Thomas Aquinas (Caputo, 1983; Stein, 2000, pp. 1-64). developed the very ontological-transcendental concepts that would later lay the foundations for phenomenology. The decisive *explosion* of the phenomenological tradition began with Edmund Husserl, who systematized the method of investigating being through intentionality, defined as the directedness of consciousness towards phenomena. Husserl's idea of the transcendental ego served as an *initial spark* (Husserl, 1980); building on Husserl, Martin Heidegger, as an existential phenomenologist, shifted the focus to the ontological movement of being and developed the existential analysis of *Dasein* as its primary frame of reference–a «universal analogy» (Czerkawski, 2024, 3930-3968).

This chapter presents a critique of Heidegger's ontology of *Dasein* as interpreted by Emmanuel Levinas. It is important to make clear from the outset that the aim here is not to reject or undermine Heidegger's ontological phenomenology. Instead, the focus is on emphasizing Levinas' justification of Husserl's transcendental phenomenology as expressed in Levinas' primary sources. In addition, I offer a detailed account of the discrepancies between Levinas and Jean Hering in relation to their immanence-to-transcendence view. In particular, I examine Hering's argument that Levinas misinterprets this view, especially in his analysis of the notion of *il y a*.[9]

Levinas claims that Heidegger's exploration of fundamental ontology neglects several key aspects of transcendental phenomenology as defined by Husserl. In particular, Levinas argues that Heidegger marginalizes the essence of being by reducing it to an entity that is ontologically self-contained. This approach, according to Levinas, overlooks not only intentionality but also the *a priori* dimensions of the self. By contrast, Heidegger claims that what is central to the existence of *Dasein* is its relationship to common sense, which exists outside of being and lies solely in its engagement with the external world—in what he calls 'being-in-the-world.'[10]

In this chapter I argue that Levinas is primarily engaged with Husserl's phenomenology and not so much with Heidegger's reductionism. The conservative/orthodox line of thought, which falsely claims that Levinas was a vehement polemicist of Husserl's transcendental idealism, urgently needs to be reoriented. On the contrary, Levinas did not intend to violate or infringe metaphysics, but aimed at developing a *new rationality* in which ethics takes precedence over ontology and transcendence over immanence (Sandford, 1999, p. 69).

Levinas' interpretation of *il y a* serves as an objection to Heidegger's insistence on the complete reduction of being to being-in-the-world or being-in-general in presence[11] as well as a critique of his complete reduction of the self to context.[12] Furthermore, I argue that Levinas blames Heidegger for failing to depart from the traditional understanding of the Cartesian *cogito*.[13] In other words, Levinas emphasizes Heidegger's inability to integrate a transcendental dialectic into his immanent criteria for being (Fagenblat, 2002, pp. 578-604). For this reason, Levinas places particular emphasis on the *il y a* by reconsidering its existential relation to being, trying to offer a reading where his *il y a* provides an alternative foundation of being to that of Heidegger's *Dasein*.[14] The categories of being outside of consciousness are not identified solely through being *qua* being (Levinas, 2003, p. 7; Panayiotou, 2024, p. 241); rather, they are understood primarily through immanence-in-transcendence, which decenters the external substance of the Cartesian *cogito* (Römer, 2019, pp. 176-180).

What is particularly interesting is that Heidegger transformed his theory on the movement of being outside its essence, presenting objections to Husserl's own transcendental idealism, and yet paid little attention to the relationship between transcendence and immanence (Drabinksi and Nelson, 2014, pp. 15-72). Heidegger avoids addressing whether intentionality is a necessary condition for *Dasein*, focusing instead on the external criteria of selfhood. For Heidegger, I am who I am not because of the representational conditions of inner intuition and cognition but because of the external, intersubjective involvement in the social world I engage with (Graham, 2009, pp. 407-413).[15]

Levinas, drawing on his expertise in both lines of thought, attempts to reinterpret the above by prioritizing transcendental idealism through the implementation of a new rationality[16] based on an ethical perspective. The case of *il y a* precedes existence within an ethical-transcendental phenomenology in which the Other transcends the categories of being through responsibility and alterity (Levinas, 1999, p. 3-38).[17]

In the second part of this study, I analyze Hering's notion of transcendentalism, arguing that his critique of Levinas is convincing. Hering shows that, at the end of *Totality and Infinity* and *Existence and Existents*, Levinas does not seem to be able to fully overcome the ontological *consensus* of the relation between Heidegger's *Dasein* and the *Other*. Hering argues in parallel that Levinas fails to overcome ontological[18] aspects from his ethical-transcendental argumentation (Boccaccini, 2016, pp. 433-448).

3. THE IL Y A BETWEEN EXISTING WITHOUT EXISTENCE AND THE ELEMENTAL

«If phenomenology is only a method of radical experience, we will find ourselves beyond phenomenology» (Levinas, 1987b, p. 54)

Before unpacking the peculiar notion of *il y a* and its relation to the ontological-transcendental debate, it is necessary to clearly explain and define some key terms. I share Philip Lawton's reading that, by saying *il y a* (*there is*), Levinas means «the *milieu* in which the self bathes before it is a self, and so before it is a subject of existence; it is the *elemental*, the indeterminate, the background of Being in which the self discovers first itself as a self and from which it thereby detaches itself to become a separate(d) being who can meet others» (Lawton, 1975, p. 477). As Lawton furthermore correctly puts it «the there is, then, is not a thing, not an object of perception or of thought, it is not approached, not intentionally constituted and not grasped by a hand or a concept. Not a thing: rather, the background of Being from which things and *I* emerge and detach themselves» (Lawton, 1975, p. 486). For Levinas, *il y a* is identical to the verb *to be*, which, in contrast to Heidegger and Hegel,[19] is anonymous, impersonal, neutral without subject or object; it is just *here-and-there* and echoes infinity without absorbing the categories of being (Sallis, 1998, p. 154). It is neither a presuppositionless, pure, indeterminate Being, as in Hegel, nor «should the relation with the *there is* be confused [*de necessitate*] with that expressed by the Heideggerian notion of *Geworfenheit*».[20] As Levinas contends, «this impersonal, anonymous, yet inextinguishable 'consummation' of being, which murmurs in the depths of nothingness itself we shall designate by the term *there is*. The *there is*, inasmuch as it resists a personal form, is 'being-in-general'» (Levinas, 1978, p. 57).

Levinas, to my mind, initiates his philosophical direction towards ethics not from Heidegger's fundamental ontology but from Husserl's transcendental phenomenology. He states that «the non-contained, which breaks the container or the forms of consciousness, thus transcends the essence or the move of knowable being which carries on its being in presence; it transcends the interestedness and simultaneity of a representable or historically reconstitutable temporality; it transcends immanence» (Levinas, 1987a, p. 166).[21] The notion of *il y a*, from my point of view, approaches particularly the Husserl's phenomenology in raising severe objections to Heidegger's *Dasein*.[22] If the reader will excuse the length of the following citation (Levinas, 1987a, p. 177),

Husserl's phenomenology has opened the possibilities. It affirms the strict solidarity between any intelligible and the psychic modalities through which and in which it is conceived: not just any meaning is accessible to any thought [...] Phenomenology is the idea that the essences of the psyche do not constitute a 'definite manifold.' It is, on the other hand, the reference of sense to the giving of sense, the Sinngebung, which animates these irreducible thoughts. Phenomenology has taught us thus not to explicate a meaning by conceiving it uniquely or principally in terms of its relations with other objective meanings for then all sense gets relativized, and every signification closed up in a system without issue. Phenomenology has taught us to make explicit or to elucidate a sense in terms of the irreducible psyche in which it is given, to thus seek it in its origin, to seek out the originating meaning.

Levinas thus emphasizes that Husserl, as the founder of phenomenology, took the pivotal first step of analyzing subjects not merely in relation to objects but in relation to themselves, introducing «a modality of the psyche leading further than negation» (p. 177). Levinas raises his objection to Heidegger's negation as «ontological difference» (Bergo, 2006) by claiming that «there [in *Time and Being*] the essential in anxiety was taken to consist in opening up the horizon of nothingness, more radically negative than that of negation, incapable of making on forget the being it negates» (Levinas, 1987a, p. 179). Later, Levinas describes *il y a* as something akin to a *forgotten* experience, which is static within a non-static framework and exists *de necessitate* both *a priori* and *a posteriori* simultaneously: «It is a deduction which is not, strictly, a deduction, of an experience which is not, strictly, an experience [...] Levinas attempts to describe, or at least to indicate, in language a deduced experience that precedes language, precedes deduction, and precedes experience» (Lawton, 1975, pp. 481-482). In Levinas' words (1969, p. 28),

These horizons endow them with a meaning – such is the essential teaching of Husserl. What does it matter if in the Husserlian phenomenology taken literally these unsuspected horizons are in their turn interpreted as thoughts aiming at objects! What counts is the idea of the overflowing of objectifying thought by a forgotten experience from which it lives. The break-up of the formal structure of thought (the noema of a noesis) into events which this structure dissimulates, but which sustain it and restore its concrete significance, constitutes a deduction – necessary and yet non-analytical. In our exposition it is indicated by expressions such as 'that is,' or 'precisely,' or 'this accomplishes that,' or 'this is produced as that.'

However, for Levinas, the time of *il y a* is not connected to Heidegger's *Being and Time* as *anxiety in presence*, and rather «comes to me prior to my freedom, from a non-present, from an immemorial» (Levinas, 1998b, p. 71).[23] Heidegger explores his argument about presence and its need to force being to move away from itself and yet return to its pure core, with influences from Aristotle and Hegel.[24] The only way for being to escape nothingness is to assume an impersonal and anonymous identity with which the subject reflects on its self-consciousness. Heidegger adopts Hegel's definition of experience and pays him great tribute. The latter disagrees with Kant's experimental dialectic of experience, since experience is not merely an intuition of the *here-and-there* that culminates in natural and experimental sciences (i.e. empiricism and positivism). Heidegger implicitly raises his objection to Kant's phenomenological intuition and at the same time endorses Hegel's concept of experience (Heidegger, 1998, pp. 20-21):[25]

> *According to the first group of [Kantian] meanings, we speak of the sciences of experience as experimental sciences [...] We must render its [subject's] essence evident as it emerges out of the relationship itself. The [Kantian] intuition which delivers the essence in this first sense, is the phenomenological intuition [...] [For Hegel] science of experience has nothing at all to do with the experimental sciences [...] Experience as denoting, both negatively and positively, undergoing an experience with something in such a way that this something is verified, experiencing it as not being what it first seemed to be, but being truly otherwise [...] It is consciousness itself that undergoes these experiences: [...] In and with itself (an ihm selbst, mit sich selbst).*

On the one hand, Heidegger argues that the only way for the *il y a* to escape nothingness is to manifest as existence itself (hypostasis), hus aligning with Hegel. On the other hand, Levinas vehemently objects to this dialectic, describing it as totalitarian (Keohane-O'Connor, 1982, p.95). Levinas therefore explores a non-conceptual transcendental ethics, which—at least at the beginning of this process—is guided by Husserl's transcendental phenomenology (1998c, pp. 91-150). Levinas in sum proposes a different hypostasis in which being is prior to all ontological attributes: firstly as responsible[26] and secondly as free to *choose* its hypostatic identity. According to Levinas, the decentering of the categories of being beyond concept and presence—as *other-than-being*—allows us to move from ontology to transcendence, an ethical dialectic that appears in relation to the Other[27] as a «pluralistic non-monadic hypostasis of the present» (Levinas, 1987b, p. 54). By 'different hypostasis,' Levinas makes references to the concept of enjoyment, which implicitly leads to being permanently distancing itself from its *egotistic* I-ness (1987b, p. 64).[28]

> *For Levinas, the argumentation regarding the movement of being—i.e., the notion of il y a in relation to subjectivity—aims at overcoming fundamental ontology. It is worth noting, however, that Levinas does not intend to violate Heidegger's ontology or fundamentally undermine it. Nevertheless, he proposes an ontology that is not entirely reduced to being-in-the-world, and a corresponding account of the self that is not entirely reduced to context (Fagenblat, 2002, p.578). The main point however is that Levinas' own texts, in contrast to Heidegger's, are highly transcendental and non-conceptual in their approach. An extract, contributed by Craig Vasey, is both highly descriptive and reflexive (Vasey, 1980, p. 469):*
>
> *In Existence and Existents this localization of consciousness is anti-Heideggerian; the subjectivity of the subject who 'is in a certain sense a substance' is considered*

to be originally non-transcendent, i.e., non-ecstatic. The relation with the other in this essay is that by which the subject – who poses itself as localized, who finds itself chained to itself in the present, and who has no time to the degree that it is solitary – has access to the transcendent, to time, and eventually, for Levinas, to the possibility of hope and of salvation.

Taking the above discussion into account we can conclude and agree with William Edelglass, who contends that Levinas' language is «(a) the entering into a relationship with the source of meaning that cannot be thematized, (b) the refusal of absolute closure, (c) the giving of signs before thematization, and (d) the opening to transcendence» (Edelglass, 2005, pp. 49, 53).

In his work *Totality and Infinity*, Levinas identifies an alternative and more sustainable term for the concept of *il y a*: the 'elemental.'[29] He compares the elemental to nothingness, yet it smoothly identifies the totality of being as an egoistic self.[30] The element, lacking an origin in any particular being, parallels Hegel's conception of pure, indeterminate being. As Levinas puts it, «indetermination here is not equivalent to the infinite surpassing limits; it precedes the distinction between the finite and the infinite» (Levinas, 1969, p. 132). The openness of being, from interiority to exteriority, is not to be understood in Heideggerian phenomenological terms. At the same time, it is an impossibility through the elemental process, as being is concerned solely with its own existence—being-there and being-in-and-for-itself rather than being-together. Levinas argues that Heidegger's insistence on fundamental ontology as the priority for the essence of being *qua* being (i.e. Heideggerian metaphysics) entangles and confines being within the limits of finitude. Specifically, according to Heidegger, non-being in the stage of nothingness finds its existence during the elemental process of creation *ex nihilo*. However, the moment the subject becomes dependent and relies solely on itself, it once again loses the ability to escape or transcend the damnation of nothingness through death and eventually returns to nothingness.[31] From a completely different perspective, for Levinas it is the encounter with the Other through language that opens up the infinite for the subject.[32] Levinas begins with Husserl's concept of phenomenality and transcends it as a mode of being to explicitly address infinity «where everything is a present absenting itself from its presence» (1969, pp. 178-179):

The face I welcome makes me pass from phenomenon to being in another sense: in discourse I expose myself to the questioning of the Other, and this urgency of the response—acuteness of the present—engenders me for responsibility [...] When I seek my final reality, I find that my existence as a 'thing in itself' begins with the presence in me of the idea of Infinity. But this relation already consists in serving the Other.

The elemental itself emerges between being and nothingness, with ontology as the starting point and transcendence as the conclusion of this process. It resides between facticity and passivity, which «returns to the heart of every negation» (Levinas, 1978, p. 61). However, it is important to emphasize that Levinas does not intend to present transcendence as something entirely outside ontology. As Levinas reminds us, «transcendence is not the fundamental movement of the ontological adventure; it is founded in the non-transcendence of position» (1978, p. 100). The crucial point we must not overlook in relation to *il y a* and nothingness is that, while Levinas engages with both Heidegger's fundamental ontology and Hegel's speculative idealism—indeed, Western idealism as a whole—in his early writings, he ultimately distances himself from both traditions for a specific and decisive reason: their exclusion of the Other. Jacques Taminiaux offers a coherent and well-reasoned explanation of this stance (Taminiaux, 1997, pp. 30-31):

> *In western idealism the sovereignty of the Ego has never been separable from the prestige of the transcendent understood as God or as Eternity but always as an existent. This is, Levinas claims, what disappears from Heidegger's ontology, which is entirely embedded in a temporality that is in no way referred to the eternal. The substitution of an ontological problem for the ontic one to which the tradition of modern idealism was devoted results, therefore, so to speak, in the closing of a window looking out onto the eternal. What is at stake [in both traditions] is the exclusion of the other to the benefit of the self.*

Therefore, it is already easy to understand that, for Levinas, subjectivity must be withdrawn from being to make room for ethical transcendence. The issue with Heidegger, according to Levinas, is that he failed to recognize that being, due to its egoistic temporality, does not escape nothingness—understood as the senses and the spatiotemporal movement of being around itself. Instead, Levinas attributed to the subject a form of ethical transcendentalism that allows to transcend the present in pursuit of infinity. In *Otherwise than Being*, Levinas implicitly critiques Heidegger's conception of subjectivity as a self-contained unity of existence and, in contrast, proposes an ethical transcendental responsibility that precedes both freedom and the senses (1991, pp. 162-163),

> *Subjectivity then shows itself as an ego, capable of a present, capable of beginning, an act of intelligence [...] But it is also posited as belonging to essence, which when assembled cannot leave anything outside, has no outside, cannot be worn away [...] The extract of the ego from being and confer upon it a transcendental status [...] It is in its ex-ception and ex-pulsion as a responsible one that a subject outside of being can be conceived.*

According to Levinas, *Dasein* as a self-contained entity reacted ontologically from the very beginning of its being, after escaping nothingness within the world and not being-together in-the-world. *Dasein* employed the concept of *il y a* without any moral or theological preconditions, attributing to it strictly some ontological and existential dimensions of being, such as self-understanding, temporality, dereliction[33] and anguish[34] (Levinas, 1996, p. 25),

Dasein understands itself from the standpoint of possibilities relating to tools, from entities within the world, and not the world itself. This is the phenomenon of the 'fall', the third characteristic of existence alongside dereliction and the project-in-draft. The fall, from which we need to detach all moral or theological recollection, is a mode of the existence of Dasein shunning its authentic existence in order to relapse into everyday life. Dasein does not understand itself in its true personality but in terms of the object it handles: it is what it does [il y a], it understands itself in virtue of the social role it professes.

Levinas, in summing up, concludes in his later writings—particularly in *Otherwise than Being*—that *Dasein* is furthermore ultimately anti-social. Levinas characterizes *Dasein* as anti-social because it is unable to transcend its strict authenticity and singularity other-than-itself. Levinas, as Chandal Bax correctly puts it, implicitly «gives a highly individualistic interpretation of Heideggerian authenticity» (Bax, 2017, p. 394), emphasizing that authenticity is falsely an achievement of *Dasein* within itself, rather than a transcendental-historical process of being-together.

4. HERING'S CRITIQUE OF LEVINAS' ETHICAL TRANSCENDENTALISM

Hering argues that Levinas' interpretation of the *il y a* is based on a misreading of *Being and Time* and on its inability to overcome fundamental ontology. I take this opportunity to demonstrate that Hering's critique of Levinas' account of *il y a* is indeed convincing.[35] Hering, before beginning his critique of Levinas, seeks first to articulate Levinas' objections to Heidegger's conception of the nature and movement of being. He notes that, according to Levinas, the problem lies in Heidegger's philosophy leaving no space for external control of the self, as all meaning is confined to the horizon of its projects and its role within the history of being. Hering then isolates an extract from Levinas' *Totality and Infinity*, aiming to refute it on the grounds that it fails to meet Heidegger's own criteria for being (Levinas, 1969, pp. 46-47; 35),

Ontology as first philosophy is a philosophy of power [...] Heideggerian ontology, which subordinates the relationship with the Other to the relation with Being in general, remains under obedience to the anonymous, and leads inevitably to another power, to imperialist domination, to tyranny.

This absolute exteriority of the metaphysical term, the irreducibility of movement to an inward play, to a simple presence of self to self, is, if not demonstrated, claimed by the word transcendent. The metaphysical movement is transcendent, and transcendence, like desire and inadequation, is necessarily a transascendence[36] *[...] Thus the metaphysician and the 'other' can not be totalized. The metaphysician is absolutely separated.*

Hering asserts that, despite Levinas' efforts to escape ontology by prioritizing exteriority, that is [ethical] transcendence, he himself inadvertently falls back into ontology due to his own skepticism. Moreover, Hering reproaches Levinas for failing to switch from an epistemological to a metaphysical account of consciousness. However, this needs further explanation as Hering is not against any epistemological account of consciousness in principle: his aim is just to show how Levinas remains burdened by a metaphysical dialectic, which he would be polemically opposed to. In sum, according to Hering, Levinas turned his own discussion of the ontology-metaphysics relation, as it is presented in *Totality and Infinity*, against himself.

Regarding consciousness in particular, Hering unequivocally asserts that the *cogito* can exist without necessarily the *κόσμος*. At first glance, Levinas appears to adopt this perspective in *Totality and Infinity*, particularly in section three, «Exteriority and the Face» (1969, pp. 187-247). Hering argues that, by invoking the *cogito* even when employing transcendental terminology, Levinas inevitably reduces ethical language to epistemological implications. This, according to Hering, is the central error in his treatment of *il y a*. Consequently, while Levinas attempts to recontextualize *il y a* by critiquing Heidegger's existential ontology, he ultimately fails to transcend fundamental ontology, as his approach remains bound by transcendental idealism. In Levinas' words reported by Hering: «We do not believe that things should have an existence independent of consciousness, since their dependence, according to our interpretation, is not conceived as the negation of transcendent existence but as its characteristic» (Levinas, 1995, p. 48). To my mind, by conceding this argument, Levinas has become ensnared in his own texts. As Daniele De Santis correctly puts it, «it is Levinas who, by thinking that he had to defend Husserl's own idealism against some of the claims made by Hering in 1926, turned the latter's discussion of the ontology-metaphysics relation into a discussion about TI [*Transcendental Idealism*] *en lui-même*» (De Santis, 2023, p. 206).

5. CONCLUSION

To conclude, Hering's critique of Levinas' ethical transcendentalism looks convincing. After Heidegger's involvement in German national socialism, a possible explanation is that Levinas strove to vehemently criticize every philosophical move in relation to ontology in order to prevent the spread of Heideggerian thought in Europe. This led Levinas to unwittingly adopt and defend Husserlian transcendental idealism, which formed the foundation of his ethics as first philosophy. However, it is evident that Levinas never succeeded in overcoming fundamental ontology, as he allegedly attempted to demonstrate in his work *Of God Who Comes to Mind* (1998a). Hering thus rightly points out that Levinas ends up asserting the impossibility of a consciousness without intentionality. In this context, he casts doubt «about how his [Levinas'] theory concerning the primacy of ontology over phenomenology in Husserl's philosophy will be received» (1932, p. 478). By criticizing Heidegger, Levinas then became entangled in a dualistic realism-idealism controversy which inevitably gave rise to several epistemological problems (Landau, 2021, pp. 460-463; Critchley, 2015). On one hand, at most one could speak of attachment to the natural attitude, that is, of postponement and weakening of transcendental reduction, but on the other hand, this does not necessarily coincide with adherence to an idealistic position that would come into contact a idealistic position in an anti-Heideggerian key. For example, Rae emphasizes Deleuze's criticism of Levinas' theory of the *face*, arguing that it is incorrect to claim that ethics takes precedence over ontology: «For Deleuze, ontology precedes ethics, because ethics emanates from and relates to the expressive vitality of the way of being of each multiplicity» (2016, p. 291). In parallel to this and in the same manner, Williams adds that «Deleuze reverses this Levinasian prioritization. For him, a truly differential ontology must underpin even the possibility of ethics» (2005, p. 44). Finally, what remains to be addressed in future research is whether Levinas can be positioned between an idealist and a realist philosophical framework.

REFERENCES

Bax, C. (2017). Otherwise than Being-with: Levinas on Heidegger and Community. *Human Studies*, *40*(3), 381–400. DOI: 10.1007/s10746-017-9432-x

Bergo, B. (2005). Ontology, Transcendence, and Immanence in Emmanuel Levinas' Philosophy. *Research in Phenomenology*, *35*(1), 141–177. DOI: 10.1163/1569164054905474

Bergo, B. (2006). Emmanuel Levinas. In Zalta, E. N., & Nodelman, U. (Eds.), *The Stanford Encyclopedia of Philosophy*.

Bergo, B. (2019). Levinas and Husserl. In Morgan, M. (Ed.), *The Oxford Handbook of Levinas* (pp. 71–102). Oxford University Press.

Boccaccini, F. (2016). De Reinach à Levinas: Hering et le réalisme phénoménologique. *Revue de Théologie et de Philosophie*, *148*, 433–448.

Caputo, J. (1977). The Question of Being and Transcendental Phenomenology: Reflections on Heidegger's Relationship to Husserl. *Research in Phenomenology*, *7*(1), 84–105. DOI: 10.1163/156916477X00077

Caputo, J. (1983). *Heidegger and Aquinas: An Essay on Overcoming Metaphysics*. Fordham University Press.

Carnap, R. (1959). The Elimination of Metaphysics through Logical Analysis of Language. In Ayer, A. J. (Ed.), *Logical Positivism* (pp. 61–80). The Free Press.

Critchley, S. (2015). *The Problem with Levinas*. Oxford University Press. DOI: 10.1093/acprof:oso/9780198738763.001.0001

Czerkawski, M. (2024). 'The soul is, in a way, all beings': Heidegger's Debts to Aristotle in *Being and Time*. *Inquiry (Oslo, Norway)*, *67*(10), 3930–3968. DOI: 10.1080/0020174X.2022.2074881

De Santis, D. (2023). *Transcendental Idealism and Metaphysics: Husserl's Critique of Heidegger*. Springer.

Dimitrova, M. (2011). *In Levinas' Trace*. Cambridge Scholars Publishing.

Dupont, C. (2015). Jean Héring and the Introduction of Husserl's Phenomenology to France. *Studia Phaenomenologica*, *15*, 129–153. DOI: 10.5840/studphaen2015158

Edelglass, W. (2005). Levinas's Language. *Analecta Husserliana*, *85*, 47–62. DOI: 10.1007/1-4020-3576-4_4

Fagenblat, M. (2019). Levinas and Heidegger. In Morgan, M. (Ed.), *The Oxford Handbook of Levinas* (pp. 103–133). Oxford University Press.

Fagenblat, M. (2002). Il y a du quotidien: Levinas and Heidegger on the Self. *Philosophy and Social Criticism*, *28*(5), 578–604. DOI: 10.1177/0191453702028005666

Gerson, L. (Ed.). (2018). *Plotinus: The Enneads*Boys-Stones, G., Trans.). Cambridge University Press.

Golob, S. (2020). What Does It Mean to 'Act in the Light of' a Norm? Heidegger and Kant on Commitments and Critique. In Burch, M., & McMullin, I. (Eds.), *Transcending Reason: Heidegger on Rationality* (pp. 79–98). Rowman & Littlefield. DOI: 10.5040/9798881816612.ch-003

Hand, S. (Ed.). (1989). *The Levinas Reader*. Basil Blackwell.

Hannan, S. (2012). To See Coming: Augustine and Heidegger on the Arising and Passing Away of Things. *Medieval Mystical Theology*, *21*(1), 75–91. DOI: 10.1558/mmt.v21i1.75

Heidegger, M. (1949). *Existence and Being*. Henry Regnery.

Heidegger, M. (1962). *Being and Time* (Macquarrie, J., & Robinson, E., Trans.). Basil Blackwell.

Heidegger, M. (1988). *Hegel's Phenomenology of Spirit* (Emad, P., & Maly, K., Trans.). Indiana University Press. DOI: 10.2307/j.ctv105b98m

Hering, J. (1932). Recension d'E. Levinas, *La théorie de l'intuition dans la phénoménologie de Husserl. Revue Philosophique de la France et de l'Étranger*, *113*(5–6), 474–481.

Hering, J. (2021). Remarks concerning Essence, Ideal Quality, and Idea. *Phenomenological Investigations*, *1*, 51–108.

Houlgate, S. (2006). *The Opening of Hegel's Logic: From Being to Infinity*. Purdue University Press.

Husserl, E. (1980). *Phenomenology and the Foundations of the Sciences (Third Book)* (Klein, T., & Pohl, W., Trans.). Martinus Nijhoff. (Original work published 1913)

Husserl, E. (1996). The Vienna Lecture and Phenomenology. In Kearney, R., & Rainwater, M. (Eds.), *The Continental Philosophy Reader* (pp. 7–22). Routledge. (Original work published 1935)

Janes, J., & Luft, S. (2020). Die angebliche Frage nach dem 'Sein des Seienden.' An Unknown Husserlian Response to Heidegger's 'Question of Being. In Burch, M., & McMullin, I. (Eds.), *Transcending Reason: Heidegger on Rationality* (pp. 237–258). Rowman & Littlefield. DOI: 10.5040/9798881816612.ch-010

Keohane-O'Connor, N. (1982). *Existence and Existents*, by Emmanuel Levinas. *JBSP : The Journal of the British Society for Phenomenology. British Society for Phenomenology, 13*(1), 94–96. DOI: 10.1080/00071773.1982.11007568

Keyes, C. D. (1972). An Evaluation of Levinas' Critique of Heidegger. *Research in Phenomenology, 2*(1), 121–142. DOI: 10.1163/156916472X00072

Landau, I. (2021). Problems in Levinas. *The European Legacy, Toward New Paradigms, 26*(5), 455–465. DOI: 10.1080/10848770.2021.1899600

Landmann, M. (1941). Socrates as a Precursor of Phenomenology. *Philosophy and Phenomenological Research, 2*(1), 15–42. DOI: 10.2307/2102672

Lawton, P. (1975). Levinas' Notion of the 'There Is'. *Tijdschrift voor Filosofie, 37*(3), 477–489.

Levinas, E. (1969). *Totality and Infinity* (Lingis, A., Trans.). Duquesne University Press. (Original work published 1961)

Levinas, E. (1978). *Existence and Existents* (Lingis, A., Trans.). Martinus Nijhoff. (Original work published 1947) DOI: 10.1007/978-94-017-1516-4

Levinas, E. (1987a). *Collected Philosophical Papers* (Lingis, A., Trans.). Martinus Nijhoff. DOI: 10.1007/978-94-009-4364-3

Levinas, E. (1987b). *Time and the Other* (Cohen, R., Trans.). Duquesne University Press. (Original work published 1947)

Levinas, E. (1989). Ethics as First Philosophy. In *The Levinas Reader* (Hand, S., Ed. & Trans.; pp. 76–87). Blackwell. (Original work published 1984)

Levinas, E. (1991). *Otherwise than Being, or Beyond Essence* (Lingis, A., Trans.). Kluwer. (Original work published 1974) DOI: 10.1007/978-94-015-7906-3

Levinas, E. (1994). *In the Time of the Nations* (Smith, M., Trans.). Indiana University Press. (Original work published 1980)

Levinas, E. (1995). *The Theory of Intuition in Husserl's Phenomenology* (Orianne, A., Trans.). Northwestern University Press. (Original work published 1963)

Levinas, E. (1996). Martin Heidegger and Ontology. *Diacritics, 26*(1), 11–32. DOI: 10.1353/dia.1996.0007

Levinas, E. (1998a). *Of God Who Comes to Mind* (Bergo, B., Trans.). Stanford University Press. (Original work published 1986) DOI: 10.1515/9781503616882

Levinas, E. (1998b). *Entre Nous: On Thinking-of-the-Other* (Smith, M., & Harshav, B., Trans.). Columbia University Press. (Original work published 1991)

Levinas, E. (1998c). *Discovering Existence with Husserl* (Cohen, R., & Smith, M., Trans.). Northwestern University Press. [1929–1977]

Levinas, E. (1999). *Alterity and Transcendence* (Smith, M., Trans.). Athlone Press. (Original work published 1995)

Levinas, E. (2003). *On Escape* (Bergo, B., Trans.). Stanford University Press. (Original work published 1935)

Lo, C. L. (2018). Time, the Image of Absolute Logos: A Comparative Analysis of the Ideas of Augustine and Husserl. *Comparative and Continental Philosophy, 10*(1), 50–61. DOI: 10.1080/17570638.2018.1465018

Marcel, G. (1952). *Metaphysical Journal* (Wall, B., Trans.). Henry Regnery. [1914–1923]

McDaniel, K. (2016). Heidegger and the 'There Is' of Being. *Philosophy and Phenomenological Research, 93*(2), 306–320. DOI: 10.1111/phpr.12257

McGrath, S. J. (2003). Heidegger and Duns Scotus on Truth and Language. *The Review of Metaphysics, 57*(2), 339–358.

Monseu, N. (2005). *Les usages de l'intentionnalité*. Éditions de l'Institut Supérieur de Philosophie.

Moran, D. (2021). Husserl and the Greeks. *JBSP : The Journal of the British Society for Phenomenology. British Society for Phenomenology, 52*(2), 98–117. DOI: 10.1080/00071773.2020.1821579

Murray, M. (1988). Husserl and Heidegger: Constructing and Reconstructing Greek Philosophy. *The Review of Metaphysics, 41*(3), 501–518.

Newman, M. (2000). Sensibility, Trauma, and the Trace: Levinas from Phenomenology to the Immemorial. In Bloechl, J. (Ed.), *The Face of the Other and the Trace of God* (pp. 90–129). Fordham University Press.

Panayiotou, S. (2024). Hegel and Levinas on the Notion of Esoteric Thought. In Pitkänen, O. (Ed.), *Philosophical Perspectives on Esotericism* (pp. 226–250). Routledge. DOI: 10.4324/9781003457046-12

Peperzak, A. (1983). Phenomenology – Ontology – Metaphysics: Levinas' Perspective on Husserl and Heidegger. *Man and World, 16*(2), 113–127. DOI: 10.1007/BF01260324

Perelsztejn, N. (2024). Escape to Judaism: Levinas's First Steps toward Becoming a Jewish Thinker. *The Journal of Jewish Thought & Philosophy*, *32*(2), 267–291. DOI: 10.1163/1477285x-12341361

Petropoulos, G. (2021). Phenomenology and Ancient Greek Philosophy: An Introduction. *JBSP : The Journal of the British Society for Phenomenology. British Society for Phenomenology*, *52*(2), 95–97. DOI: 10.1080/00071773.2021.1899053

Poleshchuk, I. (2020). Heidegger and Levinas: Metaphysics, Ontology and the Horizon of the Other. *The Indo-Pacific Journal of Phenomenology*, *10*(2), 1–10. DOI: 10.2989/IPJP.2010.10.2.4.1085

Rae, G. (2016). The Political Significance of the Face: Deleuze's Critique of Levinas. *Critical Horizons*, *17*(3–4), 279–303. DOI: 10.1080/14409917.2016.1190173

Römer, I. (2019). Levinas and Early Modern Philosophy. In Morgan, M. (Ed.), *The Oxford Handbook of Levinas* (pp. 173–194). Oxford University Press.

Russon, J. (2021). To Account for the Appearances: Phenomenology and Existential Change in Aristotle and Plato. *JBSP : The Journal of the British Society for Phenomenology. British Society for Phenomenology*, *52*(2), 155–168. DOI: 10.1080/00071773.2020.1796515

Sallis, J. (1998). Levinas and the Elemental. *Research in Phenomenology*, *28*(1), 152–159. DOI: 10.1163/156916498X00092

Sandford, S. (1999). Levinas in the Realm of the Senses: Transcendence and Intelligibility. *Angelaki*, *4*(3), 61–73. DOI: 10.1080/09697259908572056

Shockey, R. M. (2010). Heidegger's Descartes and Heidegger's Cartesianism. *European Journal of Philosophy*, *20*(2), 285–311. DOI: 10.1111/j.1468-0378.2010.00408.x

Smith, F. (2016). Gadamer and Levinas on the Topic of Sociality. *Sophia Philosophical Review*, *9*(2), 95–109.

Standish, P. (2008). Levinas and the Language of the Curriculum. In Egéa-Kuehne, D. (Ed.), *Levinas and Education* (pp. 56–66). Routledge.

Stapleton, T. J. (1983). *Husserl and Heidegger: The Question of a Phenomenological Beginning*. SUNY Press.

Stein, E. (2000). *Knowledge and Faith* (Publications, I. C. S., Trans.). (Original work published 1993)

Switzer, R. (2000). Redeeming the Elemental: Levinas on Nature. *Interdisciplinary Environmental Review*, *4*(2), 23–35. DOI: 10.1504/IER.2002.053892

Taminiaux, J. (1997). The Early Levinas's Reply to Heidegger's Fundamental Ontology. *Philosophy and Social Criticism*, *23*(6), 29–49. DOI: 10.1177/019145379702300603

Tapper, M. (1986). The Priority of Being or Consciousness for Phenomenology: Heidegger and Husserl. *Metaphilosophy*, *17*(2-3), 153–161. DOI: 10.1111/j.1467-9973.1986.tb00377.x

Vasey, C. R. (1980). *Existence and Existents*, by Emmanuel Levinas. *The Thomist*, *44*(3), 466–473. DOI: 10.1353/tho.1980.0022

Williams, E., & Standish, P. (2016). Sounds and Light: Levinas and the Elements of Thought. *Educational Philosophy and Theory*, *48*(4), 360–373. DOI: 10.1080/00131857.2015.1047820

Williams, J. (2005). *The Transversal Thought of Gilles Deleuze*. Clinamen Press.

Withy, K. (2021). Thrownness. In Wrathall, M. (Ed.), *The Cambridge Heidegger Lexicon* (pp. 753–756). Oxford University Press. DOI: 10.1017/9780511843778.204

ENDNOTES

1 Husserl, the founder of modern phenomenology, and Heidegger, Husserl's successor, were the strongest voices in the ontological-transcendental debate in contemporary phenomenology. In short, this debate centers on the question of whether transcendence precedes ontology or vice versa. Which of this line of thoughts is considered the foundation of science, the ontological or the transcendental one? Daniele De Santis, who, in my opinion, is the most comprehensive and prominent scholar on this subject, endeavors to reconstruct this debate between Husserl and Heidegger by stressing the former's contribution, «without, however, falling into any form of Husserlian apologetics». See De Santis (2023). Concerning Husserl-Heidegger phenomenology in general see also Janes and Luft (2020); Stapleton (1983); Tapper (1986, pp. 153-61); Caputo (1977, pp. 84-105).

2 The motive for this move is a reproach of the form of rationalism, plagued by positivism, that has its roots in the age of Enlightenment. Husserl strongly de-

fends the decentering of reason in the interest of modern science, by proposing a transcendental reason capable of exteriorizing intentionality as a matter of the noetic/noematic horizon. See Husserl (1996, pp. 7-22).

3 In addition, Levinas expressed his respect and loyalty to Husserlian phenomenology by dedicating considerable effort to demonstrating his admiration for Husserl's foundational work on intentionality. Levinas (1991, p. 183) highlights that «our analyses lay claim to the spirit of Husserlian philosophy, whose letter has been the return, in our age, of the permanent phenomenology restored to its rank of being a method for all philosophy. Our presentation of notions [...] remains faithful to intentional analysis».

4 Even though Levinas' philosophy owes much to Heidegger's ideas, Levinas – particularly after Heidegger's involvement with German national socialism (Fagenblat, 2019, pp. 111-123) – explores his criticism as an anti-totalitarian declaration. See Peperzak (1983, p. 123). In parallel to this, see Bergo (2005, p. 145).

5 Levinas (1969, pp. 85-86). It is worth noting that Levinas also expresses his disagreement with Husserl's interpretation of the notion of the *cogito*. Levinas critiques Husserl for falsely seeing in the cogito «a subjectivity without any support outside of itself; this cogito constitutes the idea of infinity itself and gives it to itself [merely] as an object». See Levinas (1969, p. 211). However, as this project argues, Levinas himself never succeeded in overcoming the *cogito* and intentionality.

6 Levinas refers to Parmenides multiple times in his works, being particularly impressed by Parmenides' notion of the One (monadology or henology). However, Levinas (1987b, n. 9) «attempts to both reject Parmenides (the One) and remain a philosopher». Levinas (1969, p. 104) underlines that «since Parmenides across Plotinus we have not succeeded in thinking otherwise». Additionally, R. Cohen rightly concludes in his comments in Levinas (1987b, n. 18) that the latter, on the one hand, vehemently objects to Parmenides' monadology, and on the other hand supports his criticism concerning the Nietzschean notion of non-being: «Though Levinas rejects Parmenides' philosophically decisive concept of the One, we can see in the notion of the 'there is' an acceptance of Parmenides' sanction against the path of non-being».

7 Levinas (1987b, p. 49) admires Heraclitus' existential philosophy concerning the notion of *il y a*, claiming that «if it were necessary to compare the notion of the *there is* with a great theme of classical philosophy I would think of Heraclitus».

8 There are several affinities and discrepancies between Plotinus and Levinas regarding the role of the One (*τό ν*) in relation to *il y a* as ineffable and infinite—particularly within the debate between being and nothingness. Levinas cites Plotinus' works multiple times, as he is notably an admirer of Plotinus' onto-theological aspects, especially his arguments concerning the notion of Oneness. See Levinas (1987a, pp. 105-106). Presumably, Levinas derives several ideas from the fifth *Ennead*, where Plotinus explores his argument on the conception of the *One* and his attributes against intelligibility, humans, and absolute knowledge. For instance, Levinas might agree with Plotinus' position regarding the transcendence of the One: «[The One] is beyond being. This is the requirement of negative theology» (Gerson, 2018, p. 581). However, Levinas states that intelligibility is prior to consciousness, will and freedom. It is housed between the subject and the eternal *a priori* responsibility for the *Other*. Levinas is familiar with Plotinus' texts, saying that «if you read the *Enneads*, the One doesn't even have consciousness of self, if it did have consciousness of self, it would already be multiple, as a loss of perfection. In knowledge, one is two, even when one is alone. Even when one assumes consciousness of self, there is already a split» (Levinas, 1998b, p. 112).

9 Concerning Hering's view on Husserlian phenomenology and his critique of Levinas' thought, see Dupont (2015, pp. 129-153).

10 Levinas' intention, especially in his early writings, was to escape from Heidegger's fundamental ontology. In contrast to Heidegger, as Poleshchuk (2020, p. 2) correctly puts it, «Levinas seeks for the possibility of disclosing [*il y a* through] the relation to the Other which could exceed the question of being and move to the question of the nature of subjectivity and its relation to the true Other [which is other-than-itself]». See also Perelsztejn (2024, pp. 267-291).

11 However, there are several thinkers who argue that for Heidegger this argument would be nonsense, that is, would be reduced to presence, in the sense of *Vorhandenheit*? As for Levinas partly right, and yet the sense of the presence of the il y a is complex and specified in *Existent and Existence*, and yet the question is methodological: can one take part and restitute an interpretative debate in this way in the scientific context? As for instance, Keohane-O'Connor (1982, p. 95) correctly puts it, Levinas analyses «this presence of Being-in-general as horror of the *there is*».

12 It is worth citing that analytic thinkers raises severe objections on this point: They would claim of an apparent contradiction between this position (which seems to reduce the self to the other of the world-context) and that of the related note; moreover, I am not so sure that Levinas does not consider the

Heideggerian notion of *Jemeinigket* as a form of *ipseity* as Levinas openly unleashes a vehement critique of Heidegger's ontology, as—by reducing and absorbing the other in the same—it touches on the limits of inhumanism. See Levinas (1987b, p. 109).

13 Concerning affinities and discrepancies between Heidegger and Descartes, see Shockey (2010, pp. 285-311). For the Levinas-Heidegger debate on Descartes' *cogito*, see Levinas (1969, pp. 92-93). Additionally, Levinas (1998b, p. 162) explains Heidegger's interpretation of the *cogito*: «The exchange of ideas will produce presence or representation in the unity of an utterance or an account naming or displaying a field of knowledge. It would fit within a single consciousness, within a *cogito* that remains Reason: universal Reason and egological interiority». See also Keyes (1972, p. 131) and Batnitzsky (2004, p. 14). Indeed, Heidegger (1962, p. 366), adopting the Kantian tradition of the *I think*, contends that «the *I* is a bare consciousness, accompanying all concepts [...] The *I think* is the form of apperception, which clings to every experience and precedes it».

14 Levinas, for example, opposes Heidegger's insistence on *fundamental realism*. In short, Heidegger defends that *il y a* can only be justified if it is inseparable and consistent with *Dasein*. That is, being is there if only *Dasein* is there. For a detailed analysis of this issue, see McDaniel (2016, pp. 306-320). For Levinas' critique of Heidegger's *Dasein* see also Keohane and O'Connor (1982, pp. 94-96). In fact, it is out of the question that Levinas went through the English translation to interpret the Heideggerian Dasein, so as a preliminary step a critical comparison should be made between Dasein-il y a, given that the latter notion corresponds to the literal French translation of the Heideggerian notion, or between Sein (different from entities)/existence-il y a.

15 However, much ink has spent on this issue: On the one hand, Heidegger dissociates spatial exteriority from being-in-the-world precisely by describing Dasein as transcendence, which thus defines being. If, on the other hand, one wishes to pursue this thesis, one must then specify the context of analysis on the inauthentic Dasein more precisely.

16 Levinas borrows this phrase from G. Marcel, who surprisingly critiques Heidegger's notion of self-sufficiency in relation to *Dasein*. See Marcel (1952, pp. 210-211). Levinas empowers the above argument by implicitly criticizing Heidegger being incapable of overcoming the Westernized way of thinking, that is the self-sufficiency of being *qua* being. Levinas (2003, pp. 50-51) states: «This conception of the *I* [*moi*] as self-sufficient is one of the essential marks of the bourgeois spirit and its philosophy [...] Yet this category of sufficiency

is conceived in the image of *being* such as things offer it to us. They *are* [...] And Western philosophy, in effect, has never gone beyond this. In combating the tendency to ontologize, when it did combat it, Western philosophy struggled for a better being, for a harmony between us and the world, or for the perfection of our own being».

17 However, it is difficult to convince oneself of this idea, which goes against the strongest and most widespread interpretations and, therefore, is not sufficiently gisutised by the traditional line of thought: does ethics, the transcendence of the other, not surpass existence itself, thus the being and so il y a (in *Existence and Existents* Levinas analyses existence, in its difference from the existent - thus interpreting the Heideggerian ontological difference -, through the description of the il y a)? As Wingebach (1996, p. 29) correctly puts it, Levinas criticizes Heidegger's ontology which, unlike metaphysics, «is incapable of perceiving alterity since it tends not toward otherness but toward comprehension».

18 To my mind, Boccaccini intended, on the contrary, to suggest a closeness between the Heringian interpretation of the *Stellungsnahme* and the Levinassian description of the face of the other.

19 Houlgate (2006, p. 52) correctly contends that for Hegel «the system is what emerges when thought suspends all its assumptions about itself, considers nothing but the sheer indeterminate thought of being, and holds itself open to what that thought shows itself to be». Levinas (1969, p. 36) criticizes Hegel for reducing the concept of *il y a* merely to inner thinking, that is immanence and contradiction: «Hegelian phenomenology, where self-consciousness is the distinguishing of what is not distinct, expresses the universality of the same identifying itself in the alterity of objects thought and despite the opposition of self to self». As regards the notion of *il y a* as inner thinking (esotericism) in Hegel and as exterior to being in Levinas see Panayiotou (2024, pp. 226-250). It is crucial to add at this point that Levinas' dialectic of the *il y a* is a clear and methodical move away from Hegelian thought. Levinas (1987b, p. 42) claims that «it is not a matter of traversing a series of contradictions or of reconciling them while stopping History. On the contrary it is a pluralism that does not merge to unity that I should like to make my way». See also Hand (1989, p. 35): «A presence of absence, the *there is* is beyond contradiction; it embraces and dominates its contradictory».

20 The concept of *thrownness* refers to the idea that being is cast into the world to experience its consciousness ontologically, while *Dasein* is inherently bound by time. As Withy (2021, p. 753) mentions, «Heidegger often expresses this as *Dasein's* being thrown into itself or its existential constitution. *Dasein* is

thrown into existence; the *there*». This *thrownness* is also referred to as the ontological *facticity* of *Dasein*, wherein being acquires its existence in-the-world and is oriented toward-death. Levinas, however, vehemently critiques the ontological prioritization of being's facticity, instead emphasizing transcendental passivity. J. Rolland (Levinas, 2003, p. 13) reminds us that «human existence and the positive characteristics of human finitude and nothingness [...] are defined, for Heidegger, by facticity [while Levinas prioritizes passivity]. And the understanding and interpretation of this facticity is itself the analytic ontology of *Dasein*. That is why Heidegger, and his followers define ontology as a 'hermeneutics of facticity'». In addition to this, Levinas (1987, p. 135) claims that «there is here then a passivity more passive than any passivity (and more than any facticity): filial, a pre-vious, pre-logical subjection, a one-way subjection which it would be wrong to understand on the basis of a dialogue». It is through this transcendental passivity that the *here I am* is transformed into *it is me*. I exist as a human being not because of myself but because of the Other. This passivity is purer than individuality: Pure passivity is «responsibility for the freedom of others» (p. 136).

21 See Lawton (1975, n. 10).

22 Levinas' reproach against Heidegger's concept of *il y a* is primarily related to the fact that the latter mainly insists on the question of meaning, essence and truth of being, thereby overlooking the question of the Other in the *exodus* of being from nothingness, which gives meaning to the existence of being as transcendence. See Peperzak (1997, p. 206).

23 About the immemorial past, see also Levinas (1994, pp. 76-91; 1987b, pp. 111-114). Regarding secondary literature on this issue see Dimitrova (2011, pp. 37-48); Newman (2000, pp. 90-129); Smith (2016, p. 97). Anxiety plays a decisive role in Heidegger's dialectic of being-there, which, according to Heidegger, has its starting point in nothingness (Levinas, 1987b, p. 51). Heidegger (1949, p. 230), by saying «nothing nothings», impersonalizes the activity of being and cements consciousness as a «rupture of the anonymous vigilance of the *there is* [in presence]». See Levinas (1987b, p. 51). A vehement criticism of Heidegger's term «nothing nothings» comes from Carnap (1959, p. 60-81).

24 Levinas argues that one of the most significant flaws in Heidegger's ontology is his neo-Aristotelian approach, particularly regarding the concept of being as substance. As Keohane-O'Connor (1982, p. 95) alleges «Levinas holds that Heidegger has not broken from an Aristotelian framework. Aristotle pursued questions of ontology, or metaphysics, in connection with the basic problem regarding the theory of substance».

25 Concerning the notion of *il y a* between Levinas, Hegel and Kant see Peperzak (1997, pp. 193-203).

26 Levinas (1987b, p. 104), in his intention to show that responsibility precedes freedom, characteristically claims that «responsibility for the Other signifies an original and concrete temporality, and that the universalization of presence presupposes it».

27 At this point, the later Levinas, even if he takes Husserl's transcendental phenomenology as his starting point, envisages a different kind of being in contrast to Husserl: the phenomenology of sociality, i.e. the proximity of the other as a necessary and sufficient condition for being in order to escape nothingness. For the later Levinas, the conception of *il y a* was differentiated above all when he published his *Ethics as First Philosophy* (1984). He characteristically admits that «I have attempted a 'phenomenology' of sociality starting from the face of the other person –from proximity– by understanding in its rectitude a voice that commands before all mimicry and verbal expression, in the mortality of the face from the bottom of this weakness».

28 Here, by providing the concept of enjoyment, Levinas argues against Plato, who claims that enjoyment is tautological to the categories of being and thus inseparable from being. On the contrary, Levinas (1969, p. 115) elsewhere considers enjoyment to be independent of being, because it «is the other of life, is an independence *sui generis*». Concerning the notion of enjoyment and the disagreement with Heidegger, see Golob (2020, p. 88).

29 Levinas (1969, p. 131) defines the notion of the elemental as: «Every relation or possession is situated within the non-possessable which envelops or contains without being able to be contained or enveloped. We shall call it the elemental». Concerning the transition of language from *il y a* to the elemental, see Sallis (1998, pp. 152-159); Fagenblat (2019, pp. 113-123); Williams & Standish (2016, pp. 360-373); Switzer (2000, pp. 23-35).

30 Levinas (1969, pp. 134) underlines that, attributing *Dasein*, I am «egoist without reference to the Other, I am alone without solitude, innocently egoist and alone. Not against the Others, not 'as for me ...' —but entirely deaf to the Other, outside of all communication and all refusal to communicate—without ears, like a hungry stomach».

31 However, paying particular attention to the primary text the question remains: Are we sure that it is not being left to itself, in itself and for itself, that is horrible and 'damned' rather than nothingness? Levinas, starting precisely from the notion of il y a, seems to try to show how nothingness is not an escape from

being, hence a liberation, because the latter is precisely irremediable, irremovable at the ontological level, that is, stronger than nothingness, encompassing all negation. This is of immense importance in order to answer the question of the deliberating articulation between being-non-being and and the beginning of just war in modern world.

32 Levinas aims at exteriorizing being through language in order to escape nothingness. Levinas (1969, p. 279) addresses that «the presence of the Other, or expression, source of all signification, is not contemplated as an intelligible essence [like Heidegger], but is heard as language, and thereby is effectuated exteriorly». Concerning Levinas, language and exteriority, see Panayiotou (2024, p. 232) and Standish (2008, pp. 61-63).

33 Heidegger insists that it is through dereliction that ontology and philosophy in general remain finite. Levinas vehemently rejects this position, first by exploring transcendental imperatives under Husserl's influence, and second by drawing on ethical imperatives shaped by Jewish tradition in his interpretation of *il y a*. For instance, Levinas (1996, p. 26) claims that «but, as such, [Heideggerian] philosophy is a finite possibility, determined by dereliction, by the project-in-draft and by the fall, that is, by the concrete situation of existence that philosophize».

34 Levinas (1996, p. 30) states that, according to Heidegger, «anguish must constitute the situation in which the totality of Dasein's ontological structures are individualized into a unity».

35 For a discussion of the debate between Levinas and Hering on *il y a*, being and nothingness, as well as on the central role of ontological-transcendental phenomenology in the foundation of the sciences, see Boccaccini (2016, pp. 433-448); Hering (2021, pp. 51-108; 2019; 1932, pp. 474-481); Monseu (2005).

36 By 'transascendence,' Levinas means the transcendence of transcendence, which is infinity. Hering, however, contradicts this assertion, arguing that Levinas, by transcending transcendence, returns to ontology. This is similar to Hegel's «finite infinity» or «bad infinity». See Houlgate (2006, pp, 433-435). In addition to this, as Panayiotou (2024, p. 235) puts it, «Hegel's reduction of the infinite within a finite framework automatically relegates it into an exterior realm outside of esotericism, that is of history [i.e. ontology]».

Chapter 10
Ethical and Deontological Challenges of Autonomous Weapon Systems:
A Critical Analysis of the Responsibility Gap and Dignity Arguments

Prashant Rahangdale
Symbiosis Law School, Nagpur, Symbiosis International (Deemed to be) University, Pune, India

Shashank Nande
http://orcid.org/0009-0003-0274-2251
KIIT School of Law, KIIT University, India

Ankita Kumari
Chotanagpur Law College, Namkum, India

Prachi Priyamvada
Chotanagpur Law College, Namkum, India

ABSTRACT

The moral consideration involved with autonomous weapon systems is a highly debated topic in both academic and social circles. As system development and research continue, various non-governmental bodies propose a total ban on technology. The chapter critically assesses deontological claims that prevent the use of autonomous

DOI: 10.4018/979-8-3373-6264-9.ch010

weapon systems through argument analysis. This chapter is done to find the validity and strength of the deontological case against these systems. The results show that, despite a greater catalogic abundance of influential deontological arguments against the use of autonomous weapon systems than in favour of it, the arguments against the technology are qualitatively poor in terms of tenability and relevance.

INTRODUCTION

Autonomous weapon systems (hereinafter, AWS) represent a possible paradigm shift in the field of military science and warfare. In the event of continued development, there will come systemic changes that do not merely stop at tactics and strategy but cover politics, ethics, and economic dimensions. To handle this transition well, both the Armed Forces and the Umbrella of military science must understand these systems. The shift is already evident in modern predecessors like the Bayraktar TB2 drone used in Ukraine, which means that human soldiers will no longer be needed on the immediate battlefield (Karadsheh & Sariyuce, 2022). The ethical aspects of AWS remain extremely topical. These systems were recently discussed at the United Nations in December 2021, and some parties proposed a ban on the grounds of potential adverse outcomes (Shead, 2021). Various international organisations, including "Stop Killer Robots" as well as national organisations, are proactively lobbying for a ban be implemented (Stop Killer Robots, 2022). Even though these programs highlight intuitive ethical issues, the development of weapons with self-driving features goes hand in hand (Human Rights Watch, 2022).

The debate over the ban or control of AWS raises the philosophical question of moral rectitude and what it might entail. Tales include dystopian depictions of robot oppression, as well as the possibility of soldiers abiding by the rules of war, which are resistant to human fallibility in reason. Moral philosophers are exploring these dichotomies to determine whether AWS should be developed and implemented, but no final viewpoint has emerged. AWS have provoked emotions and organised various movements aimed at prohibiting the technology. This ongoing controversy over ethical and legal issues is set against the background of consequentialist ethics and deontology. These two normative ethical theories are critical in understanding the world of research. Ronald C. Arkin (2008; 2010), one of the most famous roboticists, has been an early advocate of these systems. Arkin (2008, p. 46; 2010, p. 334) makes this argument, expressly noting that, with proper adaptation of technology and design, AWS may be more effective than human soldiers in addressing the demands of jus in bello (the conduct of war) (Arkin, 2010, p. 336). It is based on the premise that human fighters always have psychological and physiological weaknesses that machines do not. Although Arkin does not deny the difficulties posed by technology, his stance can be considered an example of a consequentialist theory, since the analysis of AWS is based on the assumption of the anticipated beneficial results of their applications (Stanford Encyclopedia of Philosophy 2020).

On the other hand, opponents employ both deontological and consequentialist viewpoints. One of the theological arguments that Sparrow (2007, pp. 74-75) has coined as foundational as the responsibility gap and subsequently has been the center of discussion is the responsibility gap. To counter this argument, it is maintained that AWS generate an ethical vacuum in which responsibility relationships are blurred or denies the existence of responsibility as more autonomy is attained. The laws of jus in bello require an actor to be legally and ethically liable for violations. The lack of AWS liability will lead to a further division of responsibility due to enhanced autonomy (Sparrow, 2007, p. 73). This represents a deontological objection, since the gap in responsibilities is considered problematic in itself, and a prohibition of such a ban without considering other possible outcomes is problematic. Unlike consequentialism, deontology focuses on the ethical goodness of actions rather than their results (Stanford Encyclopedia of Philosophy 2020).

Two prominent characteristics reflect the current discourse: 1) deontological argumentation is powerful amongst those who criticise (AWS), and 2) the discourse generally has the tendency to perceive AWS as a problem in itself, with a few exceptions of counter-example, including Arkin. The idea, promoted by Sparrow, that rivals rule the deontological sphere, seems to be a valid conclusion about the topography of research today. Nevertheless, despite the apparent might of these deontological counterarguments, the ethical debate remains. That raises an important question: Does the argument against AWS carry as much logical weight as it seems to? The available literature on the ethical problematic nature of AWS shows that most arguments have been advanced from consequentialist standpoints. Conversely, resistance is expressed in terms of the consequentialist and deontological theories. As a result, the main goal of this research paper is to critically examine the merits of the deontological argument that opponents of AWS advance. To this end, the paper examined core deontological arguments, applying a methodology of examination, critique, and response. The following Research Question is obtained as a result of this process: What is the soundness of the deontological argument on the ban of AWS, and how can it be handled? To answer this question, the study will provide a deeper understanding of the current state of research. As analyses conducted before reveal that the deontological opposition is particularly stressed, it is crucial to either prove or disprove this assumption. When it can be conclusively argued that Sparrow was either wrong or correct, the argument could be carried out in the constructive sense. In addition, the outcomes will provide a clearer picture of the ethical side should AWS's consent be granted.

BACKGROUND

The responsibility gap conceptualised in this analysis is primarily presented as a central deontological argument, even though it has aspects that can be explained in consequentialist terms. The consequentialist approach assumes that one does not have the capacity to place any agent in a legally or morally responsible position; consequently, the harm of putting AWS into practice. However, the deontological character of the argument is pegged on the intrinsic value of respect as is exhumed in the subsequent proposition: It is a scintilla of respect because war will at all be governed by morality, that one should be held responsible, or permit of being held responsible, by the decision to slay his enemy, that the deontological nature of the argument. Even the most we can give our adversary is that his life has value, that a person should dare to accept the responsibility of death on his hands (Sparrow, 2007, p. 67). This equation shows that the conclusion about the responsibility gap is determined by the rules of responsibility and respect, which are values inherent in the system and not dependent on the weapon's efficiency or the parameters of collateral damage. As a result, the gap in responsibility acts as a fundamental deontological objection in this case.

To this is added the debate over dignity, another element of deontological opposition. Consequently, Sparrow (2016, p. 111) admits Arkin's consequentialist argument that AWS could, under certain technical conditions, be more inclined to adhere to jus in bello principles than human soldiers. Nevertheless, according to Sparrow (2016, p. 112), there remains a more profound and qualitative issue: the deontological offence that creates a responsibility gap and degrades human dignity. The main idea of that argument is that a machine makes the decision to kill, and it is a serious offence against the dignity of the enemy (Sparrow, 2016, p. 109). This can be taken as confirmation of the assertion that, if deontology is given greater weight than consequentialism, the moral obligation against AWS would become stronger. Therefore, those opposed to AWS control the argument when the discussion turns to the prevalence of deontology (rights and dignity) over consequentialist ethics (outcomes and efficiency). However, arguments against these deontological arguments can be made by comparing them with existing weapon systems. Proponents use the case of precision-guided munitions to argue for an ethically acceptable technological path, namely that the so-called semi-autonomous precision weapons are much less harmful to civilians than unguided ordnance (Horowitz, 2016, p. 33). In establishing commonalities between these accepted systems and AWS, activists aim to strengthen the viability of consequentialist arguments, including the claim that AWS would fulfil the law of distinction better than human fighters do (Arkin, 2010, p. 333).

Nevertheless, Horowitz makes this comparison subtly and claims that such analogies would work when comparing AWS to the munitions level, but not when comparing AWS to the platform or operating system levels (2016, pp. 33-34). The technology will face these higher levels of autonomy, confronting the deep ethical obstacle of the responsibility gap and dignity, which can never be resolved by simply referring to the existence of precision weapons (Horowitz, 2016, pp. 33-34). Sparrow (2016, pp. 107-108) comments on the related comparative arguments that imply that, due to the use of distance and automation, the mines and the cruise missiles (which are assumed to be ethically acceptable) should not be perceived as a threat to dignity; AWS should not be treated in the same way. Sparrow (2016, p. 109) refutes this, citing empirical data indicating that the most popular view holds that the application of AWS constitutes a de facto violation of dignity. Moreover, Sparrow (2016, p. 108) questions the premise by asking whether accepted systems, such as mines, are in fact more ethically problematic than one currently admits. The matter is even more complicated as the AWS are considered in terms of jus ad bellum, the rules that determine the reasonableness of war justification before the war. One of the key consequentialist reasons to justify not using AWS is that it can be much easier to instigate a conflict (Johansson, 2018, pp. 145-146). Such an argument is based on the premise that the price of war would fall on the aggressor, as their soldiers are not put at as significant a risk. Therefore, the estimated consequence of this possible reduction of the threshold is one of the most significant adverse effects of the AWS adoption (Johansson, 2018, pp. 145-146; Horowitz, 2016, p. 33).

In addition, Linda Johansson (2018, pp. 146-147) reveals that AWS implies the threat of providing secret wars and unfair wars. In this respect, Johansson (2018, p. 146) outlines unjust war as any conflict in which one party or more employ AWS to fight against opponents who do not. In this, a financial asymmetry arises in the risks the various parties are willing to assume regarding human lives and suffering, thereby revealing injustice. Also, the consequences may be a feature of secret wars, in which continuous conflicts are difficult to characterise due to reduced transparency. Such a lack of transparency creates challenges in implementing proper ethical and legal principles (Johansson, 2018, p. 147). Other deontological reasons against AWS are given by Duncan Purves, Ryan Jenkins and Bradley Strawser (2015). The argument put forward by Purves et al. (2015, p. 851) is strong because it states that AWS cannot make decisions which do not involve moral issues. This position is premised on two aspects: first, AWS do not have the essentials to replicate the level of human ethical judgment; second, even if AWS could create such an ethical judgment, it would not be a normative decision because the right intention would be absent (Purves et al., 2015, pp. 851-852). This brings out the intuitive understanding that a system of autonomous beings cannot be a moral agent. This is the extension that justifies Sparrow's conjecture that AWS cannot be morally guilty in war (Sparrow, 2007, p. 71).

Although the AWS-related deontological issues, like the issue about dignity and the gap in ethical responsibility, raised by Sparrow, and the lack of ethical judgment, presented by Purves et al., seem to be multiple and thoroughly problematic, they can be challenged. According to Michael Robillard (2018, p. 705), AWS do not seem problematic from a deontological perspective, since the arguments presented to substantiate opposing views are based on incoherent conceptual images of the technology. According to Robillard (2018, pp. 705-706), AWS lacks goals or decision-making capacities beyond what is physically implemented in the system's software. Thus, Robillard (2018, p. 705) suggests that the arguments emphasised by Purves et al. and Sparrow are mistaken, and that the ethical discussion should treat AWS differently. Although the arguments differ, opponents of AWS have struggled to ground them on a more problematic basis, according to Michael Skerker, Duncan Purves, and Ryan Jenkins (2020, p. 197). The implication that an intuitively crucial ethical distinction exists between the death of a soldier in combat through an AWS and a human soldier has been a point that has proven challenging to articulate for opponents. It is this explanation gap that the supporters of the AWS use as the basis for their valid objections (Skerker et al., 2020, p. 197). Skerker et al. (2020, p. 197), in defence of opponents' deontological argument, propose that most ethical issues surrounding AWS can be grounded in what they call the martial contract between human combatants. From this, Skerker et al. (2020, p. 198) conclude that human soldiers are allowed to work under a silent, tacit understanding. The terms under which they engage in the battle lie in the fact that they are engaged in fighting with fellow human beings, who are enemies with a human concept of morality, and equally share in the risk of death.

Therefore, by the side of a combatant replacing his most crucial role in this martial contract with an AWS, this is a primary violation of the other party. This system can be used to develop strong deontological reasons against AWS. The proposal developed by Skerker et al. (2020, p. 198) is one means of explaining the ethical uneasiness about AWS as an intuitive, subjective form, which might be directly connected to the notions of respect and dignity, as well as to the responsibility gap mentioned above. The recent history of AWS indicates a divided terrain which does not show any (conclusively) winning arguments. Both the supporters and the detractors use consequentialist reasoning to advance their arguments. Nevertheless, only those who oppose the use of technology use deontological arguments. Although some of these deontological assertions can be logically criticised, such as the claims of Robillard and Sparrow about the lack of a sense of responsibility in Sparrow and the fact that machines cannot recreate the human ethical reasons why people make confident choices, these have done little to influence the overall discussion.

THEORETICAL FRAMEWORK

Autonomous Weapons Systems

Another salient part of this discussion on AWS is what this phenomenon actually means. By defining it too broadly, the definition may risk encompassing the latest technologies, including cruise missiles, thereby eliminating the moral novelty of the subject. On the other hand, a narrow or extreme definition leaves AWS in unrealistic, future-fantasy territory, covering up many modern systems that should be questioned. Moreover, divergent arguments often rest on different conceptualisations of AWS, which significantly shape the development of specific ethical arguments. Vast differences in definition may lead to an argument that either excludes valid systems or includes too many in toto. Simple semantic variations may lead both sides to engage in what is described as 'talking past each other,' where proponents and opponents discuss entirely different things, only to appear to be arguing over the same point. This implies that a clear definition of such definitional deviations is required to make this study valid; the discussion should be based on a conceptually shared, or at least not too far out-of-pocket, definition of AWS. As pointed out, simplistic definitions have their problems on the spur of the moment. Sparrow (2007, p. 65) puts it very plainly: "*The allegation that a weapon is autonomous entails simply that it can act on its own .... in most cases, it is an assertion that it is a fire and forget system .... Weapons of this type do not per se involve any ethical consideration to that brought about by other modern long range weapons.*" In this case, Sparrow implicitly shows that this broad definition cannot justify the problematization of AWS on grounds unique to ethics. Sparrow (2007, p. 65) presents an opposing, more rigorous definition in the exact text: To say of an agent that it is autonomous is to say that its actions are internal to it and that they serve its purposes. Moreover, in a completely free agent, these ends are ends that, in some sense, they have chosen.

The latter definition is much narrower and also covers many systems, such as cruise missiles. It, however, takes autonomy to a higher level, which can be considered cognitive thinking or agency. Although the question of philosophy of consciousness is not the subject of the proposed paper, there is a clear danger that the definitions with a tendency to focus on the notion of consciousness are going to leave the current weapon systems that form the core of the ethical discourse, and the proposed shift to radical, hypothetical cases of AWS at the same time. Scharre and Horowitz (2015, p. 16) present an alternative, functional definition of an autonomous weapon system: a weapon system that, once its function is triggered, is supposed to choose and attack the target without the human having decided to be the target. This definition defines AWS in terms of the operational action or the freedom from human intervention in single-target selection. More importantly, this argument does not, by design, establish the machine's metaphysical properties (consciousness, intelligence, and decision-making), as Scharre and Horowitz do not want to engage in pointless debates about machine intelligence and free will (2015, p. 16). Later definitions have added the aspect of meaningful human control (Sharkey, 2019, p. 75; the International Law and Disarmament Delegation of the Ministry of Foreign Affairs, 2020, p. 11). To demonstrate this, Human Rights Watch (2014) gives an example: These robots would recognise and shoot at the target without any meaningful human control, unlike any existing weapons. They would be able to decide, therefore, when to take the lives of human beings. In this respect, abstract discussions of what autonomy can be are bypassed; instead, the existence or absence of the so-called meaningful human control is used as the primary means of determining the morality and legality of the weapon system (Amoroso & Tamburrini, 2018, p. 4).

Deontology

There is a good representation of consequentialist arguments and substantial development of the debate on Autonomous Weapon Systems (AWS), as indicated in the research overview. Nevertheless, such arguments inherently cannot provide ethical responses with absolute confidence because they anticipate a consequence in itself (Amoroso & Tamburrini, 2018, p. 9). This is why the addition of deontology as a standpoint is driven by the capacity to implement the contrasting perspective to consequentialist ethics, which, by implication, would serve to balance the issue. Laid down in normative theories of deontology, the determinations of right or wrong of an action rely on the intrinsic value of that action. However, not the result, i.e. it is the meaning of deontology applied in this paper (Stanford Encyclopedia of Philosophy 2020). Such a difference can be best explained by an empirical analogy where, as in the case of a famous robber, a man requires you to give him the whereabouts of your valuables. Deontological Theory X states that one should not lie, as his/her lie is a concept that is wrong in itself, no matter what the consequences of the act are. On the other hand, under Consequentialist Theory Y, one ought to lie to deceive the robber, that is, to evade the bad event, which is that of losing his valuables. As this example depicts, deontology and consequentialist ethics offer different perspectives on what moral rightness is. Nevertheless, this does not mean that there is a universal claim that deontological theories prohibit lying or that consequentialist theories prescribe lying; such conclusions depend on the normative theories considered. An important point is that, to answer the research question correctly, the definition of deontology here is comprehensive, i.e., it concerns the nature of actions (the intrinsic intention of the action) rather than adherence to a particular, specific normative theory.

METHODOLOGY

The proposed study will use argumentation analysis as its primary methodology, namely, the deontological arguments opponents of AWS have developed. In ethical discourse, each normative camp has its own normative positions to promote through argument; thus, to critically examine and analyse these positions, argument analysis is the most suitable strategy. The arguments will serve as the primary material for this investigation, along with the scientific articles that present them. A systematic analysis using an analytical tool consisting of four questions is applied to these arguments. Dissecting arguments in a systematic way, therefore, allows outlining the exact amount of argumentation needed for advocates to be effective and refracting the broader discussion about AWS (Pietrzak, 2024). The evident benefit of argument analysis is that it provides a degree of critical examination, exposing the soundness or unsoundness of arguments, a level of analysis that a typical systematising study cannot boast. The next methodological solution would be to develop new, independent deontological arguments in support of AWS, thereby addressing the identified research gap. Though such research would add to the existing body of research, it would not be directly relevant to the current issues at hand. Overall, examining existing arguments is more valuable in this study because it offers a more user-friendly presentation and serves as a necessary starting point for further developing the discourse (Pietrzak, 2024). Moreover, this can also be a more effective strategy for producing new deontological advocacy arguments, namely by building strong objections to existing counterarguments.

The paper includes both argumentation (argument building) and argumentation analysis (criticism of existing arguments). Thus, it is rational to define how the arguments are built, then to spell out how they are assembled and digested, and finally to describe the approach to analysing and criticising them. Every argument consists of premises, middle extensions and conclusions. (Hansson, 2006, 2020). The premise or the premises are the assumptions made that require derivation to conclude. Intermediate links are used to express the logical derivation between the premises and conclusion; most notably, it is observed that some arguments do not include such links. Lastly, the conclusion is the final step in the derivation and the suggestion it aims to establish. An example of premises, intermediates, and conclusions in this set of study has the structure given below:

a) (P1; Premise 1) When you entertain yourself with regular exercise, you feel good.
b) (P2) You should not exercise when you are injured, as then you feel bad.
c) (P3) Henrik has a regular workout but feels guilty.

d) (M1; Intermediate 1) Henrik does not do anything, which can explain why he feels bad.
e) (S; Conclusion) Henrik gets wounded.

Arguing may be built in different forms under the influence of such parts, and in many cases, some premises and intermediate connections are suggested or omitted. (Hansson, 2006, 2020). To make the arguments amenable to analysis, this paper will undergo an identification, development, and supplementation process. The step of identifying involves finding and mining the particular elements of the arguments in the text (Pietrzak, 2023). The process of developing requires breaking down the argument into the parts: premises, middle and conclusion and placing them in the logical order (as shown in the example above). Supplementing is done after the development, and entails the identification of implied or omitted premises and middle links, and adding them to the logical order (Pietrzak, 2022). Such formality encourages regular handling of new arguments, that is, objections, as well as those arguments already written, which must be responded to. It thus forms one of the requirements of answering the research questions validly (Pietrzak, 2022). Nonetheless, most of the application in the study is devoted to the research of arguments. It is based on the same structural insight that the arguments ought to be analysed, formulated and supported. To advance the debate on AWS and to provide answers that will be considered valid to the research question, these steps should be carefully followed.

ANALYSIS

During the analysis stage, each argument will be examined in its own section. After a strict reconstruction, the argument becomes amenable to analysis by formulating questions. The rehabilitation process, which includes identification, translation, development, completion, reinforcement, and sharpening, can be repeated several times before the final version is submitted; the intermediate steps are stored and can be retrieved when necessary. As a result, the arguments presented are the ultimate versions of these reconstructions. A summary analysis of both arguments is provided at the end of each examination. The reconstruction of the argument on the responsibility gap made by Sparrow follows: To (P1) Be at war justly is a necessary condition that we can impose on an agent in a moral sense of the deaths we incurred against enemy combatants. However, the programmers or military commanders of an Autonomous Weapon System (AWS) can never be held to a moral standard for the resultant lethal consequences that the AWS directs at enemy combatants (P3). AWS conceptually cannot be morally responsible for its actions, including those causing the deaths of the enemy. The case of the deaths of enemy combatants caused by an AWS has no other reasonable alternative candidate to shoulder the moral responsibility. (M1'): (Assuming (P2), (P3), and (P4)) It follows that no one can be morally responsible for the death of enemy combatants whom an AWS kills.

In turn, this means that waging war with AWS is forbidden. To do so would be tantamount to verminising the enemy, like vermin are to be destroyed without consideration of morality. Is the conclusion misguided? Provided that the logical conclusion is sound, the conclusion offers a firm ground for why AWS should not be used in war. Moreover, the conclusion is, in itself, deontological, which made it suitable for the framework of this research. It proceeds from Sparrow's (2007, p. 62) thesis that it would be unethical to use such systems in war. Nonetheless, the conclusion is based on what was established beforehand, namely, a narrow definition of AWS. On the one hand, the application of Sparrow's (narrower meaning) definition allows the justification of premise (P2). This is clearly demonstrated in Sparrow's (2007, p. 65) discussion, where he writes: "*since the agent's actions were its own.*" It acted for its own ends; its actions could not be attributed to others. Conversely, the conclusion is found misplaced in the broader context of the scientific controversy at the time. The fact that Sparrow's conclusion is based on a narrow-minded conceptualisation of AWS makes weighing it against modern arguments grounded in the technology's expanded definition problematic. Such invariance is similar to the so-called straw man fallacy. The difference is that Sparrow has not already created a skewed image of AWS; the definition given is Sparrow's personal viewpoint and, as such, cannot be described as an intentional failure of argumentation. However, the issue of relevance remains underlying.

The fact that the conclusion is perceived as misplaced is an internal problem and cannot be fixed by simply refining or giving more force to the rhetoric. The excessive reliance on a narrowly defined AWS has permeated Sparrow's logical framework. Conceptualising the AWS as a full-fledged moral agent, an even narrower definition would make premise (P3) more challenging to justify, which would fail to make the argument successful. On the other hand, in case the definition is extended to exclude the fact that the actions of the agent were their own actions and based on their own purpose, it will not be possible to justify premise (P2), which in turn will lead to the disappearance of the responsibility gap. On the whole, the conclusion on the responsibility gap is quite logically sound. It is robust in advancing Sparrow's specific thesis and in supporting the deontological angle of the opposing camp in the scientific discussion. Nevertheless, the extent of this conclusion's acculturation depends on how the larger scientific community defines AWS. Seeing how the liability gap is pegged on what this study describes as a narrow definition, there is a chance of the argument and hence the conclusion being misplaced when placed within the broader scope of the discussion. Are there shifts in meaning? None of the changes in the meaning have been observed in the argument. Is the argument formulated in a valid schema?

The logical plan of the conclusion is sound. The statement that the enemy is being treated as a pest cannot be rationally traced to the premises; it is merely an opinion and not part of the conclusion. However, this rhetorical addition does not diminish the power of the analysis of the conclusion per se. Do the premises have any inappropriate or problematic ones? The assumptions are not unreasonable regarding the conclusion. Premises (P2) and (P3), however, are problematic each singly and in combination. The problem with the premise (P2) is that it assumes the inability to hold either the programmer or the commander accountable, since the responsibility cannot be passed to the commander. Although it is not a disputable opinion, it faces significant opposition (e.g., the cruise missile precedent). The counterexample is most clearly illustrated by the following: Examples of such national military AWS development would be an evil programmer. You deliberately place a backdoor code which becomes activated after 10 hours, which makes the AWS attack civilians. The strict logic of premise (P2) denies that the evil programmer can be held morally accountable, since the actions of an AWS are considered non-transferable. This example proves that premise (P2) has implausible results. This would not falsify (P2), but it would suggest that the premise needs to be reformulated or that the definition of AWS is so unreasonable in the present debate. If (P2) were revised to permit liability of the programmer or commander, the statement of the liability gap would be hard to sustain. The premise (P3) is problematic because it is based on a definition of AWS capable of making independent decisions, yet unable to be charged with the same, which is opposed by Robillard (2018, p. 708) and noted in the review of the research. Lastly, (P2) and (P3) are not a good combination for the same reason. This is (P2) to suggest that the AWS is an individual artificial actor that generates its own actions and is ambitious enough to exonerate programmers and commanders from accountability. On the other hand, (P3) states that the AWS is not liable, so it is not an agent. This creates a conceptual paradox: the AWS needs to be autonomous to be beyond the control of the human factor, yet not autonomous enough to assume its own responsibility.

Inability to Reproduce Human Ethical Judgment and Act with the Right Intention, Analysis. The thesis on the impossibility of Autonomous Weapon Systems (AWSs/AWS) partially recreating human ethical judgment and performing with the right intention is a two-sided proposal presented by Purves et al. (2015). According to the research overview, this argument can be further split into two statements: 1) AWSs cannot reproduce human terms of ethical judgment, and 2) AWSs are not able to act with the correct intention (Purves et al., 2015, pp. 851-852). As a result, the analysis below will be divided into two parts, each addressing the inability to reproduce human ethical judgment. The rational account of the case on the inability of AWS to recreate the ethical judgment of humans is crafted in the following way: (P1) Human moral judgment operates based on the anti-codifiability thesis, which states that it cannot be codified, and a definite set of rules is insufficient to account for it. Human ethical judgment requires certain mental functions: the ability to be reflective through balance sheet representations, to identify facts that constitute moral comprehension, the imagination to envision situations with moral implications, and to have moral experiences of phenomenological substance. AWS would theoretically lack all of these capabilities; consequently, it would not be able to imitate human moral judgment. (P4) The ethically justified decisions are based purely on the human ethical judgment.

Idem (M1) Based on (P3) and (P4), AWS should not be expected to make ethically correct judgments, as it lacks the necessary human ethical judgment. (M2) Since it is not possible to expect AWS to decide ethically correctly, it implies that they can be expected to decide ethically incorrectly. Indeed, based on (M1) and (M2), AWS represents an inherently ethically problematic. Is the conclusion misguided? The conclusion provides a substantive basis for why AWS should not be deployed. In this respect, the conclusion and the argument are sufficiently consistent with the research question and the scientific discussion considered. AWS is described as mainly ethically problematic, which seems to put the argument in a deontological context. The problematization of the intermediate step (M2) is, however, based on an anticipated outcome (wrong decisions), which makes it ambiguous whether the argument is somewhat deontological or consequentialist. One of the most significant aspects of the review of the conclusion is the definition of AWS. The emphasis on judgment and decision-making in the argument resembles Sparrow's more limited definition. A very slight variant of interpretation is used in Purves et al. (2015, 853): it is weapons that can be programmed to behave in such a way that not the human but the computer makes decisions and acts according to its own agenda. By virtue of this definition being more akin to the one given by Human Rights Watch (HRW) above, it is closer to the approximate research trend than Sparrow's definition of the responsibility gap.

However, such a definition is vague in its choice of words, as it decides everything and ignores the human. Deciding everything would mean that the system sets its own goals, or only makes technical decisions within an already assigned mission. In the same way, ignoring the human is indeed like HRW without meaningful human intervention, but it is different. Such confusion is not resolved further. As a result, the definition that underpinned this argument is somewhat more in line with the overall trend than Sparrow's. However, it is again subject to interpretation as having been restricted by said trend. Are there shifts in meaning? The analysis of the argument did not show any changes in meaning. Is the argument of a valid schema? Most of the argument was located in the introductory part of the article (the reconstruction process can be identified on request). Nevertheless, the reconstitution had to be supplemented with data from subsequent parts of the work, since the conclusion of the statement, which is mainly problematic from an ethical perspective, was not possible based on the briefing in the introduction. Do any of the premises seem to be inappropriate or problematic? There seems to be no premise that does not fit in connection with the conclusion. Premise (P1), however, is a pervious nodal point in the argumentation. This assumption is based on the so-called anti-codifiability thesis - in other words, the inversion of the so-called codifiability thesis - which states that ethical values are not lendable to the algorithmic codification and reduction to some list of rules; the implementation of ethical values will have to undertake an intuitive human ethical judgment (Purves et al., 2015, pp. 855-856).

In case the anti-codifiability thesis were dismissed, the argument by Purves et al. (2015) would be faulty in a significant way. Moreover, this denial may, unwillingly, support arguments in favour of AWS. The logical counter-construction would be as follows: Then there is codifiability (P1) of ethical judgment. (P2) AWSs are not vulnerable to the human shortcomings that afflict judgment (as a result of Arkin 2010, p. 336). (M1) Based on (P1) and (P2), one can say that AWSs (Autonomous Combat Platforms) may be able to implement ethical values more effectively than human soldiers. Connectedly, (S) AWSs should be used in war because they embody superior values to ethics compared to human soldiers. The so-called anti-codifiability thesis in itself is not a controversial concept. According to Purves et al. (2015, p. 857), influential philosophers of the age who have taken this stance include Kant, Ross, and McNoughton. Even a major sceptic of Purves et al. (2015), Michael Robillard (2018, p. 710), agrees with this thesis. Nonetheless, there are two camps of ethics, such as value realism, which argues that ethical values can be established. Therefore, although premise (P1) technically constitutes a possible flaw in the argument, the philosophical theory that supports it is well established.

Nevertheless, the mediating item (M2) poses a more substantial structural issue for the argument. In explaining to the reader the flaws in the inability to make morally correct decisions, Purves et al. (2015, p. 858) argue that it is much easier to program AI to execute unethical or criminal instructions than to train human soldiers to do so. Although this is an attractive argument, it may conflict with the argument for the responsibility gap. Given that here the impossibility of executing ethical judgment is problematized, the rationale is that AWS would neither attempt to find nor disobey an ethically dubious command issued by the commander; in fact, it is the commander who makes the choice. If the decision is based on the commander, then he/she should be morally accountable, and the claim that there is a gap in responsibility is thus eroded. Additionally, the rationale behind (M2) presupposes that the AWS is viewed solely as a means (a tool). As argued above, (M2) focuses on the commander's decision-making power, which directly opposes the previously defined image of AWS, in which the machine makes all choices (Purves et al., 2015, p. 853). Treating the AWS as a means to an end falls apart. This is because the argument requires focusing on the AWS as an independent decision-maker to bring out the agency's ethical shortcomings; without such a focus, the argument becomes untenable.

The reformulation of the argument of acting with the right intention by Purves et al. (2015) goes as follows: Condition number 1...(P1) It is ethically possible to have two moral rates of identical activities based on the intention of an individual that the presence of the right intention makes the activity ethically better than one that is carried out with the wrong intention. The existence of an intention means that A (the actor) has an attitude of belief or desire (or some other propositional attitude) (P2). AWS does not satisfy the properties in (P2). (M1) AWS is thus unable to do something intentionally. (S) acts of AWS (P1) and (M1) are ethically problematic since they cannot act with the right intention. Is the conclusion misguided? This conclusion provides a deontological reason to reject AWS. Since AWS is specified here in the same way as the definition above argument of Purves et al. (2015, p. 853), the components of this argument need not be defined differently. Therefore, it is not misleading to the conclusion of the estimated trend in the scientific debate of AWS. Nevertheless, the definition presented by Purves et al. (2015, p. 853) remains vague, as reflected in the analysis of the aforementioned argument. Are there shifts in meaning? No changes in meaning have been detected. Is the argument based on a legitimate schema? One main issue with the argument is that one cannot reduce the ethical problematic stated in the conclusion to the premises or the intermediate connection. One important point of contact between (P1) and (M1) is how. The mere absence of intent in an act cannot be equated with acting with the intention of doing something wrong; unless there is that particular relationship, it is hard to say there is an unhappy moral issue at hand.

This is the objection that Purves et al. (2015, p. 861) expect to receive. Their answer is to show how this issue works through an example: a sociopathic soldier who takes part in war and is only successful because of the absence of empathy. This soldier follows the order of the law of war and orders. Although he did so, as Purves et al. (2015, p. 862) stated, this instance shows that a lack of ethics can be transferred to AWS. The weakness of this defence lies in its reliance on eliciting an intuitive response from the reader rather than grounding the problem in ethical theory. Nonetheless, Purves et al. (2015, 862-863) present another justification by basing the moral question on the principle of the right intention as one of the principles of jus ad bellum. In a few words, this condition presupposes that a state can justify war only when it has the right reasons to do so (Purves et al., 2015, p. 862). They claim that right intention applies to soldiers at war and ought to apply to AWS. This is a stronger argument than the sociopathic soldier analogy. However, just as the

(M2) of the former argument is being criticised, one may wonder to what extent an AWS is merely a proxy for the commander's judgment and intention. This logical error might have been corrected in the reconstruction, even if it was accidental. Nonetheless, this supplementation would have posed the threat of undermining sustainability, because the ethical issue itself can be criticised as irrational. Therefore, this supplementation would not significantly affect the tenability of the argument.

Do the premises have any issues or problems? None of the premises has been singled out as inappropriate or problematic. Both (P1) and (P2) are based on robust theoretical frameworks: the Markowitz approach to morally worthy action and the desire-belief model (or the taking-as-a-reason model) (Purves et al., 2015, p. 861). This page offers a reconstruction of Sparrow's argument on Dignity. The discussion involving dignity, which may be credited to Sparrow, is thus re-stated as follows: (P1) Requirements of jus in bello (the laws regarding the behaviour in war), one of which is based upon the interpersonal relations between war participants. Cutting off these interpersonal relations would be a severe insult to the enemy. (P3) An AWS is necessarily unable to engage in an inter-personal relationship. We can also state that a human combatant cannot transfer his/her role to an AWS in this interpersonal relationship. Therefore, by implication of (P2) and (P3), an AWS also cannot be said to engage in an interpersonal relationship as it operates in the context of jus in bello. Thus, the application of an AWS during war is intensely offensive to an opponent.

Is the conclusion misguided? Its conclusion offers a solid deontological justification for criticising AWSs. It is explicitly deontological, which makes it non-mistaken in the context of the research environment under consideration in the current work. AWS has, in this particular argument, been defined as a type of weapon system that decides its own course of action, but remains distinct as an entire-fledged moral actor. Basically, it does not belong to either end of the conceptual spectrum: it is neither a proxy acting like a human being nor a moral agent on its own (Sparrow, 2016, p. 107). Are there shifts in meaning? There was no determination of any changes in meaning. Is the argument based on a valid schema? The argument's logic follows a lawful framework. Do any of the premises result in any inappropriateness or problem-position? Logically, the premises support the conclusion. Nonetheless, from a generally theoretical perspective, they are problematic. The premises (P1) and (P2) are based on the canonical arguments of Thomas Nagel about the moral justifications of the principles of jus in bello as proportionality and distinction (Skerker et al., 2020, p. 199). Although Nagel has a sound theoretical underpinning, objections to translating these ideas into a concrete dignity argument against AWS are easier to raise. Another major criticism of premises (P3) and (P4) is that the AWS must be considered simply as a means or tool used by a commander to achieve specific goals. Assuming that AWS is seen as a tool, the commander is still involved in the interpersonal relationship, thus depriving the force of (P3) and (P4), because a relationship between the commander and the target must exist. This criticism often argues that existing systems of autonomous operation, such as self-detonating mines or cruise missiles, show that such weapons are considered offensive. With this, Sparrow (2016, pp. 107-108) rebuts it by stating that AWS are independent of these systems, or, conversely, that the comparison between them highlights neglected ethical issues with current weapons. Finally, this objection highlights that there has always been a problem determining exactly when AWS is.

CONCLUSION

According to the deontological reasoning presented in this paper, the argument supporting the case for banning Autonomous Weapon Systems (AWS) is weak and involves both uncertainty and high in-haprobability. The arguments are considered uncertain due to sustainability issues, including the unstable theoretical relationship between ethical principles and a clear conclusion about the necessity of acting with the right intention. Additionally, the arguments can be categorised as irrelevant, since they are based on and, therefore, directed towards divergent conceptualisations of the meaning of AWS. The two essential gaps weaken the opposition's unity. Even though, in most cases, the arguments are primarily based on sound theoretical grounds, there are often sustainability problems in their applications. Other theories, like the desire-belief model and the anti-codifiability thesis, have been well supported by academic research. However, their application to the particular arguments against AWS has been found wanting. This result is multifaceted in its explanation: it may mean that the arguments were misinterpreted during the analysis or, more likely, that the authors applied these theories in inappropriate contexts. These analytical friction points are systematically analysed in this study as sustainability problems.

The deontological argument analysis also demonstrated that the most significant barrier to the argumentation is the use of conflicting definitions of AWS (Autonomous Combat Platforms). The tendency towards deviating definitions (between narrow and broad definitions) poses a major relevance problem. This problem holds that effective problematizing of AWS depends on the definition implicit or explicitly embraced in an argument. This means that arguments based on constructions of certain, limited definitions cannot be relevant to a broader range of systems. For example, the use of the term "modern cruise missiles" shows that, in most instances, these deontological arguments cannot be used to prove ethical issues with existing weapon systems, since the systems are usually beyond definition in the arguments analysing the topic. Together, that is, the relevance problem and the sustainability problems suggest that the deontological argument against AWS is not very strong. These failings directly confront Sparrow, who holds that AWS must be prohibited, especially when deontology is emphasised over consequentialism. The capacity to situate AWS within a broader moral perspective is, therefore, limited, as consequential arguments on both sides go unanalyzed. However, concerning the shown flaws in the deontological opposition, it is hardly possible that such arguments will be convincing without the reinforcement by consequentialist logic or other deontological constructions.

The four analysis questions developed by Hansson were beneficial for critiquing arguments. It should, however, be noted that these questions are not structured to highlight strengths but instead to address weaknesses. This constraint is also increased given the inability to measure ethics. In the analysis, it was clear that some of the premises depended on normative ethical theories as strong; these strong aspects were identified in the final question of the analysis merely because they did not find another more appropriate place. This is one of the validity issues to consider when interpreting the results. More so, systematic reconstruction notwithstanding, analysis is a field of interpretation per se. The most significant choice was the translation of arguments into Swedish, which affects the reliability of the analysis and the validity of the results, potentially reinforcing or undermining the results. The ethical discussion has been marked by significant ambiguity regarding the meaning of AWS. This will make the existing normative arguments more or less relevant, depending on the definition of what is and is not an AWS, which, in turn, will influence the development of future arguments. This presents an urgent requirement for the conceptual investigation. The relevance problem states that more specifications are required; in particular, it can be proposed to differentiate between weapon systems with artificial intelligence (a narrower category) and those with autonomous functions (which would help clarify matters). These categories would facilitate more specific and better arguments in each and help comprehend the general picture of ethical consequences.

Moreover, one impression from this research is that the ethical argument for AWS is skewed. The supporters and critics appear to be working with the premise that AWS is an issue that needs to be either criticised or justified. Although it is a hypothesis that cannot yet be validated, it may suggest one reason why the discussion seems dominated by opponents of AWS; doubting this premise may take the discussion in a different direction. This implies the need to perform a framing analysis of the debate or a quantitative hypothesis test using this observation as a baseline. One of the main conclusions of this paper is that the development of AWS and its application is a problem that can be viewed from a deontological perspective, which is evidently troubled by relevance issues. In the case of military professional practice, this would mean that the numerous weapon systems with autonomous functions now in operation will effectively keep such ethical arguments at bay. The use of weapon systems with simpler autonomous capabilities would involve minimal or no risk of implicating the ethical issues noted by the main deontological arguments examined in this paper. Another consideration is that the ethical debate is not entirely resolved. The argument by Sparrow, as noted in the research review, that critics of AWS were seen to triumph if deontology were considered superior to consequentialism, no longer seems to have much merit. Through these findings, it will not be possible to say that one side of the conflict is winning in the immediate future. This suggests a more profound uncertainty about the specification of future weapon systems, which will likely shape future predictions in the military field.

REFERENCES

Amoroso, D., & Tamburrini, G. (2018). The ethical and legal case against autonomy in weapons systems. *Global Jurist, 18*(1), 1–20. DOI: 10.1515/gj-2017-0012

Arkin, R. C. (2008). On the ethical quandaries of a practicing roboticist: A first-hand look. *Current Issues in Computing and Philosophy, 175*, 45–51.

Arkin, R. C. (2010). The case for ethical autonomy in unmanned systems. *Journal of Military Ethics, 9*(4), 332–341. DOI: 10.1080/15027570.2010.536402

Hansson, S. O. (2006). Safe design. *Techné (Blacksburg, Va.), 10*(1), 45–52. DOI: 10.5840/techne200610140

Hansson, S. O. (2020). Technology and mathematics. *Philosophy & Technology, 33*(1), 117–139. DOI: 10.1007/s13347-019-00348-9

Horowitz, M. C. (2016). The ethics & morality of robotic warfare: Assessing the debate over autonomous weapons. *Daedalus, 145*(4), 25–36. DOI: 10.1162/DAED_a_00409

Human Rights Watch. (2022). *Killer robots*. https://www.hrw.org/topic/arms/killer-robots

Johansson, L. (2018). Ethical aspects of military maritime and aerial autonomous systems. *Journal of Military Ethics, 17*(2-3), 140–155. DOI: 10.1080/15027570.2018.1552512

Karadsheh, J., & Sariyuce, I. (2022, April 11). *Turkish drones have become a symbol of the Ukrainian resistance*. CNN. https://edition.cnn.com/2022/04/11/middleeast/mideast-summary-04-11-2022-intl/index.html

Pietrzak, P. (2022). *Introducing the idea of ontology in statu nascendi to the broader international relations theory. International Conference Proceeding Series: International Conference on Economics and Social Sciences*. ECLSS., https://www.eclss.org/publicationsfordoi/abst11act8boo8kIE%26SS2022_antalya.pdf

Pietrzak, P. (2023). Approaching regional conflicts through the prism of ontology in statu nascendi: The new compartmentalization of the IR theory. *In Statu Nascendi – Journal of Political Philosophy and International Relations, 6*(1).

Pietrzak, P. (2024). The sixth great debate in international relations theory revolves around clarity: Ontology in statu nascendi as a cutting-edge tool for conflict management. In Pietrzak, P. (Ed.), *Dealing with regional conflicts of global importance* (pp. 1–27). IGI Global., DOI: 10.4018/978-1-6684-9467-7.ch001

Purves, D., Jenkins, R., & Strawser, B. J. (2015). Autonomous machines, moral judgment, and acting for the right reasons. *Ethical Theory and Moral Practice*, *18*(4), 851–872. DOI: 10.1007/s10677-015-9563-y

Robillard, M. (2018). No such thing as killer robots. *Journal of Applied Philosophy*, *35*(4), 705–717. DOI: 10.1111/japp.12274

Scharre, P., & Horowitz, M. C. (2015). *Autonomy in weapon systems*. Center for a New American Security. https://www.cnas.org/publications/reports/an-introduction-to-autonomy-in-weapon-systems

Sharkey, A. (2019). Autonomous weapons systems, killer robots and human dignity. *Ethics and Information Technology*, *21*(2), 75–87. DOI: 10.1007/s10676-018-9494-0

Shead, S. (2021, December 22). *UN talks to ban 'slaughterbots' collapsed*. CNBC. https://www.cnbc.com/2021/12/22/un-talks-to-ban-slaughterbots-collapsed-heres-why-that-matters.html

Skerker, M., Purves, D., & Jenkins, R. (2020). Autonomous weapons systems and the moral equality of combatants. *Ethics and Information Technology*, *22*(3), 197–209. DOI: 10.1007/s10676-020-09528-0

Sparrow, R. (2007). Killer robots. *Journal of Applied Philosophy*, *24*(1), 62–77. DOI: 10.1111/j.1468-5930.2007.00346.x

Sparrow, R. (2016). Robots and respect: Assessing the case against autonomous weapon systems. *Ethics & International Affairs*, *30*(1), 93–116. DOI: 10.1017/S0892679415000647

Stop Killer Robots. (2022). *Stop killer robots*. https://www.stopkillerrobots.org/stop-killer-robots/

ADDITIONAL READINGS

Asaro, P. M. (2012). On banning autonomous weapon systems: Human rights, automation, and the necessity of accountability. *International Review of the Red Cross*, *94*(886), 687–709. DOI: 10.1017/S1816383112000768

Champagne, M., & Tonkens, R. (2015). Bridging the responsibility gap in automated warfare. *Philosophy & Technology*, *28*(1), 125–137. DOI: 10.1007/s13347-013-0138-3

Liu, H.-Y. (2016). Refining responsibility: Differentiating two types of responsibility issues raised by autonomous weapons systems. In Bhuta, N., Beck, S., Geiss, R., Liu, H.-Y., & Kress, C. (Eds.), *Autonomous weapons systems: Law, ethics, policy* (pp. 325–344). Cambridge University Press.

Müller, V. C., & Simpson, T. W. (2016). Autonomous killer robots are probably good news. In Di Nucci, E., & Santoni de Sio, F. (Eds.), *Drones and responsibility: Legal, philosophical and socio-technical perspectives on remotely controlled weapons* (pp. 67–81). Routledge. DOI: 10.4324/9781315578187-4

Roff, H. M., & Moyes, R. (2016). *Meaningful human control, artificial intelligence and autonomous weapons*. Article 36.

Santoni de Sio, F., & van den Hoven, J. (2018). Meaningful human control as reason-responsiveness: The case of dual-mode vehicles. *Ethics and Information Technology*, *20*(3), 175–190.

Schwarz, E. (2016). Prescription drones: On the techno-biopolitical regimes of contemporary 'ethical killing'. *Security Dialogue*, *47*(1), 59–75. DOI: 10.1177/0967010615601388

KEY TERMS AND DEFINITIONS

Anti-Codifiability Thesis: The philosophical thesis that the moral judgments of humans are based on intuition and a more complicated context, and therefore cannot be completely encoded as a program in a set of rules.

Autonomous Weapon Systems (AWS): The weapon systems that, after activation, are able to pick and attack the target without any additional action by the human operator.

Consequentialism Ethical Theory: According to which the morality of an action is based only on its consequences, including the effectiveness of a strike or the reduction of overall casualties.

Deontology: A theory of ethics that determines the morality of an action by its compliance with rules and obligations (the goodness in itself) and not the effects of the action in question.

Jus in Bello: The laws of war which govern the behaviour in the conflict, namely, the obligation to comply with such principles as distinction (only fighting targets) and proportionality.

Meaningful Human Control: A theoretical standard that mandates that human beings shall retain the substantive agency and critical decision-making with respect to the precise use of lethal force.

Responsibility Gap: The ethical issue is that it is conceptually impossible to impose moral or legal responsibility on any human agent (commander or programmer) for an autonomous machine's actions.

Chapter 11
How Can AI Help in Battlefield Healthcare?

Apoorav Sharma
http://orcid.org/0009-0002-9966-7691
Lovely Professional University, India

Lovleen Marwaha
Lovely Professional University, India

Richa Singh
http://orcid.org/0009-0007-0196-4919
Lovely Professional University, India

ABSTRACT

Artificial Intelligence (AI) is quickly reshaping the healthcare on battlefield by making the use of AI better in decision-making, in bettering the management of casualties, and making healthcare staff in the high-stress, resource-constrained settings. The rapidity and precision with which medical response is provided is important in saving lives in war in modern-time. AI technologies, including predictive analytics, autonomous systems, and decision-support tools, make triage, diagnosis and treatment to be faster even in extreme conditions. As an example, AI algorithms embedded into wearable sensors may continually test the vital signs of soldiers and identify the initial signs of trauma or internal bleeding, as well as automatically notify medics or command centers. Machine learning algorithms also have the ability of analyzing large volumes of data on the battlefield in order to forecast injury patterns, evacuation routes and distribute medical resources effectively.

DOI: 10.4018/979-8-3373-6264-9.ch011

1. INTRODUCTION

One of the most problematic areas of the military activity has always been battlefield healthcare. Severe injuries, unpredictable working conditions, and lack of medical facilities are just some of the situations that soldiers are exposed to, sometimes miles away of well-equipped hospitals. In this case, diagnosis, triage, and treatment could prove to be life-saving issues in a situation like this (Kellermann *et al.,* 2023). Conventional battlefield medicine has been based on the human experience and the manual coordination which might be limited by fatigue, stress and the uncontrollable environment of a battlefield. Artificial Intelligence (AI) is the latest technology that has become one of the game changers that can improve the speed, accuracy, and efficiency of medical services in such challenging circumstances over the last few years. AI is the term which can be used to refer to the computer systems which are capable of imitating the human intelligence abilities namely to learn on the basis of the data, to identify the patterns, and make autonomous choices (Chatterjee, 2020). The technologies of AI, when implemented into the battlefield healthcare, can help medical workers do complex operations more accurately and faster. As an example, the AI-based algorithm can process real-time physiological data on wearable sensors to identify symptoms of trauma, dehydration, or internal bleeding. Such real-time identification enables the medics to focus on the treatment and provide care even before the symptoms become visible. Likewise, the predictive analytics can evaluate the pattern of the injuries throughout the battlefield and be used to distribute resources so that the medical supplies and evacuations are where they are most required (Shajari *et al.,* 2023). In addition to diagnosis and monitoring, AI is also used in the operational logistics and remote medical support. Unmanned aerial vehicles could be used to transport first aid kits or blood to isolated or unsafe locations, and robotic agents could be used to carry out more delicate tasks, which would be overseen by a remote surgeon using telemedicine teleconferencing devices. Such apps do not only increase the area of medical support but also reduce the risk to the health care staff. Nonetheless, there are other challenges in the implementation of AI in medicine on the battlefield. The problem of reliability of the data, the security of the network, and even the ethical responsibility should be handled with care to provide a safe and effective use of it (Meyer *et al.,* 2019). Even with these anxieties, the application of AI in combat medicine systems is very promising. It is also a transformation of reactive care to proactive care- in which information-based insights and intelligent automation assist in saving lives in a faster, more accurate, and more coordinated way than at any previous time. With the ongoing development of military medicine, AI will be among the key factors to determine the future of battlefield medicine, integrating human experience with technical intelligence to make unprecedented progress in survivorship and recovery rates.

2. EVOLUTION OF BATTLEFIELD MEDICINE

The practice of battlefield medicine has come to be advanced through technology after starting as a primitive first aid. The decades of innovation behind this evolution include the need to decrease mortality, enhance the accuracy of triage, and improve the medical logistics in extreme conditions. Artificial Intelligence (AI) is the most recent and most revolutionary development in combat medical systems today.

2.1 Historical Development of Combat Healthcare

The previous medical practice on the battlefield can be traced to the early organized work in the Napoleonic wars, where the first organized evacuation systems were created. Military medicine developed as a result of the military conflicts with the greatest achievements in World Wars I and II, such as antisepsis, antibiotic usage, and methods of blood transfusion. These inventions brought a new revolution in the care of wounds and prevention of infections with great changes in survival rates. Mobile surgical hospitals and quick evacuation methods like the implementation of rapid response to trauma known as the gold hour were introduced by the Vietnam and Gulf Wars. As (Biselli *et al.,* 2022) add, these developments provided the basis of contemporary battlefield healthcare with the focus on mobility, triage, and multidisciplinary cooperation. The last few decades have seen the paradigm shift in reactive and preventive and predictive strategies. Digital communication, telemedicine, and wearable sensors can help monitor in real-time and transfer medical data even in combat zones (Kalid *et al.,* 2018). This advance in technology gave the basis to AI-driven software that could automate the process of data analysis and clinical support. Therefore, the historical trend of battlefield medicine is associated with a progressive search of more rapid, intelligent, and efficient care-crowning with the modern age of the intelligent and data-enabled systems of combat healthcare.

2.2 Challenges in Traditional Battlefield Medicine

The battlefield presents special difficulties to traditional battlefield medicine because war is a chaotic event. There is the lack of access to medical equipment, slow evacuation, unpredictable type of injuries, and harsh environmental conditions that limit the possibilities to provide a timely and effective care. There is also the factor of human fatigue and stress that undermines the accuracy of judgments in high pressure situations. Research has emphasized that first-time triage and immediate stabilization are critical factors affecting survival, and they are not always possible with the use of manual assessments only (Masbi *et al.,* 2024). In addition, the use of verbal communication and manual recordkeeping has long been slowing the process of medical coordination, and also caused errors in prioritization of the casualties. In relation to (Jafar *et al.,* 2015), lack of resources and the uneven distribution of skills among the frontline medics further increase treatment delays. Loss of communication within the perimeter of a combat area complicates communication of vital medical information to the command stations. Such restrictions emphasize the criticality of automated and data-driven solutions that can help to increase the speed and reliability. Targeted at eliminating these obstacles, Artificial Intelligence offers objective and real-time decision support by means of predictive analytics, automated triage, and continuous monitoring, directly overcoming the historical obstacles to providing battlefield medicine in a traditional way.

2.3 Transition from Manual to Technology-Assisted Systems

The transition of manual care processes to ones that are aided by the use of technology is a definite epoch in the history of military healthcare. Battlefield medicine is being automated and operated on a precision level with the introduction of AI, robotics, and digital health platforms. This was preceded by early uses of telemedicine centers and handheld diagnostics, which enhanced the functioning of the field and the level of diagnostics. The synthesis of many data streams (vital signs, imaging, and environmental sensors) is now rapid, which is made possible by AI. The machine learning algorithms may be used to predict complications, optimize the allocation of resources, and aid in the process of triage (Blackbourne *et al.,* 2012). The transition likewise improves medical logistics by use of autonomous drones and robotics that move supplies and perform a remote evaluation. In addition, real time patient monitoring and transmission of physiological data to the medical command centers is now possible due to integrated battlefield networks. Such features allow the coordination of response and reduce the time spent on diagnosis and evacuation. Intelligent-assisted systems therefore make battlefield medicine a case of responsive action as opposed to a proactive action, and this greatly increases the chance of survival rates and efficiency of operations. Overall, this technological shift means a new era in war-based healthcare when an intelligent automation and human skills are combined to provide quick, safer, and more flexible medical care in war zones (Figure 1).

Figure 1. Evaluation of battlefield medicine

3. FUNDAMENTALS OF ARTIFICIAL INTELLIGENCE IN HEALTHCARE

Artificial Intelligence (AI) is transforming the healthcare sector by incorporating the ability to find solutions to intricate healthcare data, aid in the diagnosis of conditions, and improve clinical decision-making. AI can be used in the military healthcare industry to predict, optimize logistics, and enhance battlefield preparedness by automation, analysis in real time, and integration with such modern technologies as robotics and telemedicine.

3.1 Concept and Components of AI

Artificial Intelligence is a collection of computer systems that are able to mimic the cognitive processes of humans like learning, reasoning, and decision-making. Within the medical field, AI is usually divided into various technologies, such as machine learning (ML), natural language processing (NLP), or computer vision, all of which help to analyze medical data and optimize the functioning of the facility automatically. Machine learning models are trained on big datasets to detect patterns to enhance the accuracy of the diagnosis and treatment planning. NLP helps in the analysis of clinical reports and system-to-system communication, whereas computer vision can be used to interpret the images automatically in radiology and trauma assessment. These components of AI have the ability to operate synergistically to provide scalable, consistent, and rapid analysis of medical fields (Swanson *et al.*, 2023). These parts are incorporated into portable medical machines, drones, and wearable sensors in the battlefield scenario to enable real-time tracking and distance diagnosis. Between them, it is possible to mention AI-based image recognition systems that can detect internal bleeding or fractures within a few seconds using portable imaging equipment. A combination of all these factors creates the foundation of AI in combat medical care - speed, accuracy, and scalability even in disorganized conditions. Being modular, AI can be continuously learned so that the medical systems on the battlefield are constantly upgraded and updated with new data and operational requirements.

3.2 Machine Learning and Data-Driven Decision Making

Machine learning (ML), a branch of AI, entails data-driven algorithms that are not programmed. ML models are used in healthcare to analyze patient data trends and predict outcomes, prescribe treatment and identify abnormalities. These systems are incredibly useful on the battlefield (triage, prediction of injuries, and management of casualties). As (Fatima, 2024) explains, ML makes it possible to have high-performance medicine, due to the ability to integrate data of different types, including vital signs, environmental sensors, and imaging, to guide real-time decisions. Injury or disease state classification Supervised learning techniques are employed as well as unsupervised learning may be used to identify unfamiliar correlations in combat medical data. The biosensor data of soldiers in the military can be analyzed to detect dehydration, fatigue, or risk of trauma before the symptoms are manifested. ML can also be used to aid logistics by predicting the medical supplies or finding the best evacuation pathways in times of crisis. ML enhances the battlefield healthcare through ongoing data fusion whereby the reactive care is converted into proactive and predictive medical support. Such an empirical method improves survivability, minimizes human error, and precision-related decision-making in adverse operational conditions.

3.3 AI Applications in Civilian and Military Healthcare

The role of AI in civilian healthcare gives it a solid base in terms of its military use. AI helps in the diagnosis of diseases, optimal treatment, and hospital management in hospitals. As an example, deep learning neural networks have been demonstrated to be more efficient in the detection of diabetic retinopathy, skin cancer, as well as COVID-19 based on medical images. These are customized in military environments. Python-based triage systems are deployed to scan the vitals of the soldiers and injuries on the battlefield in real time and allocate medical treatment based on its severity and resources (Das *et al.,* 2022). On the same note, robotic surgical systems and autonomous drones with an AI can help in transportation of supplies or even carry out remote procedures in areas that are difficult to reach. The two-fold aspect of AI of civilian and military medicine is synergizing, therefore, increasing the speed of innovation. Military needs can be leveraged to produce more resilient and autonomous technologies that can subsequently be ruggedized to serve the battlefield, but civilian needs have been met by proven technologies in civilian hospitals. In this way, AI can provide a comprehensive solution, enabling a common ecosystem in which new developments in one sphere directly improve the other one- ultimately improving the healthcare delivery, efficiency, and safety in every setting (Table1).

Table 1. Layer-wise AI computational architecture showing sensing, edge processing, predictive inference, robotic actuation, cloud/command integration, and ethical–security compliance mechanisms

AI Layer	Input Source	Processing Unit	Core Function	Example Outcome	Constraints
Sensing Layer	Wearables, field sensors	On-device microcontrollers	Vital capture, anomaly flagging	Blood-loss alarm, heat-stress signal	Noise, hardware failure
Edge AI Layer	Vitals, portable imaging	Battlefield edge nodes	Fast ML inference, triage pre-scores	Priority tagging, complication risk	No external cloud reliance
Processing AI Layer	X-ray/USG images, maps	GPU/TPU rugged modules	CV-based trauma detection, AI routing	Fracture/bleed detection, best medevac path	Compute load, power draw
Actuation AI + Robotics	ML decisions	Autonomous bots/UAVs	Supply drop, casualty transport, tele-surgery	Kit delivery, remote surgical assist	Risk-zone navigation

continued on following page

Table 1. Continued

AI Layer	Input Source	Processing Unit	Core Function	Example Outcome	Constraints
Command/Cloud AI Layer	Aggregated mission data	AI-governed command centers	Population-level casualty prediction, resource dispatch	30% faster medevac allocation	Encryption demand
Ethical-Security AI	Medical + operational ML logs	XAI + blockchain nodes	Liability trace, bias audit, secure authentication	Transparent reasoning	Human oversight required

4. AI APPLICATIONS IN BATTLEFIELD HEALTHCARE

Artificial intelligence technologies are changing healthcare on battlefields by improving its triage, diagnosis, and logistics, as well as handling casualties. AI helps to make faster decisions, better survival rates, and use the limited resources effectively in battle scenarios through automation, data integration, and predictive modeling, which make it the foundation of next-generation military medicine.

4.1 AI-Based Injury Detection and Real-Time Monitoring

The use of AI-powered systems is very important in the process of identifying injuries and constantly checking the health of the soldiers in the battlefield. Combining wearable biosensors with AI algorithms can interpret such physiological data as heart rate, oxygen saturation, and blood loss to be able to identify trauma or internal injuries before they can manifest themselves. As stated in (Heng, 2024), machine learning algorithms have been applied in the detection of abnormal physiological patterns in real time so that medics are notified immediately. AI models that are trained using data about battlefield injuries are capable of estimating the extent of the wounds, saving man time that would otherwise be spent on the evaluation procedure. One of the subdivisions of AI is computer vision, which can analyze images taken with portable diagnostic equipment or drones and identify fractures, burns, or hemorrhages (Morse *et al.,* 2020). These systems, combined with deep learning, assist in diagnosing in complicated or low-visibility situations. Simply, AI will convert medical surveillance into an active detection tool. Repeated, robotic analysis as a means of early intervention, but also a basis of data on large-scale predictive models, can improve the health outcomes of the population at the individual, as well as the unit, level in active combat zones (Figure2).

Figure 2. AI-enabled battlefield healthcare decision loop demonstrating continuous physiological monitoring through wearables, rapid trauma recognition via computer-vision imaging, multilingual data interpretation using NLP, and machine-learning-assisted triage, operating within a human-in-the-loop command architecture for accelerated frontline medical intervention

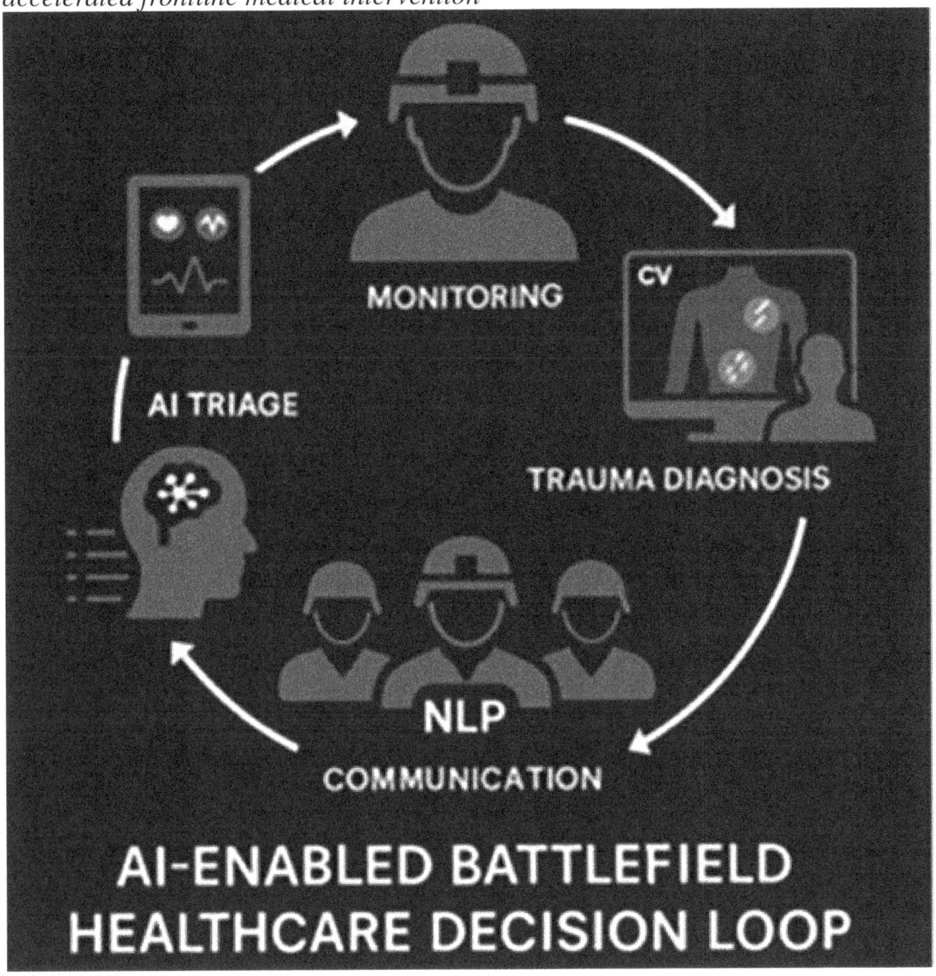

4.2 Intelligent Triage and Decision Support Systems

Triage in combat zones is traditionally based on human judgment in stress, and thus it can be inaccurate or slow. The AI-based decision-support systems (DSS) assist the medics to focus on care according to the severity of injuries, resources available and tactical circumstances. Research indicates that AI-based triage algorithms have the potential to be better than manual ones, processing multisource data, such as vitals, imaging, and environmental data. Such systems produce practical recommendations, which mean that medical staff can devote their time to treatment and not to data analysis. NLP technologies also have the benefit of supporting decisions due to their ability to summarize medical histories and combine verbal reports of various caregivers (Mahmoudi *et al.,* 2022). AI triage tools can coordinate information with teams when linked to centralized command systems to provide uniformity and less redundancy in care delivery. This smart assistance allows battlefield medics to take more accurate actions, minimizes the errors in the classification of victims, and increases the survival chances, particularly in a mass-casualty situation when every second matters.

4.3 Predictive Analytics for Casualty Management

One of the strongest uses of AI in the healthcare of the battlefield is predictive analytics. It also applies big data to predict trends on injuries, resources, and patient outcomes and make proactive decisions instead of responding to them. Using information on previous missions and current battlefield reports, AI models can forecast the pattern of the casualty count and furnish the medical teams with probable injuries. The information is useful in prepositioning of medical supplies and eviction plans to minimize delays in medical care. Algorithms that are built on AI have the capability to predict the chances of survival based on the type of injuries, physiological values, and the surrounding environment (Hunter *et al.,* 2023). This knowledge will enable the medics to concentrate scarce resources on patients that have the best chance of survival- an important aspect in high stakes battlefield situations. Predictive modeling is therefore a paradigm shift in medicine where reactive medicine is replaced by anticipatory care. Through turning raw data into foresight, AI can enable the military healthcare systems to react more quickly, more intelligently, and more efficiently.

4.4 AI in Medical Logistics and Evacuation Planning

The logistics of battlefield healthcare are also transformed by AI since evacuation routes, the distribution of resources, and managing supply chains are optimized. Based on real-time sensor, drone, and satellite feed data, AI systems dynamically evaluate the situation on a battlefield to design a safe and efficient evacuation route. The logistics systems developed by AI can shorten the evacuation time by more than 30 percent (Dehghan & Khosravian, 2024) by estimating the most efficient evacuation routes and modifying them independently. Moreover, AI-powered vehicles, drones may transport blood, medicine, or medical kits to inaccessible or remote locations.

AI logistics systems combine predictive model with real-time intelligence so that the resources are constantly adjusted to the battlefield conditions. It minimizes wastage, enhances responsiveness and helps maintain long-term medical preparedness during long-lasting operations. Comprehensively, AI-based logistics make sure that the necessary care, materials, and people are at the right place and at the right time to improve efficiency and survivability during combat medical operations (Table 2).

Table 2. Major applications of AI in battlefield healthcare

AI Technology	Application Area	Function in Battlefield Healthcare	Key Advantages
Machine Learning (ML)	Predictive Analytics & Triage	Predicts casualty risk, automates injury prioritization, and optimizes treatment sequencing	Improves survival rates, enhances decision speed
Computer Vision	Medical Imaging & Trauma Analysis	Detects fractures, bleeding, and wounds from portable imaging devices	Provides instant, accurate diagnostics
Natural Language Processing (NLP)	Communication & Data Management	Translates multilingual commands, interprets clinical notes, and improves telemedicine clarity	Reduces communication errors, accelerates coordination
AI-driven Robotics	Remote Surgery & Casualty Evacuation	Assists in surgical interventions, transport of wounded, and supply delivery in unsafe areas	Minimizes risk to medics, ensures precision care
AI + Wearable Sensors	Continuous Soldier Monitoring	Tracks vitals and detects early signs of trauma or fatigue	Enables proactive, real-time medical response
AI in Telemedicine	Remote Consultation & Diagnostics	Supports front-line medics with expert advice through real-time data interpretation	Enhances care quality in low-resource combat settings

5. INTEGRATION OF AI WITH ADVANCED TECHNOLOGIES

The real use of AI in battlefield healthcare is seen when it is combined with other technologies like robotics, telemedicine and wearable devices. Jointly, these systems make possible independent decision-making, remote diagnostics, and constant monitoring of soldiers- provide medical preparedness and accuracy even in extreme combat scenarios.

5.1 AI and Robotics in Battlefield Medicine

The field of robotics and AI is transforming the healthcare services on the battlefield through providing precision surgery, automated supply-delivery, and care of patients remotely in hazardous areas. AI is an addition to the robotic systems that facilitates real-time image analysis, adaptive control, and autonomous decision-making which are less dependent on a direct human intervention. In (Habibi *et al.,* 2025), it is stated that surgical operations and trauma patients can be kept stable by AI-assisted robots under remote control, reducing the exposure of medical staff members to risk. The robots can traverse dangerous grounds; carry victims, and even more basic tasks such as stitching wounds or bleeding. Examples of this integration include sophisticated systems such as the Trauma Pod project developed by DARPA, which is an AI-controlled robotics system with teleoperation at the battlefield to provide surgical assistance (Minopoulos *et al.,* 2023). The AI will allow the robots to understand medical images, apply design methods, and react to the changing conditions of the patients in real-time. The interactions between AI and robotics therefore push surgical and logistical ability to a whole new level beyond what humans can possibly achieve. Autonomy along with speed and precision allow AI-controlled robots to save lives even in situations that can not have human involvement due to the impossibility or delay of appearance (Figure3).

Figure 3. Interconnected autonomous battlefield medical support network integrating AI-driven real-time medevac routing optimization, UAV-enabled medical cargo deployment, secure AI-assisted telemedicine consultation, and robotic remote surgical modules, collectively forming a resilient digital-trauma care and logistics ecosystem designed for rapid casualty stabilization and controlled remote intervention

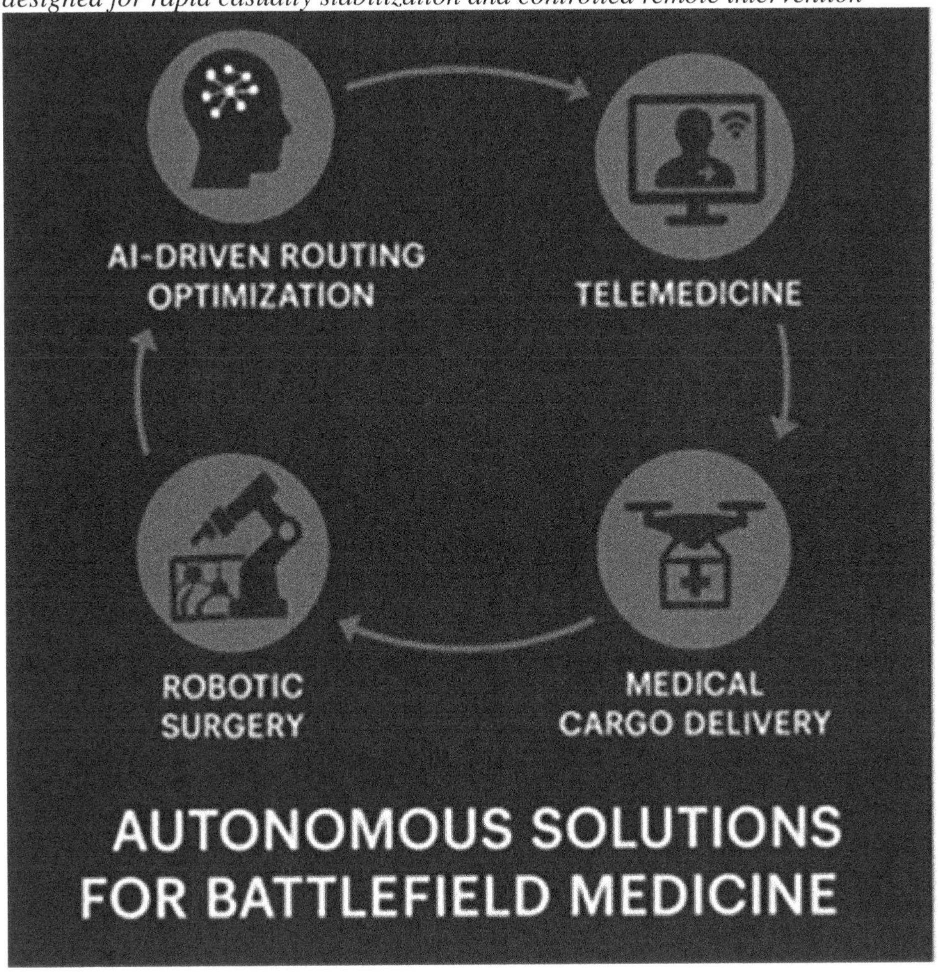

5.2 AI and Telemedicine for Remote Medical Assistance

AI-improved telemedicine will help reduce the distance between combat troops and distant specialists. Latency in communication, poor access to data, and fluctuating networks are some of the key areas of concern in battlefield healthcare- AI reduces these by compressing data, improving image quality, and automating the process of diagnosis. According to (Nazir *et al.,* 2025), AI-informed telemedicine platforms can be used to analyze the medical images passed through them, anticipate the state of patients, and propose treatment plans without engaging a specialist immediately. These systems enable an expansion and sustainability of battlefield telemedicine even under resource constraint circumstances. Non-medical staff or combat medics can also receive help in giving care under the guidance of experts with the help of AI chatbots and decision-support platforms (Sarkar *et al.,* 2024). Meanwhile, natural language processing (NLP) supports real-time translation of multi-language messages when conducting multinational military operations to enhance the comprehension and increase the reaction time. Telemedicine through AI therefore ensures continuity of care; complex medical decisions may be made remotely with high accuracy and minimal delay - which is important in the dynamic and uncertain setting of battlefield operations.

5.3 AI and Wearable Sensors for Continuous Monitoring

AI-powered wearable biosensors are transforming the real-time monitoring of the health of soldiers. These sensors constantly gather physiological information, including heart rate, blood pressure, hydration rates, etc., and send it to AI systems that process the deviation of the usual dynamics to identify fatigue, trauma, or illness at an early stage. As (Youn *et al.,* 2024) explain, physically exhausted or heat-stressed soldiers can be detected before facing the manifestation of the symptoms, and commanders can save lives by implementing the preventative measures of rest or hydration by using AI-integrated wearable devices. Deep learning algorithms capture micro-patterns in the biosignals that would not be detected using human monitoring. Such devices are capable of sending data to the command centers in combat areas to utilize it in remote triage and medical prioritization (Nazemi *et al.,* 2022). There are even more sophisticated prototypes with drug-delivery microchips, which automatically discharge the drugs when AI makes an indication of physiological distress. AI combined with wearable technology is a new dawn in the healthcare of battlefields- it will no longer be reactive care but anticipatory care. Persistent AI-aided situational awareness, soldier health maintenance, and mission success through decreasing the number of preventable injuries and medical crises all rely on situational awareness provided by continuous monitoring (Figure 4).

Figure 4. AI integration in battlefield healthcare

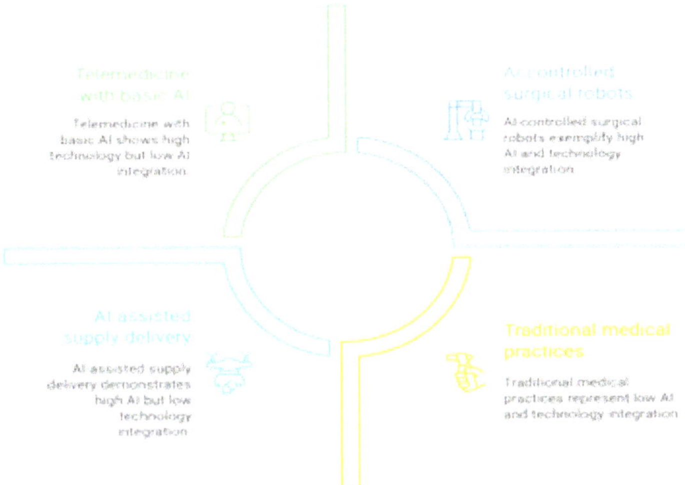

6. ETHICAL AND SECURITY CHALLENGES OF AI IN BATTLEFIELD HEALTHCARE

The introduction of AI into the healthcare of the battlefield results in the ethical, legal, and cybersecurity issues which transcend conventional medicine. These include issues of privacy, bias in algorithms, responsibility and protection of data under unhealthy conditions. In order to ensure that AI systems evolve and do not pose detrimental effects to human well-being and integrity of the operations, these issues need to be tackled.

6.1 Data Privacy and Security Risks

It relies on mass data of sensitive data of soldiers (biometric, communication networks, medical devices) to observe them in the battlefield and employ AI systems to this addiction subjects them to critical privacy and security threats. Hacks would expose the health, positions, or strategic data of troops, jeopardizing the security of both patients and missions. As (Kaviani *et al.*, 2022) point out, AI-based healthcare platforms are susceptible to adversarial attacks, small manipulations of data result in incorrect diagnosis or a misclassification, which can be deadly in battlefields. These systems should hence incorporate high levels of encryption, detection and isolation of anomalies in the network to guarantee data integrity. Additionally, remote triage and medical advice are typically provided on the battlefield by using cloud-based and edge-based AI networks. Though they are efficient, these architectures introduce several points of vulnerabilities. According to (Amin & Hossain, 2020), data transmission protocols and real-time authentication should be used to reduce the possibility of spoofing and unauthorized access to data. To reduce the risk of these threats, scientists recommend a transition to blockchain-based medical record systems and quantum-resistant encryption to guarantee the confidentiality of the data in the long term. The critical aspect of AI-assisted military medicine is that ethical data stewardship, along with robust cybersecurity design, are needed to ensure that the operational trust, as well as human safety in this domain.

6.2 Algorithmic Bias and Accountability in Decision-Making

Artificial intelligence systems are as trustworthy as the information and algorithms. The biased AI models in battlefield healthcare may result in unfair or inaccurate recommendation of triage, diagnosis, or treatment, which is a serious ethical issue. The study by revealed that AI bias is usually based on inadequate numbers of training data and this leads to unequal medical predictions between the various demographics. These biases may put soldiers in danger in military scenarios by giving preference to care that is inequitable and inappropriate interpretation of the physiological data of other ethnic and gender groups. The issue of accountability is further complicated by the fact that autonomous systems are used, and it would be difficult to determine who will be responsible in case of AI-related mistakes. According to (Zhang *et al.*, 2022), responsible AI deployment relies on the capacity of ethical governance structures, or transparency, human oversight, and traceability. One of the solutions is explainable AI (XAI), which is described as an ability to interpret and audit algorithmic decisions in real time. Healthcare providers would keep their operations and ethical responsibility through guaranteeing that AI-generated medical or tactical advice may be traced to its reasoning. Finally, to make AI-assisted battlefield care just and accountable, interdisciplinary partnership between engineers, clinicians, ethicists, and commanders is needed, that is, a human-in-the-loop approach should be established to ensure ethical control and trust of life-critical decisions.

6.3 Legal and Ethical Frameworks for Battlefield AI

The introduction of AI in military healthcare has gained quicker pace than the creation of sufficient legal and ethical frameworks of governance. The application of autonomous medical systems poses complicated issues regarding consent, liability, and adherence to the international humanitarian law. As (Defense Innovation Board,

2019) argues, the existing legal frameworks are incapable of allocating responsibility in the situations when AI behaves in ways that are unpredictable or autonomous. There are also ethical issues of the militarization of medical AI. Such technologies are dual-use and blur care and combat with a risk of a breach of the Geneva Conventions. The creation of AI ethics boards in the defense institutions can offer supervision, where the medical ethics and international law are followed. Models like the U.S. Department of Defense Ethical Principles of AI (2020) highlight five principles in responsibility, equitability, traceability, reliability, and governability and must apply to battlefield healthcare AI. Additionally, the principles of medical ethics like autonomy, beneficence, and nonmaleficence can be used to balance the military goals and the human dignity. International cooperation between technologists, ethicists, and legal professionals is essential in ensuring the future of AI in the battlefield healthcare. This will guarantee that the process of innovation is done in a responsible manner that does not jeopardize human welfare and humanitarian values even within the most complicated military settings (Table 3).

Table 3. Ethical and security challenges of AI in battlefield healthcare

Challenge Area	Description	Impact on Battlefield Healthcare	Proposed Mitigation Strategies
Data Privacy and Security	Exposure of medical and operational data through cyberattacks or unauthorized access	Risk of revealing troop health or positions; potential data manipulation	Encryption, blockchain systems, quantum-resistant security
Algorithmic Bias	Skewed predictions due to non-representative training data	Unequal triage or diagnosis accuracy among different populations	Diverse datasets, bias auditing, explainable AI
Accountability and Transparency	Difficulty identifying responsibility for AI-driven errors	Ambiguity in medical or command decision liability	Human-in-the-loop design, clear governance protocols
Legal and Ethical Ambiguity	Lack of specific laws governing AI in military medicine	Unclear standards for compliance and consent	Ethical AI frameworks, adherence to humanitarian law
Operational Reliability	AI malfunction or adversarial interference in combat zones	False readings or delayed medical response	Redundant systems, real-time auditing, secure hardware design

7. CONCLUSION

The concept of artificial intelligence (AI) has become a paradigm shifter in the healthcare on the battlefield. It improves the diagnosis, triage and treatment in war that requires speed and accuracy to survive. The AI based systems enhance efficiency of medical logistics and decision support and make it possible to make timely and accurate interventions in case of extreme pressure. Machine learning and predictive analytics can provide a toolkit to predict injuries, track the health of soldiers and distribute resources in a more efficient manner. Robotics using AI can perform remote surgeries and autonomous casualty evacuation and eliminate risks to medical staff. Artificial intelligence-based telemedicine can extend the advice of experts to the frontline medics despite communication restrictions in war-torn areas. The biosensor and AI algorithms that are installed on wearable allow round-the-clock health monitoring, including fatigue, dehydration, and internal injury. Such innovations transform the reactive medicine to proactive right at the battle line keeping combat capability intact and enhancing the recovery rates. Medical decision-making that is made using data decreases human error and increases the efficiency of operations. AI has brought ethical and security issues to the battlefield medicine, even though it promises. The problem of data privacy, algorithm bias, and responsibility still persists. Violation or discriminatory deliverables may endanger the personal rights and the security of missions. To deal with these risks, the system needs to be transparent, well encrypted, and have legal frameworks that are consistent with the international humanitarian standards. It is necessary to have interdisciplinary cooperation of engineers, physicians, ethicists, and military strategists. The use of AI in combat healthcare should not lose human supervision. This is aimed at making AI work as a complement and not a substitute to moral human decisions in life threatening scenarios. AI will further transform military medicine in the future with high levels of automation, accuracy analytics and autonomous medical solutions. Whether technology will be effective in supporting the dual goals of saving lives and upholding ethical integrity on the battlefield will depend on whether it is deployed responsibly.

REFERENCES

Amin, S. U., & Hossain, M. S. (2020). Edge intelligence and Internet of Things in healthcare: A survey. *IEEE Access : Practical Innovations, Open Solutions*, *9*, 45–59. DOI: 10.1109/ACCESS.2020.3045115

Biselli, R., Nisini, R., Lista, F., Autore, A., Lastilla, M., De Lorenzo, G., Peragallo, M. S., Stroffolini, T., & D'Amelio, R. (2022). A historical review of military medical strategies for fighting infectious diseases: From battlefields to global health. *Biomedicines*, *10*(8), 2050. DOI: 10.3390/biomedicines10082050 PMID: 36009598

Blackbourne, L. H., Baer, D. G., Eastridge, B. J., Renz, E. M., Chung, K. K., DuBose, J., Wenke, J. C., Cap, A. P., Biever, K. A., Mabry, R. L., Bailey, J., Maani, C. V., Bebarta, V., Rasmussen, T. E., Fang, R., Morrison, J., Midwinter, M. J., Cestero, R. F., & Holcomb, J. B. (2012). Military medical revolution: Deployed hospital and: en route: care. *The Journal of Trauma and Acute Care Surgery*, *73*(6), S378–S387. DOI: 10.1097/TA.0b013e3182754900 PMID: 23192059

Board, D. I. (2019). *AI principles: recommendations on the ethical use of artificial intelligence by the department of defense: supporting document*. United States Department of Defense.

Chatterjee, R. (2020). Fundamental concepts of artificial intelligence and its applications. *Journal of Mathematical Problems. Equations and Statistics*, *1*(2), 13–24.

Das, D., Biswas, S. K., & Bandyopadhyay, S. (2022). A critical review on diagnosis of diabetic retinopathy using machine learning and deep learning. *Multimedia Tools and Applications*, *81*(18), 25613–25655. DOI: 10.1007/s11042-022-12642-4 PMID: 35342328

Dehghan, M., & Khosravian, E. (2024). A review of cognitive UAVs: AI-driven situation awareness for enhanced operations. *AI and Tech in Behavioral and Social Sciences*, *2*(4), 54–65. DOI: 10.61838/kman.aitech.2.4.6

Fatima, S. (2024). Transforming healthcare with AI and machine learning: revolutionizing patient care through advanced analytics. *International Journal of Education and Science Research Review, 11*.

Habibi, S., Ivaki, N., & Barata, J. (2025). A Systematic Literature Review of Unmanned Aerial Vehicles for Healthcare and Emergency Services. *arXiv preprint arXiv:2504.08834*.

Heng, L. (2024). Strategic Overview of Applying Artificial Intelligence on the Future Battlefield.

Hunter, O. F., Perry, F., Salehi, M., Bandurski, H., Hubbard, A., Ball, C. G., & Morad Hameed, S. (2023). Science fiction or clinical reality: A review of the applications of artificial intelligence along the continuum of trauma care. *World Journal of Emergency Surgery; WJES*, *18*(1), 16. DOI: 10.1186/s13017-022-00469-1 PMID: 36879293

Jafar, A. J., Norton, I., Lecky, F., & Redmond, A. D. (2015). A literature review of medical record keeping by foreign medical teams in sudden onset disasters. *Prehospital and Disaster Medicine*, *30*(2), 216–222. DOI: 10.1017/S1049023X15000102 PMID: 25659602

Kalid, N., Zaidan, A. A., Zaidan, B. B., Salman, O. H., Hashim, M., & Muzammil, H. J. J. O. M. S. (2018). Based real time remote health monitoring systems: A review on patients prioritization and related" big data" using body sensors information and communication technology. *Journal of Medical Systems*, *42*(2), 30. DOI: 10.1007/s10916-017-0883-4 PMID: 29288419

Kaviani, S., Han, K. J., & Sohn, I. (2022). Adversarial attacks and defenses on AI in medical imaging informatics: A survey. *Expert Systems with Applications*, *198*, 116815. DOI: 10.1016/j.eswa.2022.116815

Kellermann, A. L., Kotwal, R. S., & Rasmussen, T. E. (2023). Military medicine's value to US health care and public health: Bringing battlefield lessons home. *JAMA Network Open*, *6*(9), e2335125–e2335125. DOI: 10.1001/jamanetworkopen.2023.35125 PMID: 37733341

Mahmoudi, E., Wu, W., Najarian, C., Aikens, J., Bynum, J., & Vydiswaran, V. V. (2022). Identifying caregiver availability using medical notes with rule-based natural language processing: Retrospective cohort study. *JMIR Aging*, *5*(3), e40241. DOI: 10.2196/40241 PMID: 35998328

Masbi, M., Tavakoli, N., & Dowlati, M. (2024). Challenges of providing of special care services in hospitals during emergencies and disasters: A scoping review. *BMC Emergency Medicine*, *24*(1), 238. DOI: 10.1186/s12873-024-01160-1 PMID: 39695959

Meyer, A., Cypko, M. A., Eickhoff, C., Falk, V., & Emmert, M. Y. (2019). Artificial intelligence-assisted care in medicine: A revolution or yet another blunt weapon? Potentials, challenges, and the future of implementing artificial intelligence (AI) for clinical care.

Minopoulos, G. M., Memos, V. A., Stergiou, K. D., Stergiou, C. L., & Psannis, K. E. (2023). A medical image visualization technique assisted with AI-based haptic feedback for robotic surgery and healthcare. *Applied Sciences (Basel, Switzerland), 13*(6), 3592. DOI: 10.3390/app13063592

Morse, S., Talty, K., Kuiper, P., Scioletti, M., Heymsfield, S. B., Atkinson, R. L., & Thomas, D. M. (2020). Machine learning prediction of combat basic training injury from 3D body shape images. *PLoS One*, *15*(6), e0235017. DOI: 10.1371/journal.pone.0235017 PMID: 32603356

Nazemi, M. M., Khodabandeh, A., & Hadjizadeh, A. (2022). Near-field electrospinning: Crucial parameters, challenges, and applications. *ACS Applied Bio Materials*, *5*(2), 394–412. DOI: 10.1021/acsabm.1c00944 PMID: 34995437

Nazir, A., Hussain, A., Singh, M., & Assad, A. (2025). Deep learning in medicine: Advancing healthcare with intelligent solutions and the future of holography imaging in early diagnosis. *Multimedia Tools and Applications*, *84*(17), 17677–17740. DOI: 10.1007/s11042-024-19694-8

Sarkar, S., Singh, Y. C., & Kaloiya, G. S. (2024). Psychotherapy and psychotropic drug treatment: Neurobiological and psychodynamic perspectives. *Indian Journal of Psychiatry*, *66*, S126.

Shajari, S., Kuruvinashetti, K., Komeili, A., & Sundararaj, U. (2023). The emergence of AI-based wearable sensors for digital health technology: A review. *Sensors (Basel)*, *23*(23), 9498. DOI: 10.3390/s23239498 PMID: 38067871

Swanson, K., Wu, E., Zhang, A., Alizadeh, A. A., & Zou, J. (2023). From patterns to patients: Advances in clinical machine learning for cancer diagnosis, prognosis, and treatment. *Cell*, *186*(8), 1772–1791. DOI: 10.1016/j.cell.2023.01.035 PMID: 36905928

Youn, S., Ki, M. R., Abdelhamid, M. A., & Pack, S. P. (2024). Biomimetic materials for skin tissue regeneration and electronic skin. *Biomimetics*, *9*(5), 278. DOI: 10.3390/biomimetics9050278 PMID: 38786488

Zhang, C. A., Cho, S., & Vasarhelyi, M. (2022). Explainable artificial intelligence (XAI) in auditing. *International Journal of Accounting Information Systems*, *46*, 100572. DOI: 10.1016/j.accinf.2022.100572

Chapter 12
AI-Driven Advances in Nuclear Medicine From Image Analysis to Personalized Treatment

R. Regan
University College of Engineering, Anna University, Villupuram, India

V. Nithya
http://orcid.org/0000-0002-4470-1509
University College of Engineering, Anna University, Villupuram, India

A. Thabeswar
University College of Engineering, Anna University, Villupuram, India

S. Yokeshvaran
http://orcid.org/0009-0001-3628-0839
University College of Engineering, Anna University, Villupuram, India

ABSTRACT

Artificial Intelligence (AI) is transforming nuclear medicine by boosting diagnostic precision, image analysis efficiency, and patient outcomes. AI-based techniques, especially deep learning and machine learning algorithms, have demonstrated maximum efficiency in oncology, neurology, and cardiology. Applications range from preclinical disease detection, tumor segmentation, and customized treatment planning. Constraints still exist in the form of data availability, regulatory clearance, and model interpretability. This paper is interested in AI applications in nuclear medicine imaging, model overview, discussion of datasets and evaluation metrics,

DOI: 10.4018/979-8-3373-6264-9.ch012

and summary of recent developments and challenges. Emerging research directions are explain ability, multimodal imaging integration, and ethics for enabling AI's large-scale adoption in clinics

INTRODUCTION

1. Overview of Nuclear Medicine Imaging

Nuclear medicine is a medical imaging subspecialty that is based on intravenous injection of small but well-defined quantities of radioactive tracers for imaging diagnosis and disease therapy. Unlike conventional imaging modalities, nuclear imaging provides functional data at the molecular level. PET and SPECT are widely used in oncology, neurology, and cardiology (Yang et al., 2020).

After being given to the patient, these radiopharmaceuticals may localize to particular organs or cellular receptors. Instead of depending on physical alterations in tissue anatomy, nuclear medicine may picture the degree of a disease process in the body based on cellular activity and physiology thanks to this feature of radiopharmaceuticals (Yahil et al., 2006).

Studies using nuclear medicine have the ability to detect medical issues earlier than traditional diagnostic methods in some disorders. In a way, nuclear medicine is "radiology done inside out" or "endo-radiology" since it records radiation that comes from inside the body as opposed to radiation produced by outside sources like X-rays. The use of radioactive materials in the diagnosis and treatment of illness is known as nuclear medicine and radiation therapy.

Nuclear medicine serves as a link between a specific clinical issue and an appropriate radionuclide test. It started off as a small technical instrument utilized in a few medical specialties, most notably nephrology and endocrinology. Nonetheless, it has now gained recognition as a distinct clinical field with unique training programs, specialized abilities, and a unique approach to patient care on a global scale. A solid medical education and a clinical approach to the topic are still crucial, even if practicing nuclear medicine requires a physician to acquire a lot of underlying science and technology Boukhennoufa et al. (2024). Utilizing the nuclear properties of radioactive nuclides, nuclear medicine is a branch of medicine that uses open radioactive sources to treat patients and make diagnostic assessments of their physiological and/or anatomical problems.

2. Importance of Image-Based Diagnosis

Artificial intelligence used in image analysis does away with the drawback of traditional diagnosis, thus reducing interobserver variability and enhancing diagnostic accuracy. AI can spot patterns that may be overlooked by radiologists, resulting in disease detection at an earlier stage and hence improved treatment results.

3. Role of AI in Enhancing Nuclear Medicine

AI improves nuclear imaging workflows by automating noise reduction, image reconstruction, and anomaly detection. Deep learning models such as CNNs enhance tumor detection and segmentation, enabling precise diagnostics (Bandla et al., 2023).

Figure 1. Working of nuclear medicine imaging

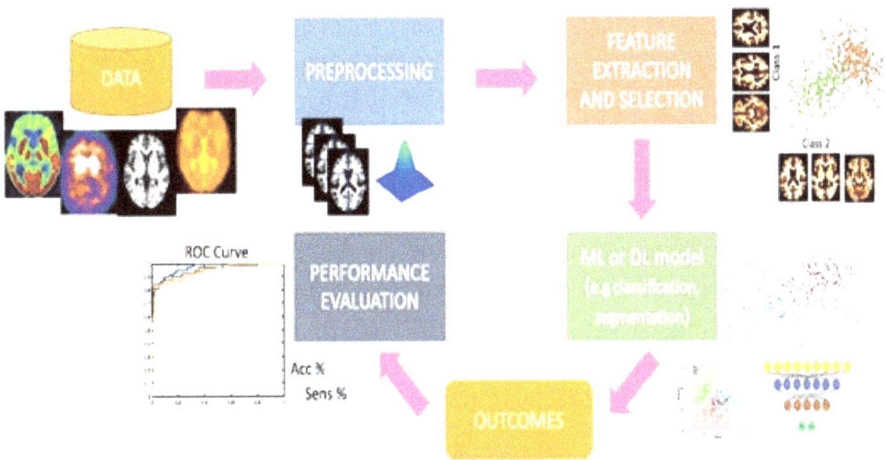

4. Challenges and Opportunities in AI Driven Diagnosis

Despite its potential, AI faces challenges such as data standardization, lack of diverse datasets, and regulatory hurdles. Ethical considerations surrounding AI transparency and patient privacy are critical (Champendal et al., 2024). AI is built on data ethics, which encompasses ownership, objectivity, transparency, informed consent, and privacy and data protection. Regretfully, these kinds of data exchanges are not unusual; the collaboration between DeepMind and the Royal Free London Foundation Trust is one example. Canadian regulations acknowledge that consumers own their private health data and that healthcare professionals are the "information custodians" of that data.

REVIEW OF LITERATURE

1. Early Research on AI in Medical Imaging

AI enhances nuclear imaging protocols with computer-aided noise reduction, image formation, and computer marking of abnormalities. Techniques such as deep learning with CNNs improve tumor detection, segmentation, and precise diagnosis.

The vast potential that artificial intelligence has to offer should be compared against data standardization issues, data sets with an over-standard, and issues related to complex regulations. Problems relating to AI systems transparency issues and protection of patient data are matters of paramount priority. Artificial intelligence applications in medical imaging started many years ago, with computer-aided diagnosis (CAD) systems emerging during the 1980s and 1990s. The first AI systems were pre-specified rule-based expert systems and hand-crafted feature extraction processes that relied pre-specified heuristics and hand-crafted feature extraction schemes. The early developments were designed mainly to aid in the identification of abnormalities on X-ray and CT images. The efficiency of such systems, however, was derailed by the hand-crafted feature issues, making them prone to inconsistency and lack of accuracy. The enhancement in statistical models, as well as computing powers, enabled applications using machine learning techniques, which consisted of the application of support vector machines (SVMs) and decision trees in nuclear medicine imaging. Although precision using such models surpassed the performance of the traditional rule-based systems, image handling and processing along with dealing with big data posed tremendous challenges. It was not until the advent of artificial neural networks (ANNs) toward the end of the 1990s that things changed as it marked a turn in utilizing artificial intelligence toward achieving better pattern recognition in medical imaging.

2. Evolution of AI Models for Nuclear Medicine

The discipline of artificial intelligence has experienced monumental progress in its application to nuclear medicine, especially with the emergence of deep learning algorithms. Convolutional neural networks (CNNs) are a technical innovation in the area of medical imaging because of their ability for learning endogenous features from SPECT and PET imaging modalities. Regular machine learning procedures do not have the ability to learn hierarchical types of images, whereas CNNs do have this facility, and that is the explanation for their usual use in the segmentation of tumors, diagnostic imaging, and the detection of lesions.

Recurrent Neural Networks (RNNs) and Long Short-Term Memory (LSTM) models have been used to make sequential imaging data more amenable to examination, especially by leveraging time-series Positron Emission Tomography (PET) scans for disease tracking. Moreover, Generative Adversarial Networks (GANs) have been helpful in nuclear medicine by improving the quality of images with the use of synthetic images and denoising methods. Transfer learning has also made pretrained artificial intelligence models usable in nuclear imaging and hence minimized the need for large amounts of labeled data.

These mentioned advancements are the result of enormous innovations in the architecture of the transformer, most notably because of the introduction of Vision Transformers (ViTs) that outperformed convolutional neural networks (CNNs). These kinds of models are pre-trained on self-attention mechanisms that help them preserve long-range dependencies of image data and thereby enhance disease diagnosis accuracy as well as the assessment of prognosis.

3. Comparisons Between Traditional and AI Based Diagnosis

Conventional nuclear imaging depends on the interpretation of a human and hence diagnosis is slow and subject to interobserver variation Perceptually Lossless Medical Image Coding. AI-based diagnosis, on the other hand, provides objective, standardized, and automated image interpretation, limiting the scope of human error and maximizing diagnostic efficiency.

Comparative analysis of traditional and AI-assisted diagnostic approaches offers several advantages related to AI systems:

- Automation and Speed: AI computers can analyze thousands of nuclear imaging scans in a matter of minutes, thereby lightening the workload of the radiologists.
- Uniformity and Consistency: Artificial intelligence models remove subjective variation in diagnosis procedures, thereby providing consistent results across various imaging centers.
- Sensitivity and Specificity: AI models have been found to be more sensitive in the identification of abnormalities compared to human radiologists in the identification of early-stage cancer.
- Artificial Intelligence Limitations: Though the advantages are provided Shen and Rangayyan (1997), artificial intelligence models sometimes experience some interpretability issues, which undermine clinicians' trust on AI-diagnoses where no evident explanations are provided.

4 Key Findings from Recent Studies

Recent studies have demonstrated the increasing application of AI in nuclear medicine. They are:

- AI-Enhanced Image Reconstruction: AI-based PET and SPECT reconstruction methods have succeeded in eliminating the imaging noise to produce high-resolution images, and it is now possible to carry out scans using reduced doses of radiation without loss of the diagnostic potential (Saboury et al., 2023).
- Segment Tumors Automatically: CNN models are superior to traditional segmentation methods in volume measurement and detection of tumor volume in oncologic image.
- Diagnosis for Alzheimer's by AI: AI programs that can read amyloid PET scans are very effective when it comes to diagnosing initial Alzheimer's in order to prescribe pre symptomatic diagnosis (Shen & Rangayyan, 1996)
- Disease Progression Forecasting Analysis: The predictive analysis of machine learning has been found to be very effective in predicting the onset of cardiovascular and neurodegenerative diseases and thus enhancing patient monitoring and therapy planning.
- The integration of multimodal (Priyadharshini et al., 2023), AI has made it possible to create advanced AI models that include imaging data from PET, MRI, and CT scans and clinical and genetic data to facilitate personalized treatment recommendations.

The result shows that artificial intelligence has potential applications to revolutionize nuclear medicine through better diagnostic accuracy, better operational efficiency, and better patient care. Availability of data, interpretability of models, and regulatory challenges are limitations for future research.

AI MODELS USED IN IMAGE DIAGNOSIS

1. Machine Learning Approaches

a) Support Vector Machines (SVM)

SVMs are widely used for medical imaging classification problems (Vaishnnave et al., 2020). They do so by building an optimal decision boundary to classify different classes of imaging data. Therefore, they can be used to distinguish between healthy and diseased tissues.

b) Random Forests

Random forests use an ensemble of decision trees for improving classification stability and accuracy in medical image analysis. Random forests are particularly utilized in identifying abnormality and making disease outcome predictions.

c) K Nearest Neighbors (KNN)

KNN is a useful and simple algorithm used in pattern recognition in nuclear imaging. The algorithm classifies points according to the majority of the class closest to them and therefore can be used to recognize outliers in image data.

2. Deep Learning Models

a) Convolutional Neural Networks (CNNs) for Image Processing

The CNNs have transformed the medical image processing because they can automatically recognize patterns and extract features. CNNs are useful in tumor detection, organ segmentation, and restoration of nuclear medicine images.

b) Recurrent Neural Networks (RNNs) for Temporal Data

RNNs handle sequential data and are applicable especially in time series medical image data analysis, for example, monitoring disease progression from a series of scans.

c) Generative Adversarial Networks (GANs) for Image Enhancement

GANs facilitate the enhancement of image quality by generating high-resolution output images from low-resolution input images. Models also assist in nuclear imaging as a denoising, contrast-enhancing, and super resolution reconstruction process.

3. Hybrid AI Strategies

Hybrid approaches integrate deep learning and machine learning approaches to improve diagnostic accuracy. Multimodality learning is the use of more than one imaging modality, for example, MRI and PET, to achieve optimal diagnostic performance.

a) Uses of Artificial Intelligence in Individual Nuclear Imaging Modalities

- **Artificial Intelligence in Positron Emission Tomography (PET)-**AI facilitates PET scan interpretation by automatic detection of lesions, improved quantitative analysis, and imaging artifact reduction.
- **Single Photon Emission Computed Tomography with Artificial Intelligence-**AI simplifies SPECT image reconstruction and diagnostic precision by removing artifacts and noise.
- **Artificial Intelligence in Magnetic Resonance Imaging (MRI)-**AI driven MRI imaging offers better image segmentation and feature extraction automatically, both of which improve disease detection.

CLINICAL APPLICATIONS AND CASE STUDIES

AI has come a long way in nuclear medicine and has led to real-world applications in medicine's multifaceted branches, including oncology, neurology, and cardiology. Real-world applications of AI-based imaging equipment have been seen to provide improved diagnostic performance, treatment planning, and monitoring of diseases. This chapter explores the uses of AI in clinical specialties and presents case studies of its effectiveness.

1. AI in Oncology

Cancer treatment and diagnosis have been greatly enhanced by AI based nuclear imaging techniques, particularly for SPECT and PET scans. AI enhances oncology imaging in the following:

- Tumor Early Detection – AI algorithms created from PETCT images can detect malignant tumors in an early stage of development, which improves prognosis and survival. Deep learning methods learn metabolic processes in PET scans to distinguish between benign and malignant lesions.
- Tumor Volume Estimation and Segmentation – AI automatically segments the tumor and helps oncologists with precise tumor growth assessment. CNN-based models and UNet models are highly accurate in tumor region segmentation of nuclear images.

Personalized Radiation Treatment Planning – AI helps with radiation therapy planning as it can forecast tumor responses to therapy. AI-driven radiomics derives features from imaging information in an attempt to tailor radiation doses and enhance treatment response.

Figure 2. PET/SPECT scan images showcasing AI-assisted tumor segmentation or disease detection

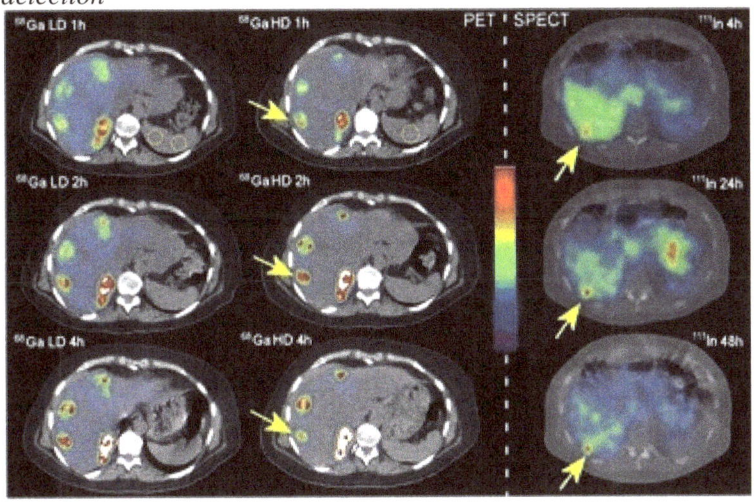

2. AI in Neurology

AI has revolutionized the early detection and management of neurodegenerative diseases by nuclear imaging Its primary uses are:

- Diagnosis of Alzheimer's Disease – AI programs analyze amyloid PET scans to diagnose early Alzheimer's even before any symptoms arise. Deep learning algorithms such as 3D CNNs can do brain image classification and detect healthy versus disease conditions.
- Monitoring Parkinson's Disease Progress – AI-assisted dopamine transporter (DaT) scans allow neurologists to monitor Parkinson's disease progress. AI algorithms are able to forecast stages of the disease and effectiveness of treatment measures.
- Ischemic Stroke Diagnosis and Prediction of Recovery – AI interprets brain perfusion scans to identify ischemic strokes early. Machine learning algorithms predict recovery and establish rehabilitation protocols.

3. AI in Cardiology

AI driven innovation has hastened nuclear cardiology for cardiac function evaluation and myocardial perfusion scans Some of the uses are:

- Coronary Artery Disease (CAD) detection – AI computer algorithms review SPECT and PET myocardial perfusion images for CAD-predictive perfusion abnormalities. AI improves diagnostic accuracy and reduces false positives
- Automated Left Ventricular Function Testing – AI-powered nuclear imaging scanners evaluate LVEF, a key cardiac performance measure. Human interpretation is replaced by AI and image-dependent variability in measurement is reduced to a minimum.
- Risk Stratification and Prognostic Modeling – Patient-specific nuclear imaging data are analyzed by AI algorithms to predict cardiovascular risks and guide personalized treatment

4. Case Studies: Real World AI Implementations in Nuclear Medicine

Various health care organizations and institutions have integrated artificial intelligence into procedures for nuclear medicine diagnosis with complete success. Such integration is facilitated by various well-documented cases as shown below:

- Case Study 1: Use of AI-Advanced PETCT for Lung Cancer Diagnosis – A leading cancer research center that utilized an AI-advanced PETCT imaging system saw a 20% rise in the rate of diagnosis of patients suffering from lung cancer. The use of artificial intelligence lowered the rates of false positives and was more effective in detecting lesions, thereby allowing healthcare providers to treat the condition at an earlier stage.
- Case Study 2: Artificial Intelligence Application to Alzheimer's Disease Diagnostic Imaging Across Centers – an international investigation utilized artificial intelligence to analyze amyloid PET scans obtained from multiple healthcare centers. The AI system diagnostic accuracy was 40% and proposed broad applicability utility in patient groups with different demographic characteristics.
- Case Study 3: Artificial Intelligence in Myocardial Perfusion Imaging in a Cardiology Clinic – Artificial intelligence application in the cardiology clinic improved myocardial perfusion imaging analysis effectiveness. Artificial intelligence greatly reduced interobserver variability and also provided rationale for the acute cardiovascular risk assessment of disease-affected patients.

The use of artificial intelligence in nuclear medicine is more pronounced today, as attested by various examples of applications and clinical case reports. This is most critical in the manner that it can facilitate the detection of disease, support the planning of treatment strategies, and ultimately improve patient outcomes.

AI DATASETS FOR NUCLEAR MEDICINE

1. Publicly Available Datasets

Accessibility of publicly available data is of significant use to the field of artificial intelligence in nuclear medicine since it enables researchers to train, validate, and create AI models.

Comprehensive used datasets include: ACRIN-FMISO is a dataset for PET imaging test development to be applied in tumor hypoxia research. It is a central element of AI-based tumor segmentation, treatment evaluation, and radiomics analysis.

The Open Access Series of Imaging Studies (OASIS) dataset is extensively utilized in Alzheimer's disease research, such as MRI and PET imaging of subjects with different degrees of cognitive impairment. Artificial intelligence models learned on OASIS data play a vital role in early diagnosis and in tracking the progression of neurodegenerative diseases.

- ADNI (Alzheimer's Disease Neuroimaging Initiative) – The database has useful imaging and clinical data that are best suited for the research of neurodegenerative diseases like Alzheimer's. Use cases of ADNI data with AI are disease diagnosis, biomarker detection, and prediction of prognosis.
- TCIA (The Cancer Imaging Archive) is an open-source data archive of imaging data for multiple cancers, including nuclear medicine imaging scans. It is widely utilized in oncology artificial intelligence applications for detection, segmentation, and radiomics analysis.
- UK Biobank is a large-scale biomedical repository of genetic, imaging, and clinical data based on a representative sample of the UK population. UK Biobank is utilized in PET and MRI data by AI researchers based on UK Biobank for any number of a variety of purposes, from cardiovascular research and neuroimaging research to clinical genetics research.
- The UK Biobank is a large biomedical database that contains genetic, imaging, and clinical data collected from a wide population sample. Researchers of artificial intelligence use PET and MRI datasets acquired from the UK Biobank to support various research studies, such as cardiovascular and neuroimaging research. The publicly available data sets are excellent tools for the

training of artificial intelligence systems but are still limited in diversity of samples, imaging protocol, and quality of annotation.

2. Hospital-Based and Proprietary Data Repositories

Though publicly available data hold promising research potential for artificial intelligence, it is significant that most advanced nuclear medicine data are built and owned by hospitals, research institutions, and private companies. Proprietary data hold much greater amounts of data, including high-resolution image data, along with clinical and genomic data. However, their potential is limited by privacy law, ethics, and institutional controls (Brown et al., 2005).

Some of the salient characteristics of hospital-based and proprietary data are:

- Institutional Imaging Databases – Many research institutions and hospitals have massive datasets of nuclear imaging, for example, PET, SPECT, and hybrid imaging methods. These are all critical to develop AI models customized to specific clinical settings (Champendal et al., 2024).
- Research Studies and Clinical Trials – Modern clinical trials produce large imaging data sets that are suitable for analysis by artificial intelligence techniques. Yet, these data sets are accessible only after rigorous ethical approvals and data-sharing agreements.
- Joint Artificial Intelligence Projects – Numerous artificial intelligence companies partner with hospitals for the formation of proprietary databases of images used for internal AI system training. These partnerships aid in research done on AI yet keep publicly funded data sets well below the scope of availability.

Efforts are being made to create secure data sharing sites that strike a balance between research requirements and patient confidentiality. Federated learning strategies, in which AI models are trained institutionally without sharing raw data, have systems for overcoming data access restrictions.

3. Issues of Data Availability and Standardization

Not with the increasing accessibility of datasets related to nuclear medicine, various obstacles impede the advancement of artificial intelligence within this discipline

- Nuclear imaging data heterogeneity is extremely prevalent between institutions because of variability in protocols used for imaging, scanner types, and patient populations. The heterogeneity poses a challenge during the construction of AI models that are going to be institutionally generalizable.

- Ethical Guidelines and Privacy Regulations – Privacy regulations, the Health Insurance Portability and Accountability Act (HIPAA) of the US, and the General Data Protection Regulation (GDPR) of the EU, impose severe limitations on the use of patient data. Researchers in the field of artificial intelligence must function in these regulation contexts to secure and use imaging datasets.
- Low Annotation of Available Datasets – High-quality AI models are created using well-annotated datasets with expert-labeled ground truth data. Manually annotating these datasets, though, is time-consuming, costly, and subject to interobserver variation.
- Data Standardization Issues – No standard imaging format, no standard annotation process, and differences in metadata standards hinder the integration of datasets acquired from various sources. Initiatives such as the DICOM (Digital Imaging and Communications in Medicine) standards attempt to address these issues, but differences remain.
- Dataset Composition Bias – The majority of datasets currently employed are developed for particular populations, and as a result, they carry embedded biases that are capable of impacting the performance of AI models. When biased datasets are employed for training AI models, their performance can be poor for other patient populations; thus, datasets should be developed that are more diverse and representative.

To counter these issues, inter-cooperation must occur among institutions, regulatory bodies, and researchers to produce standardized, open, and ethically controlled data.

Advances in synthetic data generation, federated learning, and data harmonization techniques can balance these deficits of AI research in nuclear medicine.

EVALUATING METRICS OF ARTIFICIAL INTELLIGENCE MODELS

1.Classification Performance Measurements

In artificial intelligence applications of nuclear medicine, various performance measures are used to evaluate classification models, as the objective is to quantify their predictive accuracy and reliability. Some of the most significant classification measures are:

- Accuracy – This measure is used to determine the general accuracy of predictions, in terms of the number of instances properly classified and divided by the total number of instances. Accuracy is a useful measure, but it is misleading when applied to imbalanced datasets, where one class is much larger than another.
- Sensitivity, or Recall or True Positive Rate, is the ability of the model to correctly recognize instances that have been labeled as positive, i.e., its ability to recognize the presence of a disease. High sensitivity corresponds to low false negatives, which is of paramount significance in medicine, since failure to detect a condition can be a serious consequence.
- Specificity (True Negative Rate) – Looks at the precision with which the model classifies negative cases. A specific model reduces the number of false positives, thus preventing unnecessary follow-up tests and treatments.
- Positive Predictive Value – Calculates the number of predicted positive cases that are actually positive. This is a critical measure in situations where the effect of false positives is significant, e.g., when misdiagnosing cancer.
- F1Score – A harmonic mean of recall and precision that provides a balanced measure to model performance in imbalanced datasets. Good classification performance is indicated by high F1 scores.
- Receiver Operating Characteristic Area under the Curve (ROCAUC) is a metric of the discriminative ability of the model in terms of the area under the ROC curve calculation, with the true positive rate plotted against the false positive rate on the axes and the threshold level varied. A higher value near 1 reflects higher performance (Patra et al., 2013).

2. Segmentation Metrics

Nuclear imaging AI models are likely to involve segmentation tasks, e.g., detection of tumor or anatomical structure. Segmentation model assessment does have some measures quantifying prediction and ground truth segmentations overlap.

- Dice Coefficient (F1 Score for Segmentation) – It is overlap between predicted segmentation and ground truth segmentation. It is between 0 and 1 with 1 being the best possible overlap and 0 being no overlap. It is mostly used to estimate medical image segmentation tasks.
- Intersection over Union (IoU), or Jaccard Index, calculates ground-truth and predicted segmentations overlap as intersection over ground-truth and predicted segmentations union. The greater the IoU, the more accurate the segmentation (Geng et al., 2021).

- Hausdorff Distance – It estimates the maximum distance between segmentation boundaries estimated by the model and actual boundaries and thus checks how well the model estimates actual anatomical shapes. Its value is best achieved in high-resolution medical image segmentations.

  Boundary F1 Score places greater emphasis on the agreement between boundary predictions of segmentation boundaries and actual boundaries, and therefore is of special importance to precise lesion detection in nuclear medicine imaging.

3. Robustness and Generalization of Models

The need for artificial intelligence models to display appropriate generalization properties on different imaging datasets in nuclear medicine is clear. Effective models should display enhanced performance across imaging centers, scanners, and populations. Different performance measures are employed to determine their robustness: Cross-validation is a technique that involves training and testing artificial intelligence models on multiple subsets of the data set to ensure the achievement of uniform results. The most commonly used traditional methods include k-fold cross-validation and leave-one-out cross-validation.

- External Dataset Testing – The models are tested with independent datasets derived from other organizations to determine their applicability and effectiveness in real-world situations. A good AI model should perform well when challenged with unseen data.
- Domain Adaptation Testing – Since scanner parameters and nuclear image protocols are different across institutions, testing for how models will generalize to novel domains without huge retraining efforts is crucial.

Adversarial Testing is testing the models with adversarial examples or artificially generated images to determine how robust they are against noise, artifacts, and image quality variation.

With these assessment instruments, researchers and experts in the field of artificial intelligence can ascertain that nuclear medicine models are not only precise but also trustworthy, understandable, and can generalize across different patient groups and imaging settings.

RECENT DEVELOPMENTS IN ARTIFICIAL INTELLIGENCE IN NUCLEAR MEDICINE

1. Explainable Artificial Intelligence (XAI) for Enhanced Interpretability

One of the greatest challenges in the area of AI-based nuclear medicine is the opacity of deep learning techniques, and thus there exists a transparency gap for clinicians as to the rationale behind the choices. Explainable AI (XAI) seeks to overcome this hurdle by making artificial intelligence models more transparent and interpretable.

Figure 3. Working of explainable AI (XAI)

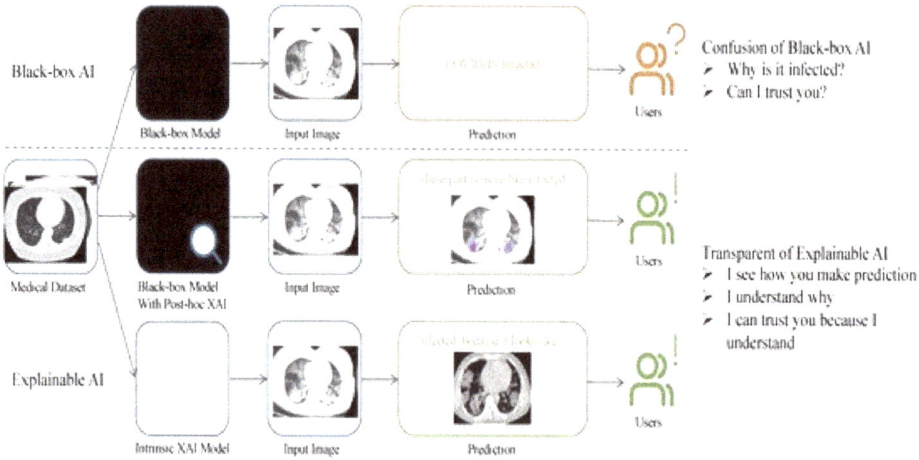

The most important breakthroughs in XAI are:

• Feature Attribution Techniques – SHAP (SHapley Additive explanations) and LIME (Local Interpretable Model-agnostic Explanations) are significant to establish the most contributing features of the medical images responsible for the decision of the artificial intelligence model.

Artificial intelligence models utilize saliency maps and heatmaps within different visualization paradigms like Gradient-weighted Class Activation Mapping (Grad-CAM) to delineate meaningful areas of nuclear scans. This allows radiologists to subsequently interpret those scans better in an attempt to provide AI-aided diagnoses (Han & Lee, 2023).

• Rule-Based AI Systems – The subsequent editions have incorporated decision-making capabilities from defined rules, where the generated output by artificial intelligence is enhanced by logical thought and medical experience and thus become more reliable.

Regulatory and Ethical Issues – As the healthcare industry becomes more reliant on artificial intelligence, regulatory agencies focus on model interpretability. Explainable AI brings AI-based nuclear medicine into compliance with accepted medical ethics and regulatory standards.

2. Federated Learning-Based Privacy-Preserving Artificial Intelligence Model

Privacy issues are among the greatest obstacles to developing AI models in nuclear medicine due to the high association of the patient's image data. Federated learning is among the primary solutions in which AI model training is feasible between institutions without sharing raw data. The primary contributions are:

- Decentralized Training of AI – Instead of pooling patient data and storing it in a single centralized database, federated learning provides for local AI model training within multiple hospitals and institutions in such a manner that model updates only are shared. This preserves the data privacy as well as being compliant with regulatory environments such as General Data Protection Regulation (GDPR) and Health Insurance Portability and Accountability Act (HIPAA).
- Block chain technology in combination with federated learning greatly enhances the security of data sharing by rendering the AI model updates tamperproof and verifiable.
- Performance and Scalability – Experimental evidence has shown that federated learning architectures can be as good in terms of performance as that of traditional AI models, without compromising data confidentiality, thus hinting at high potential for AI application in nuclear medicine Hasegawa et al. (2005) .
- The integration of medical centers is predominantly driven by high-end medical centers and research centers for artificial intelligence that wish to utilize federated learning in order to assist multicenter trials and overcome issues related to privacy.

3. Integration of Multimodal AI to Improve Diagnostics

The recent advances in artificial intelligence have led to the development of multimodal AI systems, which combine data from other sources to improve diagnostic accuracy and maximize patient care. Multimodal AI applications have several significant benefits in nuclear medicine:

- Integration of Multi-Modality Imaging – Modern AI systems integrate PET, SPECT, CT, and MRI imaging modalities in an effort to provide enhanced diagnostic information. For example, integration of PET-CT imaging provides enhanced lesion detection and tumor characterization (Jimenez-Mesa et al., 2023).
- The combination of clinical and genomic data is an area of much interest to current artificial intelligence systems, which can process not only imaging data but also patient history, lab results, and genomic data in a bid to give more personalized diagnostic and therapeutic recommendations.
- NLP for Radiology Reports – NLP-based AI models are able to read radiology reports and imaging information to enable more precise diagnoses and streamlined workflows for radiologists. Multimodal artificial intelligence can be used for prognostic and predictive modeling to predict the course of disease, treatment response, and outcomes of the patients, thus improving clinical decision-making.
- The future trends are aligned with the new role of artificial intelligence in nuclear medicine, i.e., improving interpretability, improving sensitivity to data privacy, and merging heterogeneous data sources to support more precise and efficient diagnosis processes.

CHALLENGES AND ETHICAL ISSUES

1. Data Privacy and Security Issues

The use of artificial intelligence in nuclear medicine is built on extensive medical imaging databases that contain sensitive patient data. Ensuring the privacy and security of the data is a major challenge considering

- Strong Privacy Laws – Laws like the Health Insurance Portability and Accountability Act (HIPAA) in the United States and the General Data Protection Regulation (GDPR) in the European Union provide strong laws for the handling and exchange of medical data. Developers of artificial intelligence have to comply with these legislations.
- Cyber security Vulnerabilities – The cyber security threats in Artificial Intelligence systems are unauthorized data access, ransom ware attacks (Zhou et al., 2021), and adversarial attacks that can tamper with medical im-

aging. Robust encryption protocols, access control protocols, and secure data transmission protocols must be used.

- Anonymization and Deidentification – Masking of data and synthetic data creation are a few of the techniques being tested to anonymize patient data and keep it valuable for training AI. But the issue of how to balance data utility and privacy remains a huge challenge.

2. Barriers to Adoption due to Regulations

While artificial intelligence can revolutionize nuclear medicine, regulatory frameworks for AI-based medical and diagnostic devices are yet to be developed.

Some of the regulatory challenges are amongst others:

- Validation and Approval Processes – Diagnostic machines based on artificial intelligence need to go through extensive validation and clinical trials by regulatory agencies, such as the U.S. Food and Drug Administration (FDA) and the European Medicines Agency (EMA), before approval. The time-consuming approval process is an obstacle to the implementation of AI in clinical practice (Wu et al., 2006).
- The largest hurdle to standardization is heterogeneity in imaging protocols, scanner hardware, and annotation procedures used at various institutions, and this is the largest stumbling block to the creation of AI models with good generalization ability.
- Legal and Liability Problems – Determining liability when artificial intelligence detects a medical problem incorrectly or suggests something that is in error is a major legal and ethical concern. There has to be regulation with well-defined parameters between physicians and AI so that risk is minimized or eliminated properly.

3. Bias and Fairness in AI-Driven Medical Diagnosis

- AI methods employed in nuclear medicine need to be trained on diversified data sets so that they can offer objective and impartial estimates. There are numerous causes through which AI-based diagnosis processes can develop biases
- The Representative Data Insufficiency – A significant portion of data present in the public domain is biased towards certain demographic groups; therefore, artificial intelligence models perform optimally within these groups and less than optimally in others. This disparity leads to disparity in diagnosis and treatment.

- Algorithmic Bias in AI – AI algorithms can learn and continue existing biases of training data inadvertently. For instance, a model trained on data sets with a majority of male patients will be poor for female patients in test cases.
- Ethical concerns of AI decision-making involve the significant issue of transparency. Explainable AI systems, which incorporate transparency functionalities, must ensure ethical use in healthcare industries.
- Minimizing Bias with Federated Learning and Data Augmentation – The current study is all about using federated learning, a method by which AI models are trained with decentralized data collected from different organizations, without losing objectivity and confidentiality. Complemented by this is data augmentation methods that help counter the underrepresentation of the minority patient groups in training sets.
- These challenges can be addressed through a collaborative effort from artificial intelligence researchers, medical professionals, policymakers, and regulatory bodies to ensure that AI-based nuclear medicine is safe, equitable, and of value to all patients.

FUTURE DIRECTIONS

For further advancement of advanced application of artificial intelligence in nuclear medicine, future studies must cover several various issues:

- Enhancing the explainability of AI. The XAI methods are at the core of enhancing AI system transparency and explainability. Methods like attention mechanisms, saliency maps, and feature attribution models allow radiologists to understand inferences drawn from AI, thus improving higher confidence in AI-assisted diagnosis processes.
- Multimodal Imaging Integration – The future of AI is to combine multiple imaging modalities like PET, SPECT, CT, and MRI to offer a more diagnostic image. Multimodal integration of AI can be utilized to enhance disease characterization and patient outcome (Liu et al., 2021).
- Ethical Challenges – Improvement of equity between artificial intelligence and minimizing bias should be overarching objectives to mitigate disparities in disease diagnosis. The use of federated learning and advanced techniques of data enhancement can enable more inclusive models with the guarantee of patient privacy.
- Regulatory and Clinical Validation – Regulatory systems need to be developed holistically with harmonized validation timelines to ensure that artificial intelligence models are in compliance with clinical procedures. Harmonization

of the functioning of AI researchers, healthcare professionals, and regulatory bodies is also necessary to develop trust in products based on AI-enhanced nuclear medicine technology.

- Practical Application and Regular Training – To ensure the consistent optimal functioning of artificial intelligence models, it is necessary that their databases be updated regularly with data obtained from actual operational environments. Application of adaptive learning techniques will allow these models to refine their predictions every time new clinical data becomes available (Manesh et al., 2023). With the passage of time and advancements in technology, AI will expand its horizons in the sector of nuclear medicine, and better, more efficient, and patient-centric healthcare will be realized. With the shattering of existing constraints and focus on future development, AI-based nuclear medicine will be ready to transform the sector and enhance patient care globally.

CONCLUSION

Artificial Intelligence (AI) is transforming the nuclear medicine imaging field at a revolutionary level by enhancing diagnostic accuracy, facilitating automated image analysis, and providing personalized treatment recommendations. AI-based systems have proven immense potential in the detection of diseases at an early stage, enhancing image segmentation processes, and lightening the burden of radiologists. Deep learning techniques, including convolutional neural networks (CNNs) and generative models, have enhanced image quality and interpretive value, thus making practices under nuclear medicine more efficient and accessible. In spite of these developments, a number of challenges persist (Nasir et al., 2023). Availability of data is a key challenge, as the presence of large, high-quality, annotated datasets is important for training robust AI models; however, data privacy regulations restrict data sharing across institutions. In addition, the use of artificial intelligence in hospitals is also confronted by regulation because diagnostic equipment based on AI must be well validated and certified by regulatory agencies such as the FDA and EMA.

Furthermore, the issue of model explainability is still pertinent, as a majority of AI algorithms are 'black boxes,' thereby making it challenging for clinicians to comprehend their decision-making processes. Overcoming these issues is crucial to allow artificial intelligence to achieve its full potential in nuclear medicine. Innovation in the future should focus on creating explainability through the application of explainable AI (XAI) frameworks, creating standardized datasets through federated learning methods, and enhancing collaboration between clinicians, regulatory agencies, and scientists. Artificial intelligence can revolutionize nuclear medicine with more innovation, thus enhancing patient outcomes, diagnostic accuracy, and the effectiveness of medical imaging.

REFERENCES

Bandla, M. S., Liju, A. M., Mishra, G., & Luharia, A. (2023). Artificial intelligence in healthcare with special consideration in radio imaging. In *Proceedings of the 2023 1st DMIHER International Conference on Artificial Intelligence in Education and Industry 4.0 (IDICAIEI)* (pp. 108–113). DOI: 10.1109/IDICAIEI58380.2023.10406913

Boukhennoufa, O., Comas, L., Nicod, J. M., Ungureanu, C., Zerhouni, N., & Boulahdour, H. (2024, May). Automatic detection of parathyroid glands in nuclear medicine. In *2024 IEEE International Symposium on Biomedical Imaging (ISBI)* (pp. 1-4). IEEE. DOI: 10.1109/ISBI56570.2024.10635117

Brown, R. A., Zhu, H., & Mitchell, J. R. (2005). Distributed vector processing of a new local multiscale Fourier transform for medical imaging applications. *IEEE Transactions on Medical Imaging*, *24*(5), 689–696. DOI: 10.1109/TMI.2005.845320 PMID: 15889555

Champendal, M., Ribeiro, R. S. T., Müller, H., Prior, J. O., & Dos Reis, C. S. (2024). Nuclear medicine technologists practice impacted by AI denoising applications in PET/CT images. *Radiography*, *30*(4), 1232–1239. DOI: 10.1016/j.radi.2024.06.010 PMID: 38917681

Geng, M., Meng, X., Yu, J., Zhu, L., Jin, L., Jiang, Z., Qiu, B., Li, H., Kong, H., Yuan, J., Yang, K., Shan, H., Han, H., Yang, Z., Ren, Q., & Lu, Y. (2021). Content-noise complementary learning for medical image denoising. *IEEE Transactions on Medical Imaging*, *41*(2), 407–419. DOI: 10.1109/TMI.2021.3113365 PMID: 34529565

Han, J.-W., & Lee, Y.-J. (2023). *Generative AI for radiological image data: Current trends and outlook.* In *Proceedings of the 2023 14th International Conference on Information and Communication Technology Convergence (ICTC).*

Hasegawa, T., Ishikawa, M., Maruyama, K., Inadama, N., Yoshida, E., & Murayama, H. (2005). Depth-of-interaction recognition using optical filters for nuclear medicine imaging. *IEEE Transactions on Nuclear Science*, *52*(1), 4–7. DOI: 10.1109/TNS.2004.843149

Jimenez-Mesa, C., Arco, J. E., Martinez-Murcia, F. J., Suckling, J., Ramirez, J., & Gorriz, J. M. (2023). Applications of machine learning and deep learning in SPECT and PET imaging: General overview, challenges and future prospects. *Pharmacological Research*, *197*, 106984. DOI: 10.1016/j.phrs.2023.106984 PMID: 37940064

Liu, J., Malekzadeh, M., Mirian, N., Song, T. A., Liu, C., & Dutta, J. (2021). Artificial intelligence-based image enhancement in PET imaging: Noise reduction and resolution enhancement. *PET Clinics*, *16*(4), 553–576. DOI: 10.1016/j.cpet.2021.06.005 PMID: 34537130

Manesh, A. S., Shiri, I., Khateri, M., Jenabi, E., Salimi, Y., Geramifar, P., & Zaidi, H. (2023, November). Fine-tuned BERT Language Model for Efficient Nuclear Medicine Data Retrieval. In *2023 IEEE Nuclear Science Symposium, Medical Imaging Conference and International Symposium on Room-Temperature Semiconductor Detectors (NSS MIC RTSD)* (pp. 1-1). IEEE. DOI: 10.1109/NSSMICRTSD49126.2023.10338102

Nasir, N., Alshabi, M., Al-Yateem, N., Rahman, S. A., Subu, M. A., Hijazi, H. H., . . . Saifan, A. R. (2023, June). Challenges of Artificial Intelligence in Medicine. In *2023 IEEE 47th Annual Computers, Software, and Applications Conference (COMPSAC)* (pp. 1429-1433). IEEE. DOI: 10.1109/COMPSAC57700.2023.00219

Patra, J., Moulick, H. N., & Manna, A. K. (2013). Medical image processing in nuclear medicine and bone arthroplasty. *IOSR Journal of Computer Engineering*, *13*(2), 97–110. DOI: 10.9790/0661-13297110

Priyadharshini, K., Mohan, D., & Gokulakrishnan, D. (2023). Artificial intelligence-based multimodal molecular imaging fusion framework for cancer care. In *Proceedings of the 2023 International Conference on Innovative Computing, Intelligent Communication and Smart Electrical Systems (ICSES)* (pp. 1–6). DOI: 10.1109/ICSES60034.2023.10465530

Saboury, B., Bradshaw, T., Boellaard, R., Buvat, I., Dutta, J., Hatt, M., Jha, A. K., Li, Q., Liu, C., McMeekin, H., Morris, M. A., Scott, P. J. H., Siegel, E., Sunderland, J. J., Pandit-Taskar, N., Wahl, R. L., Zuehlsdorff, S., & Rahmim, A. (2023). Artificial intelligence in nuclear medicine: Opportunities, challenges, and responsibilities toward a trustworthy ecosystem. *Journal of Nuclear Medicine*, *64*(2), 188–196. DOI: 10.2967/jnumed.121.263703 PMID: 36522184

Shen, L., & Rangayyan, R. M. (1997). A segmentation-based lossless image coding method for high-resolution medical image compression. *IEEE Transactions on Medical Imaging*, *16*(3), 301–311. DOI: 10.1109/42.585764 PMID: 9184892

Vaishnnave, M. P., Devi, K. S., & Ganeshkumar, P. (2020). Automatic method for classification of groundnut diseases using deep convolutional neural network. *Soft Computing*, *24*(21), 16347–16360. DOI: 10.1007/s00500-020-04946-0

Wu, D., Tan, D. M., Baird, M., DeCampo, J., White, C., & Wu, H. R.Hong Ren Wu. (2006). Perceptually lossless medical image coding. *IEEE Transactions on Medical Imaging*, *25*(3), 335–344. DOI: 10.1109/TMI.2006.870483 PMID: 16524089

Yahil, A., Vija, A. H., & Hawman, E. G. (2006). Enhanced feature extraction in planar nuclear medicine using Pixon minimum-complexity image processing. In *Proceedings of the 2006 IEEE Nuclear Science Symposium Conference Record* (pp. 2037–2040). DOI: 10.1109/NSSMIC.2006.354314

Yang, J., Sohn, J. H., Behr, S. C., Gullberg, G. T., & Seo, Y. (2020). CT-less direct correction of attenuation and scatter in the image space using deep learning for whole-body FDG PET: Potential benefits and pitfalls. *Radiology. Artificial Intelligence*, *3*(2), e200137. DOI: 10.1148/ryai.2020200137 PMID: 33937860

Zhou, Q., Zuley, M., Guo, Y., Yang, L., Nair, B., Vargo, A., Ghannam, S., Arefan, D., & Wu, S. (2021). A machine and human reader study on AI diagnosis model safety under attacks of adversarial images. *Nature Communications*, *12*(1), 7281. DOI: 10.1038/s41467-021-27577-x PMID: 34907229

KEY TERMS AND DEFINITIONS

Artificial Intelligence (AI): The technology known as artificial intelligence (AI) makes it possible for computers and other machines to carry out operations like learning, problem solving, and decision making that normally call for human intelligence. In essence, artificial intelligence seeks to build machines that can mimic human cognitive capacities.

Deep Learning: A branch of machine learning called "deep learning" makes use of multi-layered artificial neural networks to extract knowledge from data. It is useful in a variety of applications, including speech recognition, image identification, and natural language processing, since it is very good at identifying intricate patterns and formulating predictions.

Explainable AI (XAI): Refers to the field of artificial intelligence that focuses on making AI models and their decisions more transparent and understandable to humans. It aims to provide explanations for why a model arrives at a specific output, helping users understand the reasoning behind the AI's actions and build trust in the system.

Machine Learning: Is a field of study in artificial intelligence concerned with the development and study of statistical algorithms that can learn from data and generalize to unseen data, and thus perform tasks without explicit instructions.

Medical Image Analysis: The process of extracting meaningful information from medical images to aid in diagnosis, treatment, and patient care. It involves using various computational methods to analyze and interpret images obtained from different imaging modalities, such as X-rays, CT scans, MRI, and ultrasound.

Multimodal: Multimodal imaging combines two or more imaging techniques during a single examination to provide a more comprehensive view of a patient or subject

Nuclear Medicine: Using radioactive materials, or radiopharmaceuticals, to diagnose and cure illnesses, nuclear medicine is a branch of medicine that focuses on imaging the body's interior organs and evaluating their performance. These radioactive tracers are given to the patient, and specialized imaging equipment tracks their activity to provide information about a variety of health and disease-related topics.

Radiomics: A field that extracts and analyzes quantitative features from medical images to improve clinical decision-making. It involves using algorithms to identify and measure various aspects of images, like shape, texture, and intensity, providing data beyond what a human eye can see. This data can then be used to predict outcomes like disease classification, treatment response, and prognosis.

Chapter 13
Responsibility and Hospitality in the Doctor–Patient Relationship:
Jus in Bello and Jus Post Bellum Cases

Evangelos Ioannis Koumparoudis
http://orcid.org/0000-0002-1068-4376
Sofia University St. Kliment Ohridski, Bulgaria & Bulgarian Academy of Sciences, Bulgaria

ABSTRACT

This chapter first presents the basic models of the doctor-patient relationship (activity-passivity, clinical guidance, mutual participation, etc.) as seen through the lens of jus in bello and jus post bellum. Afterwards, it looks deeper into the philosophy of Derrida and Levinas, exposing Derrida's views of responsibility and hospitality, as well as the Levinasian ethics of asymmetry and his theory of responsibility for the Other. Their ethical theories are applied to cases of treating injured enemy soldiers and soldiers with severe PTSD and psychiatric disorders, as well as refugees, in bello and post bellum. The second part of this chapter unravels how digital health-care technologies and AI can help both in bello and post bellum by, for example, treating soldiers remotely, providing help, and collecting injured soldiers. These are approached as regards the ethics of hospitality and responsibility.

DOI: 10.4018/979-8-3373-6264-9.ch013

INTRODUCTION

This work expands on and deepens my conference presentation entitled "Doctor and Patient Relationship, through the Philosophy of Derrida and Levinas", delivered at the Derrida Today Conference in Athens in June 2024. While the original talk focused primarily on the ethical structure of the doctor-patient relationship under conditions of responsibility, hospitality, and asymmetry, this extended study situates that relationship within the extreme contexts of armed conflict and post-conflict recovery and further examines how emerging military medical technologies challenge and transform ethical relations.

War, both during and after hostilities, constitutes a condition in which ethical responsibility is subjected to intense pressure. Medical professionals make rapid and irreversible decisions under conditions of scarcity, insecurity, and uncertainty. The obligation to treat severely injured soldiers with limited resources is further complicated when those in need of care include enemy combatants, hostages, displaced civilians, or refugees. In such situations, the traditional ethical frameworks governing medical practice are strained. The doctor is confronted with decisions that cannot be fully justified by established protocols or institutional norms, revealing the tragic dimensions of ethical responsibility.

These ethical tensions are intensified by the increasing use of advanced technologies in military healthcare, including telemedicine, artificial intelligence, and unmanned systems. While these innovations expand the reach of medical assistance, they also transform the ethical structure of the doctor-patient relationship. The encounter with the patient is increasingly mediated by screens, data, and algorithms, raising fundamental questions about responsibility, proximity, and the persistence of ethical obligation when face-to-face interaction is disrupted.

The philosophies of Jacques Derrida and Emmanuel Levinas provide a particularly fertile ground for addressing these challenges. Derrida's reflections on responsibility, undecidability, and hospitality illuminate the aporetic structure of ethical decision-making, especially in contexts where no choice can be made without exclusion or sacrifice. Levinas's ethics of asymmetry and his emphasis on responsibility for the suffering Other further radicalize this inquiry. For Levinas, ethics precedes politics, and ontology challenges any attempt to reduce medical care to strategic, institutional, or technological calculations. By bringing Derrida's and Levinas's ethical philosophies into dialogue with the doctor-patient relationship in jus in bello and jus post bellum, as well as military AI, telemedical, and UAV technologies, this study moves beyond simplified or procedural models of medical ethics. Instead, it aims to reveal medical responsibility as an ongoing exposure to the Other, an exposure that persists even under conditions of war, displacement, and technological mediation.

RESEARCH METHODOLOGY

The methodology of this research is philosophical in nature. The study proceeds through a structured philosophical reading of the ethical models of the doctor-patient relationship, followed by an in-depth engagement with the ethical thought of Jacques Derrida and Emmanuel Levinas. The research unfolds in four interconnected stages. First, classical models of the doctor-patient relationship are examined within the contexts of jus in bello and jus post bellum. Second, the study conducts a philosophical analysis of Derrida's concepts of responsibility, undecidability, and hospitality, alongside Levinas's ethics of asymmetry, the face of the Other, and responsibility for suffering. Third, the concepts of Derrida and Levinas are reformulated and examined in the contexts of jus in bello and jus post bellum. Fourth, the research summarizes and presents some of the main applications of AI, UAVs, and telemedicine to military healthcare and reflects on their ethical dimensions. These technologies are not evaluated in terms of efficiency or performance but rather are examined as ethical mediations that reshape proximity, responsibility, and decision-making, based on the ideas of Derrida and Levinas.

LITERATURE REVIEW

The doctor-patient relationship has been widely examined as a fundamental ethical structure in medical practice. Early typologies by Szasz and Hollender (1956) distinguish between activity-passivity, clinical guidance, and mutual participation, highlighting different degrees of authority and patient autonomy. Emanuel and Emanuel (1992) later refine these models by introducing paternalistic, informative, interpretive, and deliberative approaches, emphasizing the ethical role of autonomy and shared decision-making. Levinas grounds ethics in an asymmetrical responsibility toward the suffering Other, revealed in the face-to-face encounter that precedes law and politics (Levinas 1969). War, for Levinas, suspends ethics by reducing individuals to forces within a totality, rendering medical care a fragile but vital interruption of violence (Hofmeyr 2024; Lau 2015). Derrida complements this view by conceiving of responsibility as emerging from undecidability and singularity rather than rule-based moral systems (Anderson 2017). Ethical action, for Derrida, is always marked by exclusion and risk. Kakoliris (2015) extends Derrida's ethics of hospitality, emphasizing the tension between unconditional ethical demands and the conditional structures of institutions and sovereignty. Bensussan (2014) emphasizes bifidity and floating ethics. Recent scholarship examines how military healthcare technologies reshape these ethical relations. Telemedicine has significantly expanded access to care in war zones and remote military settings, particularly in trauma, critical care, and mental health (Baigi et al. 2022; Mohammadi et al. 2020; Haimi 2024). The integration of artificial intelligence further enhances efficiency and clinical decision-making (Adirim and Madsen 2025).

THE DOCTOR-PATIENT RELATIONSHIP: JUS IN BELLO AND JUS POST BELLUM

The doctor-patient relationship has long been recognized as a core ethical structure in medical practice. Classical literature by Szasz and Hollender (1956), later expanded on by Emanuel and Emanuel (1992), conceptualizes the doctor-patient relationship not only as a technical interaction but also as a moral relationship. When the models are examined within the contexts of armed conflict (jus in bello) and post-war recovery (jus post bellum), their ethical significance becomes more crucial. During a war, asymmetries already present in medical practice are intensified, while post-war conditions challenge medicine to move beyond emergency to recovery and rehabilitation.

Szasz and Hollender's three-part scheme of activity-passivity, clinical guidance, and mutual participation remains one of the most influential typologies in understanding doctor-patient interactions. Each model corresponds to different clinical situations and degrees of patient capacity, but none is ethically neutral. Rather, each embeds assumptions about authority, cooperation, and moral responsibility.

The activity-passivity model is historically the oldest and is characterized by the patient's inability to meaningfully participate in decision-making. Typical examples include coma, severe trauma, anesthesia, and delirium. In such cases, the physician acts on the patient, assuming total responsibility for diagnosis and treatment. The prototype of this model is the parent-infant relationship, where care is necessarily unilateral and oriented toward protection and survival. In peacetime medicine, this model is generally treated as exceptional and temporary. In wartime medicine, however, it becomes structurally central. Large numbers of patients are suddenly rendered incapable of consent or dialogue: battlefield injuriesand other emergency situations frequently require immediate intervention without consultation. Within the framework of jus in bello, medical personnel are granted special status under international humanitarian law and are obligated to treat the wounded and sick without discrimination, including enemy combatants. In these circumstances, ethical responsibility is concentrated almost entirely in the physician's hands. The patient's identity, intentions, or political affiliation are ethically irrelevant to the medical task, even if they remain decisive in the broader conflict. This concentration of responsibility reveals the moral gravity of the activity-passivity model. Thus, war does not suspend medical ethics; rather, it exposes its most demanding form.

The second model, that of clinical guidance, presupposes a cooperative relationship between doctor and patient. The patient is conscious and capable of understanding information but remains dependent on the physician's expertise and leadership. This model is commonly applied in acute but non-catastrophic conditions and takes the parent-adolescent relationship as its prototype. In this model, parties are active participants, but the relationship is asymmetrical; the doctor guides decision-making and expects compliance. In wartime and post-conflict situations, this model appears in military hospitals, refugee camps, and emergency clinics. Patients may be able to communicate, but their autonomy is compromised due to pain, fear, displacement, and lack of alternatives. The ethical relationship here should balance beneficence and respect for emerging autonomy. The physician may guide treatment and therapy, but the patient's subjective experience of suffering and anxiety cannot be ignored. This model becomes particularly significant during the transition from jus in bello to jus post bellum. It thus serves as an intermediate ethical form between unilateral intervention and full partnership.

The third model, mutual participation, represents the most egalitarian conception of the doctor-patient relationship. It presupposes approximate equality of power, mutual interdependence, and shared responsibility for outcomes. This model is most appropriate for chronic conditions such as diabetes or heart disease, where long-term management depends on the patient's sustained engagement and self-care. Unlike the other models, mutual participation requires a more complex psychological and social organization and is not suitable for all patients or situations. After war, mutual participation gains ethical and practical significance. Survivors often face long-term disabilities and psychological trauma. Effective treatment of conditions such as post-traumatic stress disorder (PTSD) and long-term rehabilitation require continuous cooperation between doctors and patients. In these cases, medicine shifts from preservation of life to restoration of agency and social integration.

Emanuel and Emanuel's expanded typology further clarifies these ethical shifts. Their distinction between paternalistic, informative, interpretive, and deliberative models shows how different conceptions of autonomy operate within medical practice. More particularly, the deliberative model balances values and goals and is therefore especially relevant in post-conflict medicine. Because patients after war are often engaged in existential reassessment, medical decisions should consider questions of meaning, responsibility, and future orientation. Both during and after war, the doctor-patient relationship undergoes ethical transformation. War amplifies asymmetry and necessitates unilateral responsibility, while post-war recovery challenges medicine to reintroduce cooperation, dialogue, and shared decision-making.

From activity-passivity in emergency care to mutual participation in long-term recovery, medicine reflects broader transitions from survival to reconstruction. By situating these models within jus in bello and jus post bellum, we see that medical ethics is not peripheral to war but central to how societies respond to human vulnerability during and after conflict. Although these classical models are vitally important, the extreme condition of war is an asymetrical ethical situation that demands primary responsibility. What follows is an exploration of the ethical theories of Derrida and Levinas that unfolds ethical meanings beyond a simple doctor-patient relationship based on modeling. Responsibility, hospitality, and asymmetry are highlighted. After a presentation of their basic ethical theories, the doctor-patient relationship is revised and formulated both in jus in bello and jus post bellum contexts through their philosophies.

JACQUES DERRIDA ON ETHICS, RESPONSIBILITY, AND HOSPITALITY

According to Nicole Anderson (2017), Derrida does not offer an ethical theory in a traditional sense. He does not provide a system with universal principles and prescriptive accounts of right action. On the contrary, he asks how responsibility, decision, justice, and obligation are possible when the metaphysical foundations of ethics are put into question. Metaphysical ethics require stable foundations such as a self-identical subject, universal principles, rational deliberation, and rules. Derrida challenges these assumptions and asserts that ethics, like other metaphysical structures, relies on a presence of the present, the idea that meaning, truth, and moral agency can be fully present and self-identical. His method of deconstruction reveals that such presence is never pure. It is always disrupted by difference, temporality, and alterity. Deconstruction is neither ethical nor unethical as a method. It takes place within structures, inhabiting ethical systems and exposing their internal tensions. It shows that ethical systems depend on what they exclude: uncertainty, alterity, and what Derrida calls the "non-ethical opening of ethics." Ethics is ethical only because it is exposed to something other than itself, something incalculable and uncontrollable.

Anderson asserts that a central notion in Derrida's ethics is Otherness (alterity). For Derrida, ethical systems seek to regulate relations with others through general rules, but the Other always exceeds this regulation; the Other cannot be anticipated, calculated, or absorbed into a norm. This Othernessis not an opposition to ethics but its condition of possibility. Ethics exists only because it is exposed to what cannot be ethically programmed in advance. This may bring risk; the Other may bring harm, violence, or disruption. Thus, ethics is inseparable from vulnerability; one must decide without certainty, without rules that might justify a decision in advance.

For Anderson, Derrida develops his ethical insights through the concepts of justice, law, hospitality, and responsibility. He distinguishes two types of responsibility: *general,* which is governed by norms, laws, and shared principles, and *absolute,* which responds to the singularity of the Other. Absolute responsibility is not programmable; it responds to what is unique, unrepeatable, and irreplaceable. This form of responsibility cannot be justified by rules because they operate through repetition and substitution; absolute responsibility is bound to singularity and therefore resists generalization.

Anderson underlines Derrida's insistence that responsibility comes through irresponsibility. In Derrida's *The Gift of Death.* Abraham is commanded by God to sacrifice his son, thus facing the impossible decision between obeying the singular command of God and obeying the general ethical law prohibiting murder. If he obeys God, he is irresponsible to his son and community, and if he fulfills his responsibilities to his son and community, he is irresponsible to God. Responsibility is infinite, but finite agents cannot meet infinite demands. Decision always entails exclusion, violence, and betrayal; there is no pure ethical act. Therefore, ethics is founded on an insoluble contradiction between general responsibility to everyone and absolute responsibility to someone.

Anderson explains that, for Derrida, a genuine decision is only possible under a condition of undecidability; if a decision could be fully calculated according to a rule, it would merely be an application of a plan. Ethical decisions pass through uncertainty and risk, making responsibility real. One must decide without guarantees and assume responsibility for consequences, knowing that no decision can be fully justified. Thus, ethics is never purely nonviolent; every decision brings violence insofar as it excludes, prioritizes, or sacrifices others. This distinguishes Derrida from thinkers like Levinas, who seek a nonviolent ethical relation. Ethics does not eliminate violence, but it remains vigilant about it. The ethics of Derrida is not negative; it affirms the arrival of the Other, the future, and what cannot be anticipated.

Gerasimos Kakoliris (2015) situates Derrida's ethics of hospitality within current political and ethical contexts, particularly contemporary attitudes toward immigrants, refugees, and those without legal status. These are provoked by the idea that hospitality itself can be criminalized, for example, by laws that punish those who shelter undocumented migrants. For Derrida, these areexamples of political failure and crisis in the ethical foundations of Western societies.

Kakoliris asserts that at the center of Derrida's thought on hospitality is a structural aporia,a contradiction that defines ethical responsibility. Hospitality is governed by two inseparable imperatives, the law of unconditional hospitality and the conditional laws of hospitality. Responsibility emerges not from resolving this contradiction but from an endless negotiation. Derrida defines unconditional hospitality as the absolute ethical demand without any restriction, the opening of one's home to a stranger without asking for identification, justification, reciprocity, or even a name. The welcomeis prior to any determination of who the stranger is or how he behaves.

It is a radical openness to alterity, which precedes law, politics, and morality. This notion of hospitality is far different from conditional hospitality, which in Kant's view and that of others is governed by juridical, political, and moral norms that regulate borders, entry, residence, and behavior. These norms define who may enter, under what conditions, and for how long. They establish rights and duties both for the host and the guest and are an exercise of sovereignty. To offer hospitality, one must first possess a home, a territory, or a nation, and this possession necessarily involves power. Derrida emphasizes that conditional hospitality is never neutral. The host remains master of the house, retaining the authority to invite, exclude, or expel. Hospitality thus always contains an element of hostility, which Derrida calls "hostipitality."

Kakoliris's main critique of Derrida is his reliance on notions such as "pure," "true," or "absolute" hospitality. Kakoliris argues that unconditional hospitality cannot exist even as a conceptual ideal. Hospitality is always already conditioned by sovereignty, power, and exclusion. Impurity is not an external corruption of hospitality but its very condition. To posit an absolute form of hospitality risks reinstating the kind of metaphysical ideal that deconstruction sought to dismantle. Another ethical consequence of Derrida's hyperbolic hospitality is the production of perpetual guilt. Because unconditional hospitality can never be fulfilled, the subject is always ethically insufficient, always failing the Other. Derrida openly embraces this "bad conscience" as the pillar of ethical and political responsibility. One must feel remorse for those one cannot welcome, protect, or save. Kakoliris questions whether this infinite responsibility is ethically productive; while acknowledging that responsibility has no fixed limits, he suggests that equating responsibility with infinity risks paralysis and despair. Human agents are finite, situated, and limited. Ethics must therefore involve not only openness to the Other but also judgment, discrimination, and the acknowledgment of practical constraints.

According to Gérard Bensussan (2014), Derrida's ethics of the un-decisional is inseparable from the fact that every decision must be taken in relation to what is not known and not yet sensed. A decision never rests on full knowledge or certainty; it is always linked with the temporality of abandonment (prise-de-prise), when a subject holds mastery at the very moment of acting. Derrida insists that a decision is never unitary. Genuine decision, therefore, exceeds itself and obeys the rule of the "more than one toward the two," a plurality that cannot be reduced to unity. This paradoxical condition, having only one and yet not owning it, defines what Bensussan calls a floating ethic, which is without ground or fixed identity, continually movesin a "fly-sail-fly again" fashion, and is marked by irony and vulnerability. He links this ethical posture to the historical figure of the Marrano, whose existence is shaped by dissimulation, displacement, and survival under domination. In this context, decision speaks a non-language: it expresses what cannot be fully articulated within established codes. This unspoken language is the language of the Other. The response of the Other appears grammatically in the second-person imperative: "come," which Bensussan compares to Rosenzweig's "love me." Through this imperative, an order emerges without foundation; affirmation and positivity do not rely on dialectics, calculation, or prediction. Ethical meaning thus is not a meaning of a sovereign subject or a revelation but rather an exposure to alterity and hope. Thus, the ethics of the un-decisional can be understood as a messianism without a messiah, a constant opening without guarantee, fulfillment, or closure.

LEVINAS ON WAR, ETHICS, ASYMMETRY, AND THE FACE OF THE OTHER

Emmanuel Levinas's (1969) ethical philosophy provides one of the most radical counterarguments regarding war, mainly after WWII. He treats war not as a contingent political event nor as a necessity within history. He argues that war is not only an immoral practice but also a condition in which morality, responsibility, and ethical meaning simply do not exist. He proposes an ethical theory based in asymmetrical responsibility, revealed through the face-to-face encounter with the Other, which exposes violence as the ultimate negation of ethical life. In war, the primary ethical imperative,"thou shall not kill," is suspended. Killing the Other is normalized, justified, and even celebrated. War not only signifies an opposition to morality but is its temporary abolition. Ethics becomes a strategy, a survival, and victory.

Related to this, Levinasian philosophy addresses the reduction of the human being to forces within a totality. People are no longer singular moral agents; they become functions of a system, military, political, or historical. This totality is a philosophical impulse to comprehend all individuals within a unified order governed by power, causality, and necessity. Levinas asserts that Western philosophy privileges knowledge as the highest form of lucidity. To know is to relate, classify, and synthesize and to understand people as parts of a whole. This epistemology leads to totality and culminates in a vision of reality as a struggle between forces. This lucidity, according to him, is linked with a permanent possibility of war. Thinkers such as Hobbes and Nietzsche tended to an interpretation of human relations as forms of struggle rather than responsibility. War is the condition in which the truth of totality becomes visible.

Levinas's thought cannot be reduced to a simple binary between good and the evilof war; he recognizes that Being reveals itself as war, referring to a philosophical tradition from Heraclitus, Hobbes, and Nietzsche. Ontology is violent, totalizing, and competitive; ethics, on the contrary, is beyond being and linked to what Levinas calls infinity, which is not outside history but interrupts it from within; it judges each moment and each action. Ethics is fragile and exceptional, always threatened by politics, calculation, and survival instincts. Its fragility is evident in his conception of a just war. On the one hand, Levinas asserts that the suffering of the Other is immoral and never justifiable. War produces mass suffering and therefore is fundamentally unjust. On the other hand, Levinas claims that violence may be necessary when a neighbor is threatened. Justice arises with the presence of a "third" and requires judgment, comparison, and institutions. In such cases, the aggressor may no longer have a face, and violence can be legitimate. He uses the case of Palestine and Israel to support these views (Hofmeyr 2024). Levinas's notion of "useless suffering" is pivotal here; war produces suffering that cannot be justified or instrumentalized. The Holocaust, which shaped Levinas's thought, exemplifies this suffering. Any attempt to justify suffering, historically, politically, or metaphysically, is a moral outrage. What remains is only the inter human order, the responsibility awakened by the suffering of the Other (Lau 2015).

Levinas introduces the concept of the face of the Other against the totality that describes it. The face is not merely a physical appearance but the manifestation of the Other's irreducible alterity. In the encounter with the face, the self is submitted to an ethical imperative that precedes all cognition, "thou shall not kill."This relation is not ontological but metaphysical. One is responsible for the Other, without reciprocity, beforecalculation and knowledge. This responsibility is infinite, non-transferable, and cannot be discharged.

War negates this face-to-face relation. The enemy is no longer encountered as a face, but as a target, a number, or an abstract threat. Weapons, propaganda, and bureaucracy erase proximity and the ethical command of the Other. Killing becomes justified because the Other is reduced to an equal adversary in a system of violence. The asymmetrical ethical responsibility is replaced by strategic equivalence.

For Levinas, peace is neither a simple cessation of hostilities nor an equilibrium established by victory. True peace arises only from the ethical relation between singular beings. This peace is fragile and demanding. It does not erase conflict, but rather orients political and social life towards a justice far different from totality. Even when war appears unavoidable, ethical responsibility remains. The face of the Other continues to command, judging war from beyond its logic; hence Levinas allows for the possibility of distinguishing just from unjust wars, but only if politics remains subordinated to ethics rather than replacing it. This peace is not collective or institutional but rather personal, "my peace," and arises from the self's non-egotistical relation with the Other. Ethical peace resists the totalizing logic of war by affirming plurality without fusion, a "unity of plurality" grounded in infinity, not totality (Hofmeyr 2024). True peace, what he calls **"eschatological peace"**, comes from a relation with infinity, with what exceeds totality. This eschatology is not a future end of history but a judgment enacted in each moment and ethical response. Peace begins with the I: "It must be my peace." This does not mean subjective retreat but radical responsibility. The ethical subject maintains itself without egoism, exposed to the Other in love and non-indifference (Lau 2015).

THE DOCTOR-PATIENT RELATIONSHIP: JUS IN BELLO AND JUS POST BELLUM THROUGH THE PHILOSOPHIES OF DERRIDA AND LEVINAS

For Derrida, responsibility emerges only through a contradiction between general responsibility to all Others and absolute responsibility to the singular Other. Derrida generalizes this structure to all ethical decisions. Responsibility is infinite, yet human agents are finite. Every decision excludes, sacrifices, and betrays other responsibilities.

In medical practice, this aporia is pervasive. Physicians constantly navigate between general responsibility, following protocols, respecting institutional norms, coordinating with colleagues, and absolute responsibility toward the singular patient. Life-threatening cases, hard diagnoses, and invasive interventions exemplify this tension. Decisions must be made under conditions of uncertainty and risk, where no rule can fully justify the outcome. Responsibility emerges precisely where calculation fails. In jus in bello, this aporia becomes ethically tragic. Hard decisions and the treatment of enemy combatants confront physicians with undecidable choices. Saving one life may entail abandoning another. Treating a wounded enemy may indirectly enable further violence. Derrida's ethics does not resolve these dilemmas but insists that responsibility consists in deciding without guarantees and assuming the ethical burden of the decision without an alibi.

Derrida develops an ethics of the undecisional, or "floating ethics." A genuine decision can occur only in a moment of undecidability; otherwise, it would be merely the application of a rule. This undecisional moment is marked by vulnerability, exposure, a hope that the Other will come, respond, or be healed. In clinical contexts, undecidability arises when data are ambiguous, symptoms are ambiguous, or therapeutic options carry comparable risks. In both wartime and post-war medicine, such situations are frequent. Physicians must balance incomplete information, similar prognoses, and competing obligations. Ethical judgment, in this sense, is not the elimination of uncertainty but its endurance. The physician acts while knowing that the decision remains haunted by what was not chosen.

Applied to the doctor-patient relationship, Levinas's ethics radically reconfigures medical responsibility. The physician is not first a technical expert or institutional agent but a subject exposed to the suffering of the Other. The patient appears as a vulnerable face that commands care independently of identity, merit, or reciprocity.

In jus in bello, this ethical demand becomes particularly acute. Treating wounded soldiers of the opposing side exemplifies Levinasian responsibility in its purest form. Stripped of political and military identity, the enemy appears as a singular Other in pain. Medical care thus interrupts the logic of war by reinstating the face-to-face relation that war seeks to erase. Levinas insists that suffering is never justifiable. War produces what he calls "useless suffering," suffering that cannot be redeemed by historical, political, or strategic explanations. In this sense, even when violence is deemed necessary to protect a neighbor threatened by a third party, the suffering of the wounded patient remains ethically unjustifiable. The doctor-patient relationship becomes a fragile site where ethics persists within war as an interruption rather than a resolution. Medical care enacts a form of peace beyond politics, a peace grounded not in treaties or victories but in singular ethical responses.

Levinas emphasizes that ethics is constituted through speech and conversation and not from action and reaction. Genuine conversation presupposes greeting, attention, and respect for the Other. In the medical encounter, this means that healing is not reducible to technical intervention but involves proximity, listening, and acknowledgment of alterity. The physician is described by Levinas as a "hostage" to the Other, bound by responsibility beyond power, knowledge, or professional authority. In both jus in bello and jus post bellum, this conception challenges any symmetrical understanding of the doctor-patient relationship. The ethical ideal is not mutual participation or equality but an impossible orientation toward the me-Other relation. The physician strives for this relation while recognizing that it constantly withdraws, tending instead toward a me-he/she relation mediated by institutions, diagnoses, and third parties. Ethical medical practice consists in resisting this reduction without claiming to overcome it. The me-Other relationship is almost impossible, but a good doctor should strive for it ad infinitum.

APPLICATIONS OF MILITARY TECHNOLOGIES: UAVS, TELEMEDICINE, AND AI IN HEALTHCARE DELIVERY

Unmanned aerial vehicles (UAVs), originally used for military surveillance and reconnaissance, are now increasingly used for healthcare and emergency services. A systematic literature review (Baigi et al. 2022) of 15 papers from 2018-2022 and 136 primary studies published between 2007 and 2023 shows that UAVs improve emergence response times, particularly in time-critical scenarios such as cardiac arrest, with automated external defibrillators (AEDs). They are also employed in blood and vaccine delivery, disaster search and rescue, medical logistics, and telemedical support. Finally, UAVs contribute to telesurgery, enabling remote surgical support where infrastructure is compromised.

A systematic review (Mohammadi et al. 2020) examining the use and characteristics of telemedicine applications in military healthcare systems worldwide follows the PRISMA guidelines for systematic reviews. Comprehensive searches were conducted in four major scientific databases, PubMed, Scopus, Embase, and Web of Science, for English-language articles published between 2014 and 2018.The United States accounted for the majority of identified studies, reflecting its advanced infrastructure and extensive operational experience.Telemedicine was most frequently deployed in military garrisons, overseas missions, naval operations, and active war zones. Both synchronous (real-time) and asynchronous (store-and-forward) modalities were used almost equally, with a growing trend toward hybrid models that combine immediacy with flexibility. Video conferencing and email emerged as the dominant communication routes, supported by internet and satellite-based networks. From a clinical perspective, telemedicine was most commonly applied in telepsychology, telepsychiatry, teleorthopedics, tele-intensive care (tele-ICU), teledermatology, and tele-stroke services.Telepsychology was particularly prominent, addressing combat-related stress, trauma, and mental health conditions. Tele-ICU and tele-critical care applications demonstrated improved access to specialist expertise, reduced patient transfers, and enhanced decision-making in time-sensitive scenarios.

The expanding role of telemedicine in military healthcare systems enables the effective delivery of medical services to geographically dispersed and operationally constrained military populations. The author (Haimi 2024) addresses how telemedicine supports clinical decision-making, enhances access to specialized care, and improves health outcomes for service members deployed in remote, austere, or combat environments. The findings demonstrate that telemedicine has a substantial positive impact on military healthcare delivery. One of the most significant outcomes is improved access to specialist care, particularly in trauma management, critical care, mental health, dermatology, and orthopedics. Tele-ICU and tele-consultation services enable frontline clinicians to receive expert guidance in real time, reducing diagnostic uncertainty and enhancing patient safety. Telepsychiatry and telepsychology are highlighted as especially effective, addressing post-traumatic stress disorder (PTSD), anxiety, depression, and other mental health conditions prevalent among military personnel.

Artificial intelligence is emerging as a transformative enabler within the U.S. Military Health System (MHS)and is being integrated with telemedicine, which is used in expeditionary and operational military healthcare. AI-enhanced telemedicine can improve the effectiveness of remote consultations by supporting clinical decision-making, triage, patient monitoring, and diagnostic accuracy, particularly in austere or deployed environments where specialists are unavailable. Applications such as

AI-supported teleradiology, tele-critical care, and telepsychiatry are high-impact areas.AI-driven data analysis can assist clinicians by synthesizing patient histories, imaging data, and real-time physiological inputs, reducing diagnostic errors and accelerating treatment decisions. These capabilities directly enhance care delivery for deployed service members and reduce the need for medical evacuation(Adirim and Madsen 2025).

ETHICS BEYOND PROXIMITY: DERRIDA, LEVINAS, AND MILITARY TECHNOLOGIES IN CONTEMPORARY HEALTHCARE

The increasing use of telemedical practices, artificial intelligence, and military unmanned aerial vehicles in healthcare reshapes the ethical landscape of medicine. When examined through the philosophical theories of Derrida and Levinas, these developments invite a deeper ethical reflection on how responsibility, hospitality, and ethical asymmetry are preserved or threatened when the traditional face-to-face encounter is technologically mediated.

Derrida's philosophy deepens this analysis by revealing the aporetic structure of responsibility in technologically mediated care. For Derrida, responsibility is never pure or fully justified. It emerges from an irreducible contradiction between general responsibility governed by rules and absolute responsibility toward the singular Other. In medical practice, this tension is always apparent. Physicians may follow protocols, guidelines, and regulations while responding to the a singular suffering patient. Telemedicine and AI intensify this aporia. AI systems operate through generalization, pattern recognition, and probabilistic reasoning, while ethical responsibility demands attentiveness to singularity. The physician must decide when to rely on algorithmic recommendations and when to deviate from them, knowing that no decision can be fully justified in advance. In wartime and post-war contexts, Derrida's ethics of undecidability may become applicable. Telemedicine decisions are often made under conditions of incomplete information, time pressure, and competing obligations. Choosing which patient to prioritize, whether to evacuate or treat remotely, and whether to rely on AI-supported systems entails exclusion and sacrifice. Derrida insists that responsibility arises precisely here, where calculation fails. Ethical decision-making does not eliminate violence but remains vigilant toward it. Military healthcare technologies, therefore, do not absolve responsibility; they redistribute and complicate it. The physician remains responsible not only for the outcome but also for delegating aspects of care to machines. Derrida's concept of hospitality further illuminates the ethical structure of technologically mediated military healthcare. The hospital, whether physical or virtual, is a space of hospitality. Telemedicine platforms and UAV-supported care extend this hospitality across distances, yet they remain governed by access protocols, identification, security measures, and resource constraints. Derrida distinguishes unconditional hospitality, the impossible ethical demand to welcome the Other without restriction, from conditional hospitality regulated by law and power. Military healthcare exemplifies what Derrida calls "hostipitality," where care and control coexist. Telemedicine may welcome the patient while simultaneously surveilling, categorizing, and managing him within military systems. This tension is ethically productive rather than resolvable. Unconditional hospitality cannot be fully realized, but ethical responsibility consists in negotiating its impossibility without denying it. In military telemedicine, this means remaining attentive to those who cannot be reached, treated, or saved. Derrida emphasizes the importance of "bad conscience," the acknowledgment of ethical insufficiency. AI and UAV technologies promise efficiency and optimization, but Derrida's ethics resists any narrative of ethical closure. They can erase the face by replacing it with data, but they can also allow the face to appear where physical presence is impossible. They can distribute responsibility in ways that obscure accountability, but they can also enable physicians to respond to suffering they would otherwise never encounter.

Telemedicine allows physicians to treat soldiers in combat zones, remote bases, naval vessels, and post-conflict environments without physical proximity. AI systems further assist in diagnosis, monitoring, and decision-making, often under conditions of uncertainty and urgency. UAVs extend these capabilities by delivering medical supplies, blood, vaccines, and defibrillators and by supporting search, rescue, and evacuation in hostile or inaccessible environments. Levinas's philosophy forces us to ask whether ethical responsibility can survive when proximity is replaced by mediation. At first glance, telemedicine appears to weaken the Levinasian face-to-face relation. The patient's face is filtered through screens, data, images, and algorithms. AI systems synthesize patient information, sometimes without direct human encounters. UAVs deliver care without anyhuman presence. However, Levinas's ethics is not reducible to physical closeness alone. The face is not merely a visible surface; it is the manifestation of irreducible alterity and vulnerability. Telemedicine does not abolish the face, but it risks obscuring it behind technical representations. Ethical danger arises when the patient is reduced to data points, probabilities, or logistical objectives. In military healthcare, this danger is intensified by operational priorities such as readiness, efficiency, and survival. Nevertheless, these technologies can also serve as ethical interruptions of war. Telemedicine enables physicians to respond to suffering across enemy lines and borders and in inaccessible zones, reenabling responsibility where war would have erased it. Treating wounded enemy soldiers remotely exemplifies Levinasian asymmetry: care is offered without reciprocity, merit, or political justification. The ethical command of the face persists even when mediated by technology.

CONCLUSIONS

The doctor-patient relationship, when situated in jus in bello and jus post bellum, exposes responsibility in its most demanding and fragile form. Responsibility arises precisely where calculation fails. Every medical decision, especially in conditions of scarcity or danger, entails exclusion, sacrifice, and an ethical remainder. Derrida's notion of hospitality further illuminates the structure of medical care. Hospitals, clinics, and even telemedical platforms are spaces of conditional hospitality, governed by rules, identification, and power, yet unconditional hospitality persists as an impossible ethical demand that haunts every act of care. In military medicine, this tension appears with particular clarity: care is offered within systems of control, surveillance, and strategic priority. Ethical responsibility consists not in resolving this tension, but in remaining attentive to it, sustaining what Derrida describes as a necessary "bad conscience" regarding those who cannot be fully welcomed, treated, or saved.

Levinas complements and radicalizes this perspective by grounding ethics in the face of the suffering Other. War, for Levinas, suspends ethics by erasing the face and reducing human beings to targets or forces within a totality, yet medical care interrupts this suspension. Treating wounded enemy soldiers, refugees, and traumatized civilians reinstates the face-to-face relation that war would abolish. Responsibility here is asymmetrical, infinite, and non-reciprocal.The suffering of the Other is never justifiable, even when violence is deemed necessary to protect a third party. Medicine thus becomes a fragile site where ethics persists within war, not as reconciliation but as interruption.

These ethical insights become even more complex in the context of military medical technologies such as telemedicine, artificial intelligence, and UAVs. These technologies extend care beyond physical proximity, raising the question of whether responsibility can survive mediation. Derrida's ethics suggests that technology does not eliminate responsibility but redistributes and complicates it. Delegating care to AI systems or UAVs does not absolve the physician; it introduces new sites of undecidability and new forms of ethical risk. Levinas, meanwhile, warns against the reduction of the patient to data, images, or logistical objectives, yet proximity, in Levinas's sense, is not merely spatial. The face can still command responsibility through mediation, provided that singular suffering is not absorbed into abstraction. Ethics in wartime and post-war medicine must be understood as an ongoing exposure rather than a solved problem. Responsibility exceeds proximity, extends beyond certainty, and resists technological closure. Whether in direct care, remote treatment, or AI-supported decision-making, the physician remains answerable to the suffering Other, without alibi, and without final justification.

REFERENCES

Adirim, T., & Madsen, C. (2025). Artificial intelligence in the U.S. military health system: Forging a new frontier for clinical care and efficiency. *Military Medicine*, *190*(7–8), 199–205. DOI: 10.1093/milmed/usae428 PMID: 39292525

Anderson, N. (2017). Ethics. In Colebrook, C. (Ed.), *Derrida: Key concepts* (pp. 153–166). Acumen.

Baigi, S. F. M., Kimiafar, K., Sarbaz, M., Abbaszadeh, A., & Mousavi, A. S. (2022). Effect of telemedicine in military medicine: A literature review. *Paramedical Sciences and Military Health*, *17*(2), 77–90.

Bensussan, G. (2014). Title of chapter. In Cohen-Levinas, D., & Crépon, M. (Eds.), *Levinas–Derrida* (pp. 249–xx). Hermann.

Emanuel, E. J., & Emanuel, L. L. (1992). The physician–patient relationship. *Journal of the American Medical Association*, *267*(16), 2221–2226. DOI: 10.1001/jama.1992.03480160079038 PMID: 1556799

Haimi, M. (2024). Telemedicine in war zones: Prospects, barriers, and meeting the needs of special populations. *Frontiers in Medicine*, *11*, 1417025. Advance online publication. DOI: 10.3389/fmed.2024.1417025 PMID: 39497840

Hofmeyr, A. B. (2024). The challenge that war poses to Levinas's thought. *Philosophia: International Journal of Philosophy*, *25*(1), 26–44.

Kakoliris, G. (2015). Jacques Derrida on the ethics of hospitality. In Imafidon, E. (Ed.), *The ethics of subjectivity: Perspectives since the dawn of modernity* (pp. 144–156). Palgrave Macmillan. DOI: 10.1057/9781137472427_9

Lau, K.-y. (2015). War, peace and love: The logic of Lévinas. *Social Imaginaries*, *1*(2), 103–126. DOI: 10.5840/si20151219

Levinas, E. (1969). *Totality and infinity: An essay on exteriority* (Lingis, A., Trans.). Duquesne University Press. (Original work published 1961)

Mohammadi, R., Tabanejad, Z., Abhari, S., Honarvar, B., Lazem, M., Maleki, M., & Garavand, A. (2020). A systematic review of the use of telemedicine in the military forces worldwide. *Shiraz E Medical Journal*, *21*(11), e99343. Advance online publication. DOI: 10.5812/semj.99343

Szasz, T. S., & Hollender, M. H. (1956). The basic models of the doctor–patient relationship. *Archives of Internal Medicine*, *97*, 585–592. PMID: 13312700

KEY TERMS AND DEFINITIONS

Absolute and General Responsibility: This general ethical paradox is elucidated by the paradox with which Abraham, in the biblical story, is confronted: God commands Abraham to sacrifice on Mount Moriah, as a burnt offering, his son Isaac (Genesis 22: 1–19). Abraham is faced with the harrowing choice of whether to fulfil his responsibility towards the singular, in this case God, or to obey the general ethical laws of his society, and fulfil his responsibility towards all others (Isaac, Sarah, his friends, family and community). By singular or singularity, Derrida is referring not just to God, but to that which is irreducible, unrepeatable, heterogeneous, and idiosyncratic (another person or event, for instance).

Artificial Intelligence: The study of [intelligent] agents that receive precepts from the environment and take action. Each such agent is implemented by a function that maps percepts to actions, and we cover different ways to represent these functions, such as production systems, reactive agents, logical planners, neural networks, and decision-theoretic systems" source: Russell, Stuart J., and Peter Norvig. 2010. Artificial intelligence: a modern approach, 3rd ed. Upper Saddle River, N.J.: Prentice Hall. https://www.worldcat.org/oclc/688385283. ISBN 9780132071482.

Ethics of Asymmetry: The meeting is based on a radical asymmetry; I am always more responsible than the Other. The face of the Other, as in a manner the I, occupies the position of the Other in his fragility and nudity of the face, is the responsibility for the Other, as "the impossibility for the other man, the impossibility of leaving him alone in the mystery of death". This relation is based on love; this love is not love or eros in general, nor a reduction to altruism, the goodness of a generous nature, but it is rather an an-archic bond between the subject and the good that comes from the outside. Source: Emmanuel Levinas, "Humanism and An-Archy," *Revue Internatinal de la Philoso- phie*, no. 85 (1968): 65-82.

Telemedicine: the use of electronic information and communications technologies to provide and support health care when distance separates the participants. Institute of Medicine (US) Committee on Evaluating Clinical Applications of Telemedicine; Field MJ, editor. Telemedicine: A Guide to Assessing Telecommunications in Health Care. Washington (DC): National Academies Press (US); 1996. 1, Introduction and Background. Available from: https://www.ncbi.nlm.nih.gov/books/NBK45440/

Unmanned Autonomous Vehicles: Fully autonomous vehicles which do not require a driver at all. Source: Christoph Bartneck, Christoph Lütge, Alan Wagner, Sean Welsh, *An Introduction to Ethics and AI*, Charm: Springer, 2021, p.40.

Compilation of References

Adirim, T., & Madsen, C. (2025). Artificial intelligence in the U.S. military health system: Forging a new frontier for clinical care and efficiency. *Military Medicine*, *190*(7–8), 199–205. DOI: 10.1093/milmed/usae428 PMID: 39292525

Alderman, R. (2015). Looking toward 7th and 8th generation warfare. Military Embedded Systems, 7 July. Available at: https://militaryembedded.com/unmanned/isr/looking-toward-7th-and-8th-generation-warfare (Accessed: 14.04.2026).

Allen, G. C. (2023). China's Pursuit of Defense Technologies: Implications for U.S. and Multilateral Export Control and Investment Screening Regimes. https://www.uscc.gov/sites/default/files/2023-04/Gregory_Allen_Testimony.pdf

Allen, G. C. (2024, October 15). Blocking China's Access to AI Chips Matters to U.S. National Security. Center for Strategic and International Studies. https://www.csis.org/analysis/blocking-chinas-access-ai-chips-matters-us-national-security

Althobiti, M., Nhung, T. T. T., Verma, S., Albugami, R. R., & Kumar, R. (2025). Artificial intelligence and biosensors: Transforming cancer diagnostics. *Medicine in Novel Technology and Devices*, *27*, 100378. DOI: 10.1016/j.medntd.2025.100378

Altmann, J., & Sauer, F. (2017). Autonomous weapon systems and strategic stability. *Survival*, *59*(5), 117–142. DOI: 10.1080/00396338.2017.1375263

Álvarez-Machancoses, Ó., Galiana, E. J. D., Cernea, A., De La Viña, J. F. S., & Fernández-Martínez, J. L. (2020). <p>On the Role of Artificial Intelligence in Genomics to Enhance Precision Medicine</p>. *Pharmacogenomics and Personalized Medicine*, *13*, 105–119. DOI: 10.2147/PGPM.S205082 PMID: 32256101

Amilakhvari, L., & Baghaturia, O. (2024). Russia's Militaristic Rhetoric, Imperialism, and Expansion: Wars in Georgia and Ukraine. In Pietrzak, P. (Ed.), *Dealing With Regional Conflicts of Global Importance* (pp. 161–181). IGI Global Scientific Publishing., DOI: 10.4018/978-1-6684-9467-7.ch008

Amin, S. U., & Hossain, M. S. (2020). Edge intelligence and Internet of Things in healthcare: A survey. *IEEE Access : Practical Innovations, Open Solutions*, *9*, 45–59. DOI: 10.1109/ACCESS.2020.3045115

Amoroso, D., & Guglielmo, T. (2018). The ethical and legal case against autonomy in weapons systems. *Global Jurist, 17*(3).

Amoroso, D., & Tamburrini, G. (2018). The ethical and legal case against autonomy in weapons systems. *Global Jurist*, *18*(1), 1–20. DOI: 10.1515/gj-2017-0012

Amri, M. M., & Abed, S. A. (2023). The Data-Driven Future of Healthcare: A review. *Mesopotamian Journal of Big Data*, *2023*, 68–74. DOI: 10.58496/MJBD/2023/010

Anderson, K., & Waxman, M. (2013). *Law and ethics for autonomous weapon systems: Why a ban won't work and how the laws of war can.* The Hoover Institution, Stanford University, Jean Perkins Task Force on National Security and Law Essay.

Anderson, N. (2017). Ethics. In Colebrook, C. (Ed.), *Derrida: Key concepts* (pp. 153–166). Acumen.

Anduril. (2022). *Autonomy for every mission: Lattice is an open operating system for defense that brings autonomy to the toughest missions.* https://www.anduril.com/lattice/

Antebi, L. (2021). *Artificial intelligence and national security in Israel.* Institute for National Security Studies (INSS).

Arkin, R. (2010). The case for ethical autonomy in unmanned systems. *Journal of Military Ethics*, *9*(4), 332–341. DOI: 10.1080/15027570.2010.536402

Arkin, R. C. (2008). On the ethical quandaries of a practicing roboticist: A first-hand look. *Current Issues in Computing and Philosophy*, *175*, 45–51.

Asaro, P. (2017). Moral and ethical perspectives. In R. Acheson, M. Bolton, & E. Minor (Eds.), *The humanitarian impact of drones* (pp. 141–149). Reaching Critical Will.

Asaro, P. (2020). Autonomous weapons and the ethics of artificial intelligence. In *Ethics of Artificial Intelligence* (pp. 212–236). online edn, Oxford Academic. https://doi.org/DOI: 10.1093/oso/9780190905033.003.0008

Asaro, P. (2013). On banning autonomous weapon systems: Human rights, automation, and the dehumanization of lethal decision-making. *International Review of the Red Cross*, *94*(886), 687–709. DOI: 10.1017/S1816383112000768

Australasian Legal Information Institute. (2022). *Queensland consolidated acts.* Retrieved March 3, 2024, from http://www5.austlii.edu.au/au/legis/qld/consol_act/cc189994/s27.html#:~:text=(1)%20A%20person%20is%20not,to%20control%20the%20person's%20actions

Ayoub, K., & Payne, K. (2016). Strategy in the age of artificial intelligence. *The Journal of Strategic Studies*, *39*(5-6), 793–819. DOI: 10.1080/01402390.2015.1088838

Backes, O., & Swab, A. (2019). *Cognitive Warfare: The Russian Threat to Election Integrity in the Baltic States.*

Backes, O., & Swab, A. (2019). *Cognitive Warfare. The Russian Threat to Election Integrity in the Baltic States.* Belfer Center for Science and International Affairs.

Baigi, S. F. M., Kimiafar, K., Sarbaz, M., Abbaszadeh, A., & Mousavi, A. S. (2022). Effect of telemedicine in military medicine: A literature review. *Paramedical Sciences and Military Health*, *17*(2), 77–90.

Bai, Y. (2025). Moral judgment in the age of artificial intelligence: How humans perceive ethical decisions made by machines. *Humanistic Management Journal*, *10*(3), 361–385. Advance online publication. DOI: 10.1007/s41463-025-00215-0

Bandla, M. S., Liju, A. M., Mishra, G., & Luharia, A. (2023). Artificial intelligence in healthcare with special consideration in radio imaging. In *Proceedings of the 2023 1st DMIHER International Conference on Artificial Intelligence in Education and Industry 4.0 (IDICAIEI)* (pp. 108–113). DOI: 10.1109/IDICAIEI58380.2023.10406913

Bankov, B. (2024). Hybrid Warfare: How to Escape the Conceptual Gray-Zone. *Journal of Strategic Security*, *17*(1), 1–23. https://digitalcommons.usf.edu/jss/vol17/iss1/1. DOI: 10.5038/1944-0472.17.1.2118

Bax, C. (2017). Otherwise than Being-with: Levinas on Heidegger and Community. *Human Studies*, *40*(3), 381–400. DOI: 10.1007/s10746-017-9432-x

Belkin, P., Bowen, A. S., Nelson, R. M., & Welt, C. (2024, December 23). Russia's war against Ukraine: U.S. policy and the role of Congress. Congressional Research Service.

Bender, E. M., Gebru, T., McMillan-Major, A., & Shmitchell, S. (2021). On the dangers of stochastic parrots: Can language models be too big? *Proceedings of the 2021 ACM Conference on Fairness, Accountability, and Transparency*, 610–623. DOI: 10.1145/3442188.3445922

Ben-Israel, G. M. (2025). Deconstructing Warfare: Redefining Superpowers, Battlefields, Wars, and Armies. In Pietrzak, P. (Ed.), *International Relations Theory and Philosophical Political Insights Into Conflict Management* (pp. 175–208). IGI Global Scientific Publishing., DOI: 10.4018/979-8-3693-9626-1.ch008

Bensussan, G. (2014). Title of chapter. In Cohen-Levinas, D., & Crépon, M. (Eds.), *Levinas–Derrida* (pp. 249–xx). Hermann.

Bergo, B. (2005). Ontology, Transcendence, and Immanence in Emmanuel Levinas' Philosophy. *Research in Phenomenology*, *35*(1), 141–177. DOI: 10.1163/1569164054905474

Bergo, B. (2006). Emmanuel Levinas. In Zalta, E. N., & Nodelman, U. (Eds.), *The Stanford Encyclopedia of Philosophy*.

Bergo, B. (2019). Levinas and Husserl. In Morgan, M. (Ed.), *The Oxford Handbook of Levinas* (pp. 71–102). Oxford University Press.

Bernstein, D. L. (2022). *Woke antisemitism: How a Progressive Ideology Harms Jews*. Wicked Son.

Bhavana, S., & Mithra, S. G. S. (2025). Revolutionizing healthcare with an in-depth analysis of AI's transformative impact in India. *Social Sciences & Humanities Open*, *11*, 101467. DOI: 10.1016/j.ssaho.2025.101467

Bigman, Y. E., & Gray, K. (2018). People are averse to machines making moral decisions. *Cognition*, *181*, 21–34. DOI: 10.1016/j.cognition.2018.08.003 PMID: 30107256

Bilotta, M., Rocca, R., & Alcaro, S. (2025). Next-Generation Drug Discovery: The AI revolution in pharmaceutical research. *Artificial Intelligence in the Life Sciences*, *100149*, 100149. Advance online publication. DOI: 10.1016/j.ailsci.2025.100149

Birch, M. (2024). Who did that? AI assisted targeting and the lowering of thresholds in Gaza. *Medicine, Conflict, and Survival*, *40*(2), 97–100. DOI: 10.1080/13623699.2024.2364937 PMID: 38915179

Biselli, R., Nisini, R., Lista, F., Autore, A., Lastilla, M., De Lorenzo, G., Peragallo, M. S., Stroffolini, T., & D'Amelio, R. (2022). A historical review of military medical strategies for fighting infectious diseases: From battlefields to global health. *Biomedicines*, *10*(8), 2050. DOI: 10.3390/biomedicines10082050 PMID: 36009598

Bitko, G. (2025, May 15). Powering Tomorrow: How Strategic AI infrastructure Policy Can Secure America's technological edge. *Forbes*. https://www.forbes.com/sites/gordonbitko/2025/05/15/powering-tomorrow-how-strategic-ai-infrastructure-policy-can-secure-americas-technological-edge/

Blackbourne, L. H., Baer, D. G., Eastridge, B. J., Renz, E. M., Chung, K. K., DuBose, J., Wenke, J. C., Cap, A. P., Biever, K. A., Mabry, R. L., Bailey, J., Maani, C. V., Bebarta, V., Rasmussen, T. E., Fang, R., Morrison, J., Midwinter, M. J., Cestero, R. F., & Holcomb, J. B. (2012). Military medical revolution: Deployed hospital and: en route: care. *The Journal of Trauma and Acute Care Surgery*, *73*(6), S378–S387. DOI: 10.1097/TA.0b013e3182754900 PMID: 23192059

Board, D. I. (2019). *AI principles: recommendations on the ethical use of artificial intelligence by the department of defense: supporting document*. United States Department of Defense.

Boccaccini, F. (2016). De Reinach à Levinas: Hering et le réalisme phénoménologique. *Revue de Théologie et de Philosophie*, *148*, 433–448.

Bornstein, J. (2015, March 24–26). *DOD autonomy roadmap: Autonomy community of interest* [Conference session]. National Defense Industrial Association 16th Annual Science and Engineering Conference/Defense Tech Exposition, Springfield, VA, United States. https://ndiastorage.blob.core.usgovcloudapi.net/ndia/2015/SET/WedBornstein.pdf

Bostrom, N. (2014). *Superintelligence: Paths, dangers, strategies*. Oxford University Press.

Bouderhem, R. (2024). Shaping the future of AI in healthcare through ethics and governance. *Humanities & Social Sciences Communications*, *11*(1), 416. Advance online publication. DOI: 10.1057/s41599-024-02894-w

Boukhennoufa, O., Comas, L., Nicod, J. M., Ungureanu, C., Zerhouni, N., & Boulahdour, H. (2024, May). Automatic detection of parathyroid glands in nuclear medicine. In *2024 IEEE International Symposium on Biomedical Imaging (ISBI)* (pp. 1-4). IEEE. DOI: 10.1109/ISBI56570.2024.10635117

Boulanin, V., & Verbruggen, M. (2017). *Mapping the development of autonomy in weapon systems*. Stockholm International Peace Research Institute (SIPRI).

Boyle, M. (2015). The legal and ethical implications of drone warfare. *International Journal of Human Rights*, *19*(2), 105–126. DOI: 10.1080/13642987.2014.991210

Brandao, P. R. (2025). The impact of artificial intelligence on modern society. *AI*, *6*(8), 190. DOI: 10.3390/ai6080190

Braverman, H. (1974). *Labor and monopoly capital.* Monthly Review Press. DOI: 10.14452/MR-026-03-1974-07_1

Brooks, R. (2014). Drones and the international rule of law. *Ethics & International Affairs, 28*(1), 83–103. DOI: 10.1017/S0892679414000070

Brown, D. (2011). Proportionality in modern just war theory: A tort-based approach. *Journal of Military Ethics, 10*(3), 213–229. DOI: 10.1080/15027570.2011.608496

Brown, R. A., Zhu, H., & Mitchell, J. R. (2005). Distributed vector processing of a new local multiscale Fourier transform for medical imaging applications. *IEEE Transactions on Medical Imaging, 24*(5), 689–696. DOI: 10.1109/TMI.2005.845320 PMID: 15889555

Broyles, D., Crichton, R., Jolliffe, B., Sæbø, J. I., & Dixon, B. E. (2016). Shared longitudinal health records for clinical and population health. In *Elsevier eBooks* (pp. 149–162). https://doi.org/DOI: 10.1016/B978-0-12-803135-3.00010-4

Brunstetter, D., & Braun, M. (2011). The implications of drones on the just war tradition. *Ethics & International Affairs, 25*(3), 337–358. DOI: 10.1017/S0892679411000281

Buchanan, A., & Keohane, R. O. (2004). The preventive use of force: A cosmopolitan institutional proposal. *Ethics & International Affairs, 18*(1), 1–22. DOI: 10.1111/j.1747-7093.2004.tb00447.x

Buonocore, F., Annosi, M. C., De Gennaro, D., & Riemma, F. (2024). Digital transformation and social change: Leadership strategies for responsible innovation. *Journal of Engineering and Technology Management, 74*, 101843. DOI: 10.1016/j.jengtecman.2024.101843

Burston, D., & Jacobsen, K. (2025). *Authoritarianism in All its Guises: Perspectives on Right, Left, and Center*. Taylor & Francis. DOI: 10.4324/9781003389927

Buzan, B. (2015). The English School: A neglected approach to International Security Studies. *Security Dialogue, 46*(2), 126–143. DOI: 10.1177/0967010614555944

Campaign to Stop Killer Robots. (2018). *Country views on killer robots*. https://www.stopkillerrobots.org/wp-content/uploads/2018/11/KRC_CountryViews22Nov2018.pdf

Campaign to Stop Killer Robots. (n.d.). *Campaign to Stop Killer Robots*. https://www.stopkillerrobots.org/

Campbell, J. (2014, November 18). U.S. policy to counter Nigeria's Boko Haram. *Council on Foreign Relations*. https://www.cfr.org/reports/us-policy-counter-nigerias-boko-haram

Caputo, J. (1977). The Question of Being and Transcendental Phenomenology: Reflections on Heidegger's Relationship to Husserl. *Research in Phenomenology*, *7*(1), 84–105. DOI: 10.1163/156916477X00077

Caputo, J. (1983). *Heidegger and Aquinas: An Essay on Overcoming Metaphysics*. Fordham University Press.

Carnap, R. (1959). The Elimination of Metaphysics through Logical Analysis of Language. In Ayer, A. J. (Ed.), *Logical Positivism* (pp. 61–80). The Free Press.

Carter, B., Van Nuys, M., & Albayrak, C. (2020). State sponsorship of religiously motivated terrorism: A deadly combination. *Democracy and Security*, *17*(2), 181–209. DOI: 10.1080/17419166.2020.1848558

Cellan-Jones, R. (2014, December 2). *Stephen Hawking warns artificial intelligence could end mankind*. BBC. https://www.bbc.com/news/technology-30290540

Center for Strategic & International Studies. (2023, April 18*). Testify with Gregory C. Allen: China's Pursuit of Defense Technologies* [Video]. YouTube. https://www.youtube.com/watch?v=eSudeO55On0

Center for Strategic & International Studies. (2023, September 25). Testify with Greg Allen: Advanced Technology - Examining Threats to National Security [Video]. YouTube. https://www.youtube.com/watch?v=fL_8WmGoo_o

Chambers, G., Smith, C., & Rees, L. (2017). *Command and control in modern warfare: The importance of talent, experience and expertise*. https://researchcentre.army.gov.au/sites/default/files/command_control_b5.pdf

Champendal, M., Ribeiro, R. S. T., Müller, H., Prior, J. O., & Dos Reis, C. S. (2024). Nuclear medicine technologists practice impacted by AI denoising applications in PET/CT images. *Radiography*, *30*(4), 1232–1239. DOI: 10.1016/j.radi.2024.06.010 PMID: 38917681

Chandrakar, M. (2024). Telehealth and digital tools enhancing healthcare access in rural systems. *Discover Public Health*, *21*(1), 144. Advance online publication. DOI: 10.1186/s12982-024-00271-1

Chan, H. Y. (2018). *Advance directives: Rethinking regulation, autonomy & healthcare decision-making*. Springer. DOI: 10.1007/978-3-030-00976-2

Chatterjee, R. (2020). Fundamental concepts of artificial intelligence and its applications. *Journal of Mathematical Problems. Equations and Statistics, 1*(2), 13–24.

Chaudhari, N., Bhade, M., Havelikar, U., & Patel, V. P. (2025). Artificial intelligence and robotics in intensive care units (ICUs): A review of critical care innovations. *Intelligent Hospital, 1*(2), 100010. DOI: 10.1016/j.inhs.2025.100010

Chávez, K., & Swed, O. (2021). The proliferation of drones to violent nonstate actors. *Defence Studies, 21*(1), 1–24. DOI: 10.1080/14702436.2020.1848426

Chayka, K. (2022, March 3). Ukraine becomes the world's "First TikTok war." The New Yorker. https://www.newyorker.com/culture/infinite-scroll/watching-the-worlds-first-tiktok-war

Chen, S. (2018, July 1). China military develops robotic submarines to launch a new era of sea power. *South China Morning Post.* https://www.scmp.com/news/china/society/article/2156361/china-developing-unmanned-ai-submarines-launch-new-era-sea-power

Chen, E., Prakash, S., Reddi, V. J., Kim, D., & Rajpurkar, P. (2023). A framework for integrating artificial intelligence for clinical care with continuous therapeutic monitoring. *Nature Biomedical Engineering, 9*(4), 445–454. DOI: 10.1038/s41551-023-01115-0 PMID: 37932379

Chettri, S. K., Deka, R. K., & Saikia, M. J. (2025). Bridging the gap in the adoption of trustworthy AI in Indian healthcare: Challenges and opportunities. *AI, 6*(1), 10. DOI: 10.3390/ai6010010

Cheung, T., Anderson, E., & Yang, F. (2017). *Chinese defense industry reforms and their implications for US–China military technological competition. Study of Innovation and Technology in China (SITC).* Research Brief.

Coeckelbergh, M. (2022). *The political philosophy of AI.* Polity Press.

Cole, R., Simmons, S., Duncan, J., Cole, J., & Peacock, J. G. (2025). Readying military medicine for AI-enabled warfare. *Military Medicine*, usaf460. Advance online publication. DOI: 10.1093/milmed/usaf460 PMID: 41003652

Colther, C., & Doussoulin, J. P. (2024). Artificial intelligence: Driving force in the evolution of human knowledge. *Journal of Innovation & Knowledge, 9*(4), 100625. DOI: 10.1016/j.jik.2024.100625

Coombs, C., Hislop, D., Taneva, S. K., & Barnard, S. (2020). The strategic impacts of Intelligent Automation for knowledge and service work: An interdisciplinary review. *The Journal of Strategic Information Systems*, *29*(4), 101600. DOI: 10.1016/j.jsis.2020.101600

Cordova-Pozo, K., & Rouwette, E. A. (2023). Types of scenario planning and their effectiveness: A review of reviews. *Futures*, *149*, 103153. DOI: 10.1016/j.futures.2023.103153

Council on Foreign Relations. (2025, November 3). Boko Haram. https://www.cfr.org/keywords/boko-haram

Crawford, N. (2013). *Accountability for killing: Moral responsibility for collateral damage in America's post-9/11 wars*. Oxford University Press. DOI: 10.1093/acprof:oso/9780199981724.001.0001

Critchley, S. (2015). *The Problem with Levinas*. Oxford University Press. DOI: 10.1093/acprof:oso/9780198738763.001.0001

Cummings, M. L. (2021). Rethinking the maturity of artificial intelligence in safety-critical settings. *AI Magazine, 42*(1). Obtenido de https://ojs.aaai.org/aimagazine/index.php/aimagazine/article/view/7394

Cummings, M. (2017). *Artificial intelligence and the future of warfare*. International Security Department and US and the Americas Programme, Chatham House.

Curts, R. J., & Campbell, D. E. (2005). *Building an ontology for command & control*. 10th International Command and Control Research and Technology Symposium: The Future of C2. DOI: 10.21236/ADA464304

Czerkawski, M. (2024). 'The soul is, in a way, all beings': Heidegger's Debts to Aristotle in *Being and Time. Inquiry (Oslo, Norway)*, *67*(10), 3930–3968. DOI: 10.1080/0020174X.2022.2074881

Dahlin, E. (2023). Think Differently We must! An AI Manifesto for the Future. *AI & Society*, *39*(3), 1423–1426. DOI: 10.1007/s00146-022-01620-x

Das, D., Biswas, S. K., & Bandyopadhyay, S. (2022). A critical review on diagnosis of diabetic retinopathy using machine learning and deep learning. *Multimedia Tools and Applications*, *81*(18), 25613–25655. DOI: 10.1007/s11042-022-12642-4 PMID: 35342328

Das, S. K., Dasgupta, R. K., Roy, S. D., & Shil, D. (2024). AI in Indian healthcare: From roadmap to reality. *Intelligent Pharmacy*, *2*(3), 329–334. DOI: 10.1016/j.ipha.2024.02.005

De Ágreda, Á. G. (2020). Ethics of autonomous weapons systems and its applicability to any AI systems. *Telecommunications Policy*, *44*(6), 101953. DOI: 10.1016/j.telpol.2020.101953

De Santis, D. (2023). *Transcendental Idealism and Metaphysics: Husserl's Critique of Heidegger*. Springer.

De Swielande, T. S., Orinx, K., & Peiffer, S. (2024). La guerre cognitive au coeur de la stratégie chinoise de socialisation. *Ingénierie Cognitique*, *7*(2). Advance online publication. DOI: 10.21494/ISTE.OP.2024.1153

Decker, S. E., Matthieu, M. M., Smith, B. N., & Landes, S. J. (2023). Barriers and Facilitators to Dialectical Behavior Therapy Skills Groups in the Veterans Health Administration. *Military Medicine*, *189*(5-6), 1055–1063. DOI: 10.1093/milmed/usad123 PMID: 37104810

Dégallier-Rochat, S., Kurpicz-Briki, M., Endrissat, N., & Yatsenko, O. (2022). Human augmentation, not replacement: A research agenda for AI and robotics in the industry. *Frontiers in Robotics and AI*, *9*, 997386. DOI: 10.3389/frobt.2022.997386 PMID: 36267424

Dehais, F., Lafont, A., Roy, R., & Fairclough, S. (2020). A neuroergonomics approach to mental workload, engagement and human performance. *Frontiers in Neuroscience*, *14*, 268. DOI: 10.3389/fnins.2020.00268 PMID: 32317914

Dehghan, M., & Khosravian, E. (2024). A review of cognitive UAVs: AI-driven situation awareness for enhanced operations. *AI and Tech in Behavioral and Social Sciences*, *2*(4), 54–65. DOI: 10.61838/kman.aitech.2.4.6

Department of Defence. (2023, March 6). *The first Ocius Bluebottle uncrewed surface vessels for the ADF*. https://www.minister.defence.gov.au/media-releases/2023–03–06/first-ocius-bluebottle-uncrewed-surface-vessels-adf

Department of Defense Responsible AI Working Council. (2022). *Responsible artificial intelligence strategy and implementation pathway*. U.S. Department of Defense. https://media.defense.gov/2022/Jun/22/2003022604/-1/-1/0/Department-of-Defense-Responsible-Artificial-Intelligence-Strategy-and-Implementation-Pathway.PDF

Department of Defense. (1978). *Managing weapon system software: Progress and problems* (Unclassified Digest of a Classified Report PSAD-78–112 B-16305e). https://www.gao.gov/products/psad-78–112

Department of Defense. (2012). *Autonomy in weapon systems* (Directive 3000.09). U.S. Department of Defense. https://www.esd.whs.mil/portals/54/documents/dd/issuances/dodd/300009p.pdf

Department of Defense. (2021). *Implementing responsible artificial intelligence in the Department of Defense*. Obtenido de Office of the Secretary of Defense: https://media.defense.gov/2021/May/27/2002730593/-1/-1/0/IMPLEMENTING-RESPONSIBLE-ARTIFICIAL-INTELLIGENCE-IN-THE-DEPARTMENT-OF-DEFENSE.PDF?utm_source=chatgpt.com

Department of Defense. (2023). *Autonomy in weapon systems* (DOD Directive 3000.09). https://www.esd.whs.mil/Portals/54/Documents/DD/issuances/dodd/300009p.PDF

Der Derian, J., & Rollo, S. (2023). Quantum warfare. In *Routledge handbook of the future of warfare* (pp. 319–329). Routledge. DOI: 10.4324/9781003299011-34

Detert, J. R., Treviño, L. K., & Sweitzer, V. L. (2008). Moral disengagement in ethical decision making: A study of antecedents and outcomes. *The Journal of Applied Psychology*, *93*(2), 374–391. DOI: 10.1037/0021-9010.93.2.374 PMID: 18361639

Devitt, S. K., Gan, M., Scholz, J., & Bolia, R. S. (2021). *A method for ethical AI in defence* (DSTG-TR-3786). Defence Science and Technology Group. https://www.dst.defence.gov.au/publication/ethical-ai

Devitt, S. K. (2023). Bad, mad and cooked apples: Responsibility for unlawful targeting in human-AI military teams. In Schraagen, J. M. (Ed.), *Responsible use of AI in military systems*. CRC Press., https://arxiv.org/abs/2211.06326

Dhillon, N., Jeon, N., Gurkan, U., Gupta, A. S., Bonomo, R., Drummy, L., Zhang, M., & Chance, M. (2023). Military medicine and medical research as a source of inspiration and innovation to solve national security and health challenges in the 21st century. *Pathogens & Immunity*, *8*(1), 51–63. DOI: 10.20411/pai.v8i1.596 PMID: 37799210

Dimitrova, M. (2011). *In Levinas' Trace*. Cambridge Scholars Publishing.

Dixon, D., Sattar, H., Moros, N., Kesireddy, S. R., Ahsan, H., Lakkimsetti, M., Fatima, M., Doshi, D., Sadhu, K., & Hassan, M. J. (2024). Unveiling the Influence of AI Predictive analytics on patient Outcomes: A Comprehensive Narrative review. *Cureus*, *16*(5), e59954. DOI: 10.7759/cureus.59954 PMID: 38854327

Dodig-Crnkovic, G., & Çürüklü, B. (2012). Robots: Ethical by design. *Ethics and Information Technology*, *14*(1), 61–71. DOI: 10.1007/s10676-011-9278-2

Donelle, L., Comer, L., Hiebert, B., Hall, J., Shelley, J. J., Smith, M. J., Kothari, A., Burkell, J., Stranges, S., Cooke, T., Shelley, J. M., Gilliland, J., Ngole, M., & Facca, D. (2023). Use of digital technologies for public health surveillance during the COVID-19 pandemic: A scoping review. *Digital Health*, *9*, 20552076231173220. DOI: 10.1177/20552076231173220 PMID: 37214658

Donnelly, S. B. (2005, December 4). Long-distance warriors. *Time*. https://content.time.com/time/subscriber/article/0,33009,1137653,00.html

Dorbolo, J. (2001). *Just war theory*. Oregon State University. http://oregonstate.edu/instruct/phl201/modules/just_war_theory/criteria_intro.html

Doyle, M. (1995). On democratic peace. *International Security*, *19*(4), 164–184. DOI: 10.2307/2539124

Dupont, C. (2015). Jean Héring and the Introduction of Husserl's Phenomenology to France. *Studia Phaenomenologica*, *15*, 129–153. DOI: 10.5840/studphaen2015158

Dutt, K. (2018, October 11). *World leaders in legal education and research - Legal Wires*. Legal Wires - World Leaders in Legal Education and Research. https://legal-wires.com/columns/necessity-and-proportionality-in-international-humanitarian-law-legal-boundaries-in-armed-conflict/

Economist Intelligence Unit. (2017). *The Economist Intelligence Unit's Democracy Index*. https://infographics.economist.com/2019/DemocracyIndex/

Edelglass, W. (2005). Levinas's Language. *Analecta Husserliana*, *85*, 47–62. DOI: 10.1007/1-4020-3576-4_4

Elish, M. C. (2019). Moral crumple zones: Cautionary tales in human-robot interaction. *Engaging Science, Technology, and Society*, *5*, 40–60. DOI: 10.17351/ests2019.260

Emanuel, E. J., & Emanuel, L. L. (1992). The physician–patient relationship. *Journal of the American Medical Association*, *267*(16), 2221–2226. DOI: 10.1001/jama.1992.03480160079038 PMID: 1556799

Endsley, M. R. (2023). Supporting human-AI teams: Transparency, explainability, and situation awareness. *Computers in Human Behavior*, *140*, 107574. DOI: 10.1016/j.chb.2022.107574

Enemark, C. (2019). Drones, risk, and moral injury. *Critical Military Studies*, *5*(2), 150–167. DOI: 10.1080/23337486.2017.1384979

Entous, A. (2025, March 30). The secret history of America's involvement in the Ukraine War. The New York Times. https://web.archive.org/web/20250330021832/https://www.nytimes.com/interactive/2025/03/29/world/europe/us-ukraine-military-war-wiesbaden.html

European Parliament. (2018). *Resolution on autonomous weapon systems (2018/2752(RSP))*. http://www.europarl.europa.eu/doceo/document/TA-8-2018-0341_EN.html?redirect

European People's Party. (2019, February 20). *New deal is a huge victory for security and defence in Europe*. https://www.eppgroup.eu/newsroom/news/huge-victory-for-security-and-defence-in-europe

Fagenblat, M. (2002). Il y a du quotidien: Levinas and Heidegger on the Self. *Philosophy and Social Criticism*, *28*(5), 578–604. DOI: 10.1177/0191453702028005666

Fairclough, N. (2003). *Analyzing discourse: Textual analysis for social research*. Routledge. DOI: 10.4324/9780203697078

Faiyazuddin, M., Rahman, S. J. Q., Anand, G., Siddiqui, R. K., Mehta, R., Khatib, M. N., Gaidhane, S., Zahiruddin, Q. S., Hussain, A., & Sah, R. (2025). The Impact of Artificial Intelligence on Healthcare: A comprehensive review of advancements in diagnostics, treatment, and operational efficiency. *Health Science Reports*, *8*(1), e70312. DOI: 10.1002/hsr2.70312 PMID: 39763580

Fann, K., & Bursi, C. (2023). *Protecting the Porcupine: Why Taiwan Matters*. Blog. Center for International Affairs at Havard University.

Farhud, D. D., & Zokaei, S. (2021). Ethical issues of artificial intelligence in medicine and healthcare. *Iranian Journal of Public Health*, *50*(11), i–v. DOI: 10.18502/ijph.v50i11.7600 PMID: 35223619

Farooq, U. (2019, May 14). *The second drone age: How Turkey and the US became a killer drone power*. The Intercept. https://theintercept.com/2019/05/14/turkey-second-drone-age/

Fatima, S. (2024). Transforming healthcare with AI and machine learning: revolutionizing patient care through advanced analytics. *International Journal of Education and Science Research Review, 11*.

Ferguson, S. M., & Langer, L. J. (2021). The U.S. National Institutes of Health – founding a national biomedical "Innovation Ecosystem.". *Journal of Commercial Biotechnology*, *26*(1), 72–82. DOI: 10.5912/jcb972 PMID: 33859535

Finlan, A. (2021). The shape of warfare to come: A Swedish perspective 2020–2045. *Defense and Security Analysis*, *37*(4), 472–491. DOI: 10.1080/14751798.2021.1995976

Fitz-Gibbon, S. (2006). *Not mentioned in despatches.*

Floridi, L., & Sanders, J. (2024). Minds and Machines. *On the morality of artificial agents, 14*(3), 349–379. DOI: 10.1023/B:MIND.0000035461.63578.9d

Fontes, R., & Kamminga, J. (2023, March 24). Ukraine A Living Lab for AI Warfare. https://www.nationaldefensemagazine.org/articles/2023/3/24/ukraine-a-living-lab-for-ai-warfare

Fontes, C., Hohma, E., Corrigan, C. C., & Lütge, C. (2022). AI-powered public surveillance systems: Why we (might) need them and how we want them. *Technology in Society*, *71*, 102137. DOI: 10.1016/j.techsoc.2022.102137

Forden, G., Podvig, P., & Postol, T. A. (2000). False alarm, nuclear danger. *IEEE Spectrum*, *37*(3), 31–39. DOI: 10.1109/6.825657

Fordham, B., & Walker, T. (2005). Kantian liberalism, regime type, and military resource allocation: Do democracies spend less? *International Studies Quarterly*, *49*(1), 141–157. DOI: 10.1111/j.0020-8833.2005.00338.x

Freedberg, S. J., Jr. (2020, November 9). Meet the Army's future family of robot tanks: RCV. *Breaking Defense*. https://breakingdefense.com/2020/11/meet-the-armys-future-family-of-robot-tanks-rcv/

Freiberger, E. (2013, July 18). Just War Theory and the Ethics of Drone Warfare. *E-International Relations*. https://www.e-ir.info/2013/07/18/just-war-theory-and-the-ethics-of-drone-warfare/

Funaiole, M. P., & Hart, B. (2025, September 2). China's Military in 10 Charts. Center for Strategic and International Studies. https://www.csis.org/analysis/chinas-military-10-charts

Fuster, V., Swirski, F. K., & Nadkarni, G. N. (2025). Conventional and precision medicine: Opposites or complementary ends? *Frontiers in Science*, *3*, 1701495. Advance online publication. DOI: 10.3389/fsci.2025.1701495

Future of Life Institute. (2015). *Autonomous weapons: An open letter from AI & robotics researchers*. https://futureoflife.org/open-letter-autonomous-weapons/

Galeotti, M. (2018). The mythical 'Gerasimov Doctrine' and the language of threat. *Critical Studies on Security*, *7*(2), 157–161. DOI: 10.1080/21624887.2018.1441623

Gebru, T., & Mitchell, M. (2022, June 17). Opinion: We warned Google that people might believe AI was sentient. Now it's happening. *The Washington Post*. https://www.washingtonpost.com/opinions/2022/06/17/google-ai-ethics-sentient-lemoine-warning/

Geist, E. (2016). It's already too late to stop the AI arms race—We must manage it instead. *The Bulletin of the Atomic Scientists*, *72*(5), 318–321. DOI: 10.1080/00963402.2016.1216672

Geng, M., Meng, X., Yu, J., Zhu, L., Jin, L., Jiang, Z., Qiu, B., Li, H., Kong, H., Yuan, J., Yang, K., Shan, H., Han, H., Yang, Z., Ren, Q., & Lu, Y. (2021). Content-noise complementary learning for medical image denoising. *IEEE Transactions on Medical Imaging*, *41*(2), 407–419. DOI: 10.1109/TMI.2021.3113365 PMID: 34529565

Gentry, C. E. (2014). Epistemic bias: Legitimate authority and politically violent nonstate actors. In Lang, A. F.Jr, O'Driscoll, C., & Williams, J. (Eds.), *The future of just war* (pp. 17–29). University of Georgia Press., DOI: 10.2307/j.ctt46nbn3.4

Gerson, L. (Ed.). (2018). *Plotinus: The Enneads*Boys-Stones, G., Trans.). Cambridge University Press.

Ghosh, B., & Thompson, M. (2009, June 1). The CIA's silent war in Pakistan. *Time*. https://content.time.com/time/subscriber/article/0,33009,1900248-2,00.html

Ghosh, A., Saini, A., & Barad, H. (2025). Artificial intelligence in governance: Recent trends, risks, challenges, innovative frameworks and future directions. *AI & Society*, *40*(7), 5685–5707. DOI: 10.1007/s00146-025-02312-y

Gignac, G. E., & Szodorai, E. T. (2024). Defining intelligence: Bridging the gap between human and artificial perspectives. *Intelligence*, *104*, 101832. DOI: 10.1016/j.intell.2024.101832

Githens-Mazer, J. (2012). The rhetoric and reality: Radicalization and political discourse. *International Political Science Review*, *33*(5), 556–567. DOI: 10.1177/0192512112454416

Golob, S. (2020). What Does It Mean to 'Act in the Light of' a Norm? Heidegger and Kant on Commitments and Critique. In Burch, M., & McMullin, I. (Eds.), *Transcending Reason: Heidegger on Rationality* (pp. 79–98). Rowman & Littlefield. DOI: 10.5040/9798881816612.ch-003

Gore, M. N., & Olawade, D. B. (2024). Harnessing AI for public health: India's roadmap. *Frontiers in Public Health*, *12*, 1417568. DOI: 10.3389/fpubh.2024.1417568 PMID: 39399702

Gray, C. S. (1999). Strategic culture as context: The first generation of theory strikes back. *Review of International Studies, 25*(1), 49–69. DOI: 10.1017/S0260210599000492

Griffin, B. J., Purcell, N., Burkman, K., Litz, B. T., Bryan, C. J., Schmitz, M., Villierme, C., Walsh, J., & Maguen, S. (2019). Moral injury: An integrative review. *Journal of Traumatic Stress, 32*(3), 350–362. DOI: 10.1002/jts.22362 PMID: 30688367

Gu, D., Andreev, K., & Dupre, M. E. (2021). Major trends in population growth around the world. *China CDC Weekly, 3*(28), 604–613. DOI: 10.46234/ccdcw2021.160 PMID: 34594946

Gu, Y., Goez, J. C., Guajardo, M., & Wallace, S. W. (2021). Autonomous vessels: State of the art and potential opportunities in logistics. *International Transactions in Operational Research, 28*(4), 1706–1739. DOI: 10.1111/itor.12785

Habibi, S., Ivaki, N., & Barata, J. (2025). A Systematic Literature Review of Unmanned Aerial Vehicles for Healthcare and Emergency Services. *arXiv preprint arXiv:2504.08834.*

Hackman, J. R., & Oldham, G. R. (1976). Motivation through the design of work: Test of a theory. *Organizational Behavior and Human Performance, 16*(2), 250–279. DOI: 10.1016/0030-5073(76)90016-7

Hadlington, L., Binder, J., Gardner, S., Karanika-Murray, M., & Knight, S. (2023). The use of artificial intelligence in a military context: Development of the attitudes toward AI in defense (AAID) scale. *Frontiers in Psychology, 14*, 1164810. DOI: 10.3389/fpsyg.2023.1164810 PMID: 37213385

Haerem, T., Kuvaas, B., Bakken, B. T., & Karlsen, T. (2011). Do military decision makers behave as predicted by prospect theory? *Journal of Behavioral Decision Making, 24*(5), 482–497. DOI: 10.1002/bdm.704

Hagen, R. A. (2024, February 16). Cognitive Warfare, Cybersecurity, and the AI Challenge. https://www.linkedin.com/pulse/cognitive-warfare-cybersecurity-ai-challenge-raymond-andr%C3%A8-hagen-fvsef

Haggard, P. (2017). Sense of agency in the human brain. *Nature Reviews. Neuroscience, 18*(4), 196–207. DOI: 10.1038/nrn.2017.14 PMID: 28251993

Haggard, P., Clark, S., & Kalogeras, J. (2002). Voluntary action and conscious awareness. *Nature Neuroscience, 5*(4), 382–385. DOI: 10.1038/nn827 PMID: 11896397

Haimi, M. (2024). Telemedicine in war zones: Prospects, barriers, and meeting the needs of special populations. *Frontiers in Medicine*, *11*, 1417025. Advance online publication. DOI: 10.3389/fmed.2024.1417025 PMID: 39497840

Hajjar, H. (2019). The counterterrorism war paradigm versus international humanitarian law: The legal contradictions and global consequences of the US 'War on Terror'. *Law & Social Inquiry*, *44*(4), 922–956. DOI: 10.1017/lsi.2018.26

Han, J.-W., & Lee, Y.-J. (2023). *Generative AI for radiological image data: Current trends and outlook.* In *Proceedings of the 2023 14th International Conference on Information and Communication Technology Convergence (ICTC).*

Hancock, P. A., Vincenzi, D. A., Wise, J. A., & Mouloua, M. (2008). *Human factors in simulation and training*. CRC Press. DOI: 10.1201/9781420072846

Hand, S. (Ed.). (1989). *The Levinas Reader*. Basil Blackwell.

Hannan, S. (2012). To See Coming: Augustine and Heidegger on the Arising and Passing Away of Things. *Medieval Mystical Theology*, *21*(1), 75–91. DOI: 10.1558/mmt.v21i1.75

Hansson, S. O. (2006). Safe design. *Techné (Blacksburg, Va.)*, *10*(1), 45–52. DOI: 10.5840/techne200610140

Hansson, S. O. (2020). Technology and mathematics. *Philosophy & Technology*, *33*(1), 117–139. DOI: 10.1007/s13347-019-00348-9

Hartley, D. S. (2021). *An ontology of modern conflict: including conventional combat and unconventional conflict* (1st ed. 2021). Springer., DOI: 10.1007/978-3-030-53214-7

Hartzell, C. L. (2023, July 10). Future weapons technology of 2040. Army University Press. https://www.armyupress.army.mil/Journals/NCO-Journal/Archives/2023/July/Future-Weapons-Technology-of-2040/

Hasegawa, T., Ishikawa, M., Maruyama, K., Inadama, N., Yoshida, E., & Murayama, H. (2005). Depth-of-interaction recognition using optical filters for nuclear medicine imaging. *IEEE Transactions on Nuclear Science*, *52*(1), 4–7. DOI: 10.1109/TNS.2004.843149

Hassan, M., Borycki, E. M., & Kushniruk, A. W. (2024). Artificial intelligence governance framework for healthcare. *Healthcare Management Forum*, *38*(2), 125–130. DOI: 10.1177/08404704241291226 PMID: 39470044

Hassen, M. Z. (2025). The risk of moral outsourcing: Why artificial intelligence cannot and should not make our ethical decisions. *EthAIca*, *4*, 428. DOI: 10.56294/ai2025428

Hasson, J., & Slama, A. (2023). *IHL's reasonable military commander standard and culture: Applying the lessons of ICL and IHRL*. Tulsa Law Review.

Heidegger, M. (1949). *Existence and Being*. Henry Regnery.

Heidegger, M. (1962). *Being and Time* (Macquarrie, J., & Robinson, E., Trans.). Basil Blackwell.

Heidegger, M. (1988). *Hegel's Phenomenology of Spirit* (Emad, P., & Maly, K., Trans.). Indiana University Press. DOI: 10.2307/j.ctv105b98m

Helkala, K. M., Lucas, G., Barrett, E., & Syse, H. (2023). Editorial: Ethical challenges in AI-enhanced military operations. *Frontiers in Big Data*, *6*, 1229252. Advance online publication. DOI: 10.3389/fdata.2023.1229252 PMID: 37405063

Helmore, E. (2009, August 23). US now trains more drone operators than pilots. *The Guardian*. https://www.theguardian.com/world/2009/aug/23/drones-air-force-robot-planes

Heng, L. (2024). Strategic Overview of Applying Artificial Intelligence on the Future Battlefield.

Hering, J. (1932). Recension d'E. Levinas, *La théorie de l'intuition dans la phénoménologie de Husserl. Revue Philosophique de la France et de l'Étranger*, *113*(5–6), 474–481.

Hering, J. (2021). Remarks concerning Essence, Ideal Quality, and Idea. *Phenomenological Investigations*, *1*, 51–108.

Herschinger, E., Bozay, K., Von Drachenfels, M., Decker, O., & Joppke, C. (2020). A Threat to Open Societies? Conceptualizing the Radicalization of Society. *DOAJ*. DOI: 10.4119/ijcv-3807

Hirani, R., Noruzi, K., Khuram, H., Hussaini, A. S., Aifuwa, E. I., Ely, K. E., Lewis, J. M., Gabr, A. E., Smiley, A., Tiwari, R. K., & Etienne, M. (2024). Artificial Intelligence and Healthcare: A Journey through History, Present Innovations, and Future Possibilities. *Life (Basel, Switzerland)*, *14*(5), 557. DOI: 10.3390/life14050557 PMID: 38792579

Hofmeyr, A. B. (2024). The challenge that war poses to Levinas's thought. *Philosophia: International Journal of Philosophy*, *25*(1), 26–44.

Holland Michel, A. (2020). *The black box, unlocked: Predictability and understandability in military AI*. United Nations Institute for Disarmament Research (UNIDIR).

Horowitz, M. (2018). Artificial intelligence, international competition, and the balance of power. *Texas National Security Review, 1*(3).

Horowitz, M. (2010). *The diffusion of military power: Causes and consequences for international politics*. Princeton University Press.

Horowitz, M. C. (2016). The ethics & morality of robotic warfare: Assessing the debate over autonomous weapons. *Daedalus*, *145*(4), 25–36. DOI: 10.1162/DAED_a_00409

Horowitz, M., Kreps, S., & Fuhrman, M. (2016). Separating fact from fiction in the debate over drone proliferation. *International Security*, *41*(2), 7–42. DOI: 10.1162/ISEC_a_00257

Hosang, J. B. (2020). *Rules of engagement and the international law of military operations*. Oxford University Press. DOI: 10.1093/oso/9780198853886.001.0001

Houlgate, S. (2006). *The Opening of Hegel's Logic: From Being to Infinity*. Purdue University Press.

Hughes, D. J. (Ed.). (1993). *Moltke on the art of war: Selected writings*. Random House Publishing Group.

Human Rights Watch. (2012). *Losing humanity: The case against killer robots*. https://www.hrw.org/report/2012/11/19/losing-humanity/case-against-killer-robots

Human Rights Watch. (2014, May 13). *12 key arguments for a preemptive ban on fully autonomous weapons*. https://www.hrw.org/news/2014/05/13/advancing-debate-killer-robots

Human Rights Watch. (2016, April 11). *Killer robots and the concept of meaningful human control*. https://www.hrw.org/news/2016/04/11/killer-robots-and-concept-meaningful-human-control

Human Rights Watch. (2022). *Killer robots*. https://www.hrw.org/topic/arms/killer-robots

Hung, T., & Hung, T. (2022). How China's Cognitive Warfare works: A Frontline Perspective of Taiwan's Anti-Disinformation Wars. *Journal of Global Security Studies*, *7*(4), ogac016. Advance online publication. DOI: 10.1093/jogss/ogac016

Hunter, A. (2017). *Defense acquisition trends, 2016: The end of the contracting drawdown*. Center for Strategic and International Studies.

Hunter, O. F., Perry, F., Salehi, M., Bandurski, H., Hubbard, A., Ball, C. G., & Morad Hameed, S. (2023). Science fiction or clinical reality: A review of the applications of artificial intelligence along the continuum of trauma care. *World Journal of Emergency Surgery; WJES, 18*(1), 16. DOI: 10.1186/s13017-022-00469-1 PMID: 36879293

Hurley, M. E., Lang, B. H., Kostick-Quenet, K. M., Smith, J. N., & Blumenthal-Barby, J. (2024). Patient consent and the right to notice and explanation of AI systems used in health care. *The American Journal of Bioethics : AJOB, 25*(3), 102–114. DOI: 10.1080/15265161.2024.2399828 PMID: 39288291

Husain, A. (2021). AI is shaping the future of war. *Prism, 9*(3), 50–61.

Husserl, E. (1980). *Phenomenology and the Foundations of the Sciences (Third Book)* (Klein, T., & Pohl, W., Trans.). Martinus Nijhoff. (Original work published 1913)

Husserl, E. (1996). The Vienna Lecture and Phenomenology. In Kearney, R., & Rainwater, M. (Eds.), *The Continental Philosophy Reader* (pp. 7–22). Routledge. (Original work published 1935)

Hynek, N., & Solovyeva, A. (2022). *Militarizing artificial intelligence*. Routledge. DOI: 10.4324/9781003045489

Inamdar, A. V., Kale, M., Deogaonkar, A., Dariya, S. S., & Dey, A. K. (2025). A unified health ecosystem in India: Exploring the Ayushman Bharat digital mission. *International Journal of Community Medicine and Public Health, 12*(8), 3813–3819. DOI: 10.18203/2394-6040.ijcmph20252497

Innes, J. M., & Morrison, B. W. (2021). Artificial intelligence and psychology. In *The Routledge social science handbook of AI* (pp. 30–57). Routledge. DOI: 10.4324/9780429198533-4

International Committee for Robot Arms Control. (n.d.). *International Committee for Robot Arms Control*. https://www.icrac.net/

International Committee of the Red Cross. (2006). A guide to the legal review of new weapons, means and methods of warfare: Measures to implement Article 36 of Additional Protocol I of 1977. *International Review of the Red Cross, 88*(864). https://www.icrc.org/eng/assets/files/other/irrc_864_icrc_geneva.pdf

International Committee of the Red Cross. (2014). *Autonomous weapon systems: Technical, military, legal and humanitarian aspects*. https://www.icrc.org/en/document/report-icrc-meeting-autonomous-weapon-systems-26-28-march-2014

International Committee of the Red Cross. (2016). *Autonomous weapon systems: Implications of increasing autonomy in the critical functions of weapons.* https://www.icrc.org/en/publication/4283-autonomous-weapons-systems

International Committee of the Red Cross. (2021). *ICRC position on autonomous weapons systems.* https://www.icrc.org/en/document/icrc-position-autonomous-weapon-systems

International Committee of the Red Cross. (2022). *What you need to know about autonomous weapons.* https://www.icrc.org/en/document/what-you-need-know-about-autonomous-weapons

International Committee of the Red Cross. (n.d.). *Treaties, states parties and commentaries.* https://ihl-databases.icrc.org/applic/ihl/ihl.nsf/ART/150-110029?OpenDocument

International Institute for Strategic Studies. (2011). Unmanned aerial vehicles: Emerging lessons and technologies. In *The Military Balance 2011* (pp. 20–26).

Ivanov, I., & Dronzina, T. (2025). In ISIS We Trust: Why Does Radicalization Still Matter? In Pietrzak, P. (Ed.), *Exploring the Implications of Local and Regional Conflicts* (pp. 127–150). IGI Global Scientific Publishing., DOI: 10.4018/979-8-3693-2877-4.ch005

Jacovi, A., Marasović, A., Miller, T., & Goldberg, Y. (2021). Formalizing trust in artificial intelligence: Prerequisites, causes and goals of human trust in AI. *Proceedings of the 2021 ACM Conference on Fairness, Accountability, and Transparency*, 624–635. DOI: 10.1145/3442188.3445923

Jafar, A. J., Norton, I., Lecky, F., & Redmond, A. D. (2015). A literature review of medical record keeping by foreign medical teams in sudden onset disasters. *Prehospital and Disaster Medicine*, *30*(2), 216–222. DOI: 10.1017/S1049023X15000102 PMID: 25659602

Jasko, K., LaFree, G., Piazza, J., & Becker, M. H. (2022). A comparison of political violence by left-wing, right-wing, and Islamist extremists in the United States and the world. *Proceedings of the National Academy of Sciences of the United States of America*, *119*(30), e2122593119. DOI: 10.1073/pnas.2122593119 PMID: 35858413

Jauhiainen, J., & Erbsen, H. (2023, April 3). Multilevel governance in the temporal protection and integration of Ukrainians within the European Union: The case of Estonia. *Journal of European Integration*, *45*(3), 413–430. DOI: 10.1080/07036337.2023.2190109

Jensen, P. A. (2021, January 1). *Intro.* blog.biocomm.ai. https://blog.biocomm.ai/2021/01/01/national-artificial-intelligence-initiative-overseeing-and-implementing-the-united-states-national-ai-strategy/t

Jervis, R. (2017). *Perception and misperception in international politics.* Center for International Affairs.

Jiao, A. (2026, January 15). China AI Strategy: Policy, Regulation & Global. *Impact in, 2025,* •••. https://ashleydudarenok.com/china-ai-strategy/

Jimenez, G., Tan, W. S., Virk, A. K., Low, C. K., Car, J., & Ho, A. H. Y. (2018). Overview of systematic reviews of advance care planning: Summary of evidence and global lessons. *Journal of Pain and Symptom Management, 56*(3), 436–459.e425. DOI: 10.1016/j.jpainsymman.2018.05.016 PMID: 29807158

Jimenez-Mesa, C., Arco, J. E., Martinez-Murcia, F. J., Suckling, J., Ramirez, J., & Gorriz, J. M. (2023). Applications of machine learning and deep learning in SPECT and PET imaging: General overview, challenges and future prospects. *Pharmacological Research, 197,* 106984. DOI: 10.1016/j.phrs.2023.106984 PMID: 37940064

Johansson, L. (2015). *Äkta robotar.* Fri Tanke.

Johansson, L. (2018). Ethical aspects of military maritime and aerial autonomous systems. *Journal of Military Ethics, 17*(2-3), 140–155. DOI: 10.1080/15027570.2018.1552512

Johnson, J. (2022). The AI Commander Problem: Ethical, political, and psychological Dilemmas of Human-Machine Interactions in AI-enabled Warfare. *Journal of Military Ethics, 21*(3–4), 246–271. DOI: 10.1080/15027570.2023.2175887

Johnson, J. (2023). Automating the OODA loop in the age of intelligent machines: Reaffirming the role of humans in command-and-control decision-making in the digital age. *Defence Studies, 23*(1), 43–67. DOI: 10.1080/14702436.2022.2102486

Johnson, M., & Axinn, S. (2013). The morality of autonomous robots. *Journal of Military Ethics, 12*(2), 129–141. DOI: 10.1080/15027570.2013.818399

Jonze, S. (Director). (2013). *Her* [Film]. Warner Bros. Pictures.

Jose, B., Cherian, J., Verghis, A. M., Varghise, S. M., S, M., & Joseph, S. (2025). The cognitive paradox of AI in education: Between enhancement and erosion. *Frontiers in Psychology, 16,* 1550621. DOI: 10.3389/fpsyg.2025.1550621 PMID: 40297599

Joshi, K. (2024). Artificial General Intelligence (AGI): A Comprehensive review. *Journal of the Epidemiology Foundation of India, 2*(3), 93–96. DOI: 10.56450/JEFI.2024.v2i03.004

K, S. (2024, October 3). *AI: Simulating Human Intelligence for Real-World Tasks | InfoSecured.ai*. InfoSecured.ai. https://www.infosecured.ai/i/ai-definitions/artificial-intelligence-simulating-human-intelligence/

Kaag, J., & Kreps, S. (2013). Drones and democratic peace. *The Brown Journal of World Affairs*, *19*(11).

Kahl, C. H. (2026, February 6). The myth of the AI race: Neither America nor China can achieve true tech dominance. *Foreign Affairs*.

Kahneman, D. (2011). *Thinking, fast and slow*. Farrar, Straus and Giroux.

Kahneman, D., & Tversky, A. (1979). Prospect theory: An analysis of decision under risk. *Econometrica*, *47*(2), 263–292. DOI: 10.2307/1914185

Kakoliris, G. (2015). Jacques Derrida on the ethics of hospitality. In Imafidon, E. (Ed.), *The ethics of subjectivity: Perspectives since the dawn of modernity* (pp. 144–156). Palgrave Macmillan. DOI: 10.1057/9781137472427_9

Kalani, M., & Anjankar, A. (2024). Revolutionizing Neurology: The role of Artificial intelligence in advancing diagnosis and treatment. *Cureus*, *16*(6), e61706. DOI: 10.7759/cureus.61706 PMID: 38975469

Kalid, N., Zaidan, A. A., Zaidan, B. B., Salman, O. H., Hashim, M., & Muzammil, H. J. J. O. M. S. (2018). Based real time remote health monitoring systems: A review on patients prioritization and related" big data" using body sensors information and communication technology. *Journal of Medical Systems*, *42*(2), 30. DOI: 10.1007/s10916-017-0883-4 PMID: 29288419

Kania, E. B. (2019). Chinese military innovation in the AI revolution. *RUSI Journal*, *164*(5–6), 26–34. DOI: 10.1080/03071847.2019.1693803

Kant, I. (2011). *Immanuel Kant: Groundwork of the metaphysics of morals*. Cambridge University Press., DOI: 10.1017/CBO9780511973741

Kapoor, N., Sanjana, S. N., Davalagi, S. B., Balu, P. S., & Sethia, S. (2024). AI Horizons in Indian Healthcare: A vision for transformation and equity. *Indian Journal of Community Medicine : Official Publication of Indian Association of Preventive & Social Medicine*, *49*(Suppl 2), S210–S216. DOI: 10.4103/ijcm.ijcm_806_24 PMID: 40124859

Karadsheh, J., & Sariyuce, I. (2022, April 11). *Turkish drones have become a symbol of the Ukrainian resistance*. CNN. https://edition.cnn.com/2022/04/11/middleeast/mideast-summary-04-11-2022-intl/index.html

Karenxcheng. (2022, August). *Created the first ever AI cover for Cosmopolitan magazine! #shorts* [Video]. YouTube. https://youtu.be/8fthDHDshvg

Kaviani, S., Han, K. J., & Sohn, I. (2022). Adversarial attacks and defenses on AI in medical imaging informatics: A survey. *Expert Systems with Applications, 198*, 116815. DOI: 10.1016/j.eswa.2022.116815

Keegan, J. (2004). *Intelligence in war: Knowledge of the enemy from Napoleon to al-Qaeda*. Random House.

Kellermann, A. L., Kotwal, R. S., & Rasmussen, T. E. (2023). Military medicine's value to US health care and public health: Bringing battlefield lessons home. *JAMA Network Open*, *6*(9), e2335125–e2335125. DOI: 10.1001/jamanetworkopen.2023.35125 PMID: 37733341

Keohane-O'Connor, N. (1982). *Existence and Existents*, by Emmanuel Levinas. *JBSP : The Journal of the British Society for Phenomenology. British Society for Phenomenology*, *13*(1), 94–96. DOI: 10.1080/00071773.1982.11007568

Keyes, C. D. (1972). An Evaluation of Levinas' Critique of Heidegger. *Research in Phenomenology*, *2*(1), 121–142. DOI: 10.1163/156916472X00072

Khurshid, S., Friedman, S. F., Al-Alusi, M. A., Kany, S., Sommers, T., Anderson, C. D., Ho, J. E., McManus, D. D., Borowsky, L. H., Ashburner, J. M., Lubitz, S. A., Atlas, S. J., Maddah, M., Singer, D. E., Ellinor, P. T., Khurshid, S., Friedman, S. F., Al-Alusi, M. A., Kany, S., & Ellinor, P. T. (2025). Artificial intelligence-enabled analysis of handheld single-lead electrocardiograms to predict incident atrial fibrillation: An analysis of the VITAL-AF randomized trial. *NPJ Digital Medicine*, *8*(1), 776. Advance online publication. DOI: 10.1038/s41746-025-02164-2 PMID: 41299008

Khurshudyan, I., Ilyushina, M., & Khudov, K. (2022, December 2). Ukraine and Russia are fighting the "first full-scale drone war." *The Washington Post*. https://www.washingtonpost.com/world/2022/12/02/drones-russia-ukraine-air-war/

Kirilova, N. V. (2024). Regional Conflict Prevention and Perceived Power Competition: Six Elements of Power. In Pietrzak, P. (Ed.), *Analyzing Global Responses to Contemporary Regional Conflicts* (pp. 252–274). IGI Global., DOI: 10.4018/979-8-3693-2837-8.ch012

Kissinger, H., Schmidt, E., & Huttenlocher, D. (2021). *The age of AI: And our human future*. John Murray.

Konert, A., & Balcerzak, T. (2021). Military autonomous drones (UAVs) - from fantasy to reality. Legal and Ethical implications. *Transportation Research Procedia*, *59*, 292–299. DOI: 10.1016/j.trpro.2021.11.121

Korac, M. (2018). Feminists against sexual violence in war: The question of perpetrators and victims revisited. *Social Sciences (Basel, Switzerland)*, *7*(10), 182. DOI: 10.3390/socsci7100182

Korać, S. (2018). Depersonalisation of killing: Towards a 21st century use of force "beyond good and evil"? *Philosophy and Society*, *29*(1), 1–152. DOI: 10.2298/FID1801049K

Koumparoudis, E. I. (2024). Thou Shall Not Kill: The Ethics of AI in Contemporary Warfare. In Pietrzak, P. (Ed.), *Dealing with Regional Conflicts of Global Importance* (pp. 299–313). IGI Global Scientific Publishing., DOI: 10.4018/978-1-6684-9467-7.ch015

Krishnan, A., & Devarajan, Y. (2025). Sustainable wastewater treatment with AI: A path to enhanced hygiene and environmental protection. *Process Safety and Environmental Protection*, *201*, 107487. DOI: 10.1016/j.psep.2025.107487

Krishna, V. V. (2023). A I and contemporary challenges: The good, bad and the scary. *Journal of Open Innovation*, *10*(1), 100178. DOI: 10.1016/j.joitmc.2023.100178

Kumar, A. (2023). The transformation of the Indian healthcare system. *Cureus*, *15*(5), e39079. DOI: 10.7759/cureus.39079 PMID: 37378105

Kumar, G., Baharia, R., Singh, K., Gupta, S. K., Joy, S., Sharma, A., & Rahi, M. (2024). Addressing challenges in vector control: A review of current strategies and the imperative for novel tools in India's combat against vector-borne diseases. *BMJ Public Health*, *2*(1), e000342. DOI: 10.1136/bmjph-2023-000342 PMID: 40018116

Kuttan, N., Pundkar, A., Gadkari, C., Patel, A., & Kumar, A. (2025). Transforming emergency medicine with artificial intelligence: From triage to clinical decision support. *Multidisciplinary Reviews*, *8*(10), 2025285. Advance online publication. DOI: 10.31893/multirev.2025285

Kuyer, P., & Gordijn, B. (2023). Nudge in perspective: A systematic literature review on the ethical issues with nudging. *Rationality and Society*, *35*(2), 191–230. DOI: 10.1177/10434631231155005

Lachow, I. (2017). The upside and downside of swarming drones. *The Bulletin of the Atomic Scientists*, *73*(2), 96–101. DOI: 10.1080/00963402.2017.1290879

Landau, I. (2021). Problems in Levinas. *The European Legacy, Toward New Paradigms*, *26*(5), 455–465. DOI: 10.1080/10848770.2021.1899600

Landmann, M. (1941). Socrates as a Precursor of Phenomenology. *Philosophy and Phenomenological Research*, *2*(1), 15–42. DOI: 10.2307/2102672

Latif, S. D., Hazrin, N. B., Koo, C. H., Ng, J. L., Chaplot, B., Huang, Y. F., El-Shafie, A., & Ahmed, A. N. (2023). Assessing rainfall prediction models: Exploring the advantages of machine learning and remote sensing approaches. *Alexandria Engineering Journal*, *82*, 16–25. DOI: 10.1016/j.aej.2023.09.060

Lau, K.-y. (2015). War, peace and love: The logic of Lévinas. *Social Imaginaries*, *1*(2), 103–126. DOI: 10.5840/si20151219

Lawton, P. (1975). Levinas' Notion of the 'There Is'. *Tijdschrift voor Filosofie*, *37*(3), 477–489.

Lazar, S. (2016). War. In Zalta, E. N. (Ed.), *The Stanford encyclopedia of philosophy* (Winter 2016 ed.). Stanford University., https://plato.stanford.edu/entries/war/

Lee, A. R., Koo, D., Kim, I. K., Lee, E., Kim, H. H., Yoo, S., Kim, J., Choi, E. K., & Lee, H. (2023). Identifying facilitators of and barriers to the adoption of dynamic consent in digital health ecosystems: A scoping review. *BMC Medical Ethics*, *24*(1), 107. DOI: 10.1186/s12910-023-00988-9 PMID: 38041034

Lepenies, R., & Malecka, M. (2018). The ethics of behavioural public policy. In *The Routledge handbook of ethics and public policy*. Routledge. DOI: 10.4324/9781315461731-41

Levinas, E. (1969). *Totality and Infinity* (Lingis, A., Trans.). Duquesne University Press. (Original work published 1961)

Levinas, E. (1969). *Totality and infinity: An essay on exteriority* (Lingis, A., Trans.). Duquesne University Press. (Original work published 1961)

Levinas, E. (1978). *Existence and Existents* (Lingis, A., Trans.). Martinus Nijhoff. (Original work published 1947) DOI: 10.1007/978-94-017-1516-4

Levinas, E. (1987a). *Collected Philosophical Papers* (Lingis, A., Trans.). Martinus Nijhoff. DOI: 10.1007/978-94-009-4364-3

Levinas, E. (1987b). *Time and the Other* (Cohen, R., Trans.). Duquesne University Press. (Original work published 1947)

Levinas, E. (1991). *Otherwise than Being, or Beyond Essence* (Lingis, A., Trans.). Kluwer. (Original work published 1974) DOI: 10.1007/978-94-015-7906-3

Levinas, E. (1994). *In the Time of the Nations* (Smith, M., Trans.). Indiana University Press. (Original work published 1980)

Levinas, E. (1995). *The Theory of Intuition in Husserl's Phenomenology* (Orianne, A., Trans.). Northwestern University Press. (Original work published 1963)

Levinas, E. (1996). Martin Heidegger and Ontology. *Diacritics*, *26*(1), 11–32. DOI: 10.1353/dia.1996.0007

Levinas, E. (1998a). *Of God Who Comes to Mind* (Bergo, B., Trans.). Stanford University Press. (Original work published 1986) DOI: 10.1515/9781503616882

Levinas, E. (1998b). *Entre Nous: On Thinking-of-the-Other* (Smith, M., & Harshav, B., Trans.). Columbia University Press. (Original work published 1991)

Levinas, E. (1998c). *Discovering Existence with Husserl* (Cohen, R., & Smith, M., Trans.). Northwestern University Press. [1929–1977]

Levinas, E. (1999). *Alterity and Transcendence* (Smith, M., Trans.). Athlone Press. (Original work published 1995)

Levinas, E. (2003). *On Escape* (Bergo, B., Trans.). Stanford University Press. (Original work published 1935)

Levy, J. (1988). Domestic politics and war. *The Journal of Interdisciplinary History*, *18*(4), 653–673. DOI: 10.2307/204819

Liang, H., & Xue, Y. (2009). Avoidance of information technology threats: A theoretical perspective. *Management Information Systems Quarterly*, *33*(1), 71–90. DOI: 10.2307/20650279

Lidskog, R. (2020). Samhället utmanat? Artificiell intelligens och sociologisk kunskap [Society challenged? Artificial intelligence and sociological knowledge]. *Sociologisk Forskning*, *57*(2), 99–120. DOI: 10.37062/sf.57.19591

Li, E., Clarke, J., Ashrafian, H., Darzi, A., & Neves, A. L. (2022). The Impact of Electronic Health record interoperability on safety and Quality of Care in High-Income Countries: Systematic review. *Journal of Medical Internet Research*, *24*(9), e38144. DOI: 10.2196/38144 PMID: 36107486

Lieberman, H. R., Bathalon, G. P., Falco, C. M., Morgan, C. A., Niro, P. J., & Tharion, W. J. (2005). The fog of war: Decrements in cognitive performance and mood associated with combat-like stress. *Aviation, Space, and Environmental Medicine*, *76*(7), C7–C14. PMID: 16018323

Liivoja, R., Massingham, E., & McKenzie, S. (2022). The legal requirement for command and the future of autonomous military platforms. *International Law Studies*, *99*(1), 27.

Liu, J., Malekzadeh, M., Mirian, N., Song, T. A., Liu, C., & Dutta, J. (2021). Artificial intelligence-based image enhancement in PET imaging: Noise reduction and resolution enhancement. *PET Clinics*, *16*(4), 553–576. DOI: 10.1016/j.cpet.2021.06.005 PMID: 34537130

Li, Y., Yi, X., Fu, J., Yang, Y., Duan, C., & Wang, J. (2025). Reducing misdiagnosis in AI-driven medical diagnostics: A multidimensional framework for technical, ethical, and policy solutions. *Frontiers in Medicine*, *12*, 1594450. DOI: 10.3389/fmed.2025.1594450 PMID: 41244777

Lo, C. L. (2018). Time, the Image of Absolute Logos: A Comparative Analysis of the Ideas of Augustine and Husserl. *Comparative and Continental Philosophy*, *10*(1), 50–61. DOI: 10.1080/17570638.2018.1465018

Lohr, K. D. (2025). Digital Literacy and Access: Equity from a global and local perspective. *New Directions for Adult and Continuing Education*, *2025*(185), 39–43. DOI: 10.1002/ace.20559

Lokhorst, G.-J., & van den Hoven, J. (2012). Responsibility for military robots. In Lin, P., Abney, K., & Bekey, G. A. (Eds.), *Robot ethics: The ethical and social implications of robotics* (pp. 145–156). MIT Press.

Luckett, T., Bhattarai, P., Phillips, J., Agar, M., Currow, D., Krastev, Y., & Davidson, P. M. (2015). Advance care planning in 21st century Australia: A systematic review and appraisal of online advance care directive templates against national framework criteria. *Australian Health Review : A Publication of the Australian Hospital Association*, *39*(5), 552–560. DOI: 10.1071/AH14187 PMID: 26027723

Lukyanenko, R., Maass, W., & Storey, V. C. (2022). Trust in artificial intelligence: From a foundational trust framework to emerging research opportunities. *Electronic Markets*, *32*(4), 1993–2020. DOI: 10.1007/s12525-022-00605-4

Lushenko, P. (2021). Review of the special issue "robotics autonomous systems and warfare," Small Wars and Insurgencies 31, 4 June 2020. *Small Wars & Insurgencies*, *32*(3), 550–570. DOI: 10.1080/09592318.2021.1891618

Luttwak, E. N. (1996). A post-heroic military policy. *Foreign Affairs*, *75*(4), 33–44. DOI: 10.2307/20047657

Maan, A. K., & Cobaugh, P. (2018). *Introduction to narrative warfare: A primer and study guide*. Narrative Strategies LLC.

Maas, M. (2019). How viable is international arms control for military artificial intelligence? Three lessons from nuclear weapons. *Contemporary Security Policy*, *40*(3), 285–311. DOI: 10.1080/13523260.2019.1576464

Madhiraju, P. (2025, June 9). *How AI is reshaping the role of pathologists and radiologists in oncology*. Business-News-Today.com. https://business-news-today.com/how-ai-is-reshaping-the-role-of-pathologists-and-radiologists-in-oncology/

Mahmoudi, E., Wu, W., Najarian, C., Aikens, J., Bynum, J., & Vydiswaran, V. V. (2022). Identifying caregiver availability using medical notes with rule-based natural language processing: Retrospective cohort study. *JMIR Aging*, *5*(3), e40241. DOI: 10.2196/40241 PMID: 35998328

Majumdar, S., Pendleton, B., & Gupta, A. (2025). Red Teaming AI Red Teaming. https://doi.org//arXiv.2507.05538DOI: 10.48550

Manesh, A. S., Shiri, I., Khateri, M., Jenabi, E., Salimi, Y., Geramifar, P., & Zaidi, H. (2023, November). Fine-tuned BERT Language Model for Efficient Nuclear Medicine Data Retrieval. In *2023 IEEE Nuclear Science Symposium, Medical Imaging Conference and International Symposium on Room-Temperature Semiconductor Detectors (NSS MIC RTSD)* (pp. 1-1). IEEE. DOI: 10.1109/NSSMICRTSD49126.2023.10338102

Mansfield, E. D., & Snyder, J. (2002). Democratic transitions, institutional strength, and war. *International Organization*, *56*(2), 297–337. DOI: 10.1162/002081802320005496

Marcel, G. (1952). *Metaphysical Journal* (Wall, B., Trans.). Henry Regnery. [1914–1923]

Marshall, A. (2016). From civil war to proxy war: Past history and current dilemmas. *Small Wars & Insurgencies*, *27*(2), 183–195. DOI: 10.1080/09592318.2015.1129172

Marsili, M. (2023). Guerre à la Carte: Cyber, Information, Cognitive Warfare and the Metaverse. *Applied Cybersecurity & Internet Governance*, *2*(1), 1–11. DOI: 10.60097/ACIG/162861

Martin, K. D., & Zimmermann, J. (2024). Artificial intelligence and its implications for data privacy. *Current Opinion in Psychology*, *58*, 101829. DOI: 10.1016/j.copsyc.2024.101829 PMID: 38954851

Masbi, M., Tavakoli, N., & Dowlati, M. (2024). Challenges of providing of special care services in hospitals during emergencies and disasters: A scoping review. *BMC Emergency Medicine*, *24*(1), 238. DOI: 10.1186/s12873-024-01160-1 PMID: 39695959

McCauley, C., & Moskalenko, S. (2008). Mechanisms of Political Radicalization: Pathways toward Terrorism. *Terrorism and Political Violence*, *20*(3), 415–433. DOI: 10.1080/09546550802073367

McConoly. R. (2021, June 21). "What is Network-Centric Warfare? Naval Post-Naval News and Information" ttps://navalpost.com/what-is-network-centric-warfare.

McDaniel, K. (2016). Heidegger and the 'There Is' of Being. *Philosophy and Phenomenological Research*, *93*(2), 306–320. DOI: 10.1111/phpr.12257

McGowan, C. R., Baxter, L., DuBois, M., Sheather, J., Khondaker, R., Cummings, R., & Watkins, K. (2020). Preparing humanitarians to address ethical problems. *Conflict and Health*, *14*(1), 72. DOI: 10.1186/s13031-020-00319-4 PMID: 33292441

McGrath, S. J. (2003). Heidegger and Duns Scotus on Truth and Language. *The Review of Metaphysics*, *57*(2), 339–358.

McKenzie, S. (2020). When is a ship a ship? Use by state armed forces of uncrewed maritime vehicles and the United Nations Convention on the Law of the Sea. *Melbourne Journal of International Law*, *21*(2), 373–402.

McKenzie, S. (2021). Autonomous technology and dynamic obligations: Uncrewed maritime vehicles and the regulation of maritime military surveillance in the exclusive economic zone. *Asian Journal of International Law*, *11*(1), 146–175. DOI: 10.1017/S2044251321000011

Meadows, A. J., Oppenheim, B., Guerrero, J., Ash, B., Badker, R., Lam, C. K., Pardee, C., Ngoon, C., Savage, P. T., Sridharan, V., Madhav, N. K., & Stephenson, N. (2022). Infectious disease underreporting is predicted by Country-Level preparedness, politics, and pathogen severity. *Health Security*, *20*(4), 331–338. DOI: 10.1089/hs.2021.0197 PMID: 35925788

Mearsheimer, J. J. (2001). *The Tragedy of Great Power Politics*. W. W. Norton.

Mettraux, G. (2009). *The law of command responsibility*. Oxford University Press. DOI: 10.1093/acprof:oso/9780199559329.001.0001

Meyer, A., Cypko, M. A., Eickhoff, C., Falk, V., & Emmert, M. Y. (2019). Artificial intelligence-assisted care in medicine: A revolution or yet another blunt weapon? Potentials, challenges, and the future of implementing artificial intelligence (AI) for clinical care.

Miller, S. (2023, September 1). Cognitive warfare: An ethical analysis. *Ethics and Information Technology*, *25*(3), 46. Advance online publication. DOI: 10.1007/s10676-023-09717-7

Mill, J. S. (1860). *On liberty* (2nd ed.). John W. Parker & Son.

Mills, D. (2013, May 24). Obama's speech on drone policy. *The New York Times*. https://www.nytimes.com/2013/05/24/us/politics/transcript-of-obamas-speech-on-drone-policy.html

Minopoulos, G. M., Memos, V. A., Stergiou, K. D., Stergiou, C. L., & Psannis, K. E. (2023). A medical image visualization technique assisted with AI-based haptic feedback for robotic surgery and healthcare. *Applied Sciences (Basel, Switzerland)*, *13*(6), 3592. DOI: 10.3390/app13063592

Mishra, U. S., Yadav, S., & Joe, W. (2024). The Ayushman Bharat Digital Mission of India: An assessment. *Health Systems and Reform*, *10*(2), 2392290. DOI: 10.1080/23288604.2024.2392290 PMID: 39437234

Mohammadi, R., Tabanejad, Z., Abhari, S., Honarvar, B., Lazem, M., Maleki, M., & Garavand, A. (2020). A systematic review of the use of telemedicine in the military forces worldwide. *Shiraz E Medical Journal*, *21*(11), e99343. Advance online publication. DOI: 10.5812/semj.99343

Monseu, N. (2005). *Les usages de l'intentionnalité*. Éditions de l'Institut Supérieur de Philosophie.

Montemayor, C., Halpern, J., & Fairweather, A. (2021). In principle obstacles for empathic AI: Why we can't replace human empathy in healthcare. *AI & Society*, *37*(4), 1353–1359. DOI: 10.1007/s00146-021-01230-z PMID: 34054228

Moran, D. (2021). Husserl and the Greeks. *JBSP : The Journal of the British Society for Phenomenology. British Society for Phenomenology*, *52*(2), 98–117. DOI: 10.1080/00071773.2020.1821579

Morgenthau, H. J. (1978). *Politics Among Nations: The Struggle of Power and Peace* (5th ed.). Alfred A. Knopf.

Morrison, S. R. (2020). Advance directives/care planning: Clear, simple, and wrong. *Journal of Palliative Medicine*, *23*(7), 878–879. DOI: 10.1089/jpm.2020.0272 PMID: 32453620

Morse, S., Talty, K., Kuiper, P., Scioletti, M., Heymsfield, S. B., Atkinson, R. L., & Thomas, D. M. (2020). Machine learning prediction of combat basic training injury from 3D body shape images. *PLoS One*, *15*(6), e0235017. DOI: 10.1371/journal.pone.0235017 PMID: 32603356

Moseley, A. (2004). Just war theory. *Internet Encyclopedia of Philosophy*. https://iep.utm.edu/justwar/

Mulchandani, N., & Shanahan, J. (2022). *Software-Defined Warfare: Architecting the DOD's Transition to the Digital Age.* https://www.csis.org/analysis/software-defined-warfare-architecting-dods-transition-digital-age

Mulchandani, N., & Shanahan, L. G. J. N. (2022). *Software-Defined Warfare: Architecting the DOD's transition to the digital age.* https://www.csis.org/analysis/software-defined-warfare-architecting-dods-transition-digital-age

Müller, H. (2004). The antinomy of democratic peace. *International Politics, 41*(4), 494–520. DOI: 10.1057/palgrave.ip.8800089

Murray, M. (1988). Husserl and Heidegger: Constructing and Reconstructing Greek Philosophy. *The Review of Metaphysics, 41*(3), 501–518.

Nancholas, B. (2023, October 13). *Narrow artificial intelligence: advantages, disadvantages, and the future of AI - University of Wolverhampton.* University of Wolverhampton. https://online.wlv.ac.uk/narrow-artificial-intelligence-advantages-disadvantages-and-the-future-of-ai/

Narayan, A., Bhushan, I., & Schulman, K. (2024a). India's evolving digital health strategy. *NPJ Digital Medicine, 7*(1), 284. DOI: 10.1038/s41746-024-01279-2 PMID: 39414924

Nasir, N., Alshabi, M., Al-Yateem, N., Rahman, S. A., Subu, M. A., Hijazi, H. H., . . . Saifan, A. R. (2023, June). Challenges of Artificial Intelligence in Medicine. In *2023 IEEE 47th Annual Computers, Software, and Applications Conference (COMPSAC)* (pp. 1429-1433). IEEE. DOI: 10.1109/COMPSAC57700.2023.00219

National Institute of Standards and Technology. (2025). *Outline: Proposed zero draft for a standard on AI TEVV (Test, Evaluation, Validation, and Verification).* Obtenido de https://www.nist.gov/system/files/documents/2025/07/15/Outline_%20Proposed%20Zero%20Draft%20for%20a%20Standard%20on%20AI%20TEVV-for-web.pdf

National Science and Technology Council. (2016). *The National Artificial Intelligence Research and Development Strategic Plan.* https://www.nitrd.gov/news/national_ai_rd_strategic_plan.aspx

National Transport Commission. (2018). *Changing driving laws to support automated vehicles.* https://www.ntc.gov.au/transport-reform/ntc-projects/changing-driving-laws-support-AVs

NATO. (2021, October 22). *NATO releases first-ever strategy for artificial intelligence.* https://www.nato.int/cps/en/natohq/news_187934.htm

Nazemi, M. M., Khodabandeh, A., & Hadjizadeh, A. (2022). Near-field electrospinning: Crucial parameters, challenges, and applications. *ACS Applied Bio Materials*, *5*(2), 394–412. DOI: 10.1021/acsabm.1c00944 PMID: 34995437

Nazir, A., Hussain, A., Singh, M., & Assad, A. (2025). Deep learning in medicine: Advancing healthcare with intelligent solutions and the future of holography imaging in early diagnosis. *Multimedia Tools and Applications*, *84*(17), 17677–17740. DOI: 10.1007/s11042-024-19694-8

Neacsu, M. (2010). *Hans J. Morgenthau's Theory of International Relation, Disenchantment and Re-enchantment*. Palgrave Macmillan. DOI: 10.1057/9780230245761

Neumann, P. R. (2013). The trouble with radicalization. *International Affairs*, *89*(4), 873–893. DOI: 10.1111/1468-2346.12049

New America. (n.d.). *The drone war in Pakistan*. https://www.newamerica.org/international-security/reports/americas-counterterrorism-wars/the-drone-war-in-pakistan/

Newman, M. (2000). Sensibility, Trauma, and the Trace: Levinas from Phenomenology to the Immemorial. In Bloechl, J. (Ed.), *The Face of the Other and the Trace of God* (pp. 90–129). Fordham University Press.

Nichols, R., Horstman, J., & Nitz, I. (2023). Army Health Systems Doctrine and Training in Relation to Antibiotics: A Systematic Review. *Military Medicine*, *188*(7-8), 1649–1655. DOI: 10.1093/milmed/usac210 PMID: 35830415

Norris, K. (2020). The drone threat to just war theory: Responding to Braun. *International Relations*, *34*(4), 603–607. DOI: 10.1177/0047117820967983

Nyholm, S., & Smids, J. (2016). The ethics of accident-algorithms for self-driving cars: An applied trolley problem? *Ethical Theory and Moral Practice*, *19*(5), 1275–1289. DOI: 10.1007/s10677-016-9745-2

Obama, B. (2012, October 22). Second presidential debate [Video]. YouTube. https://www.youtube.com/watch?v=ZDDJLuIREAQ

OECD AI Policy Observatory Portal. (2025). https://oecd.ai/en/

Office of the Surgeon General. (2006). *Mental health advisory team (MHAT) IV Operation Iraqi Freedom 05-07*. National Technical Reports Library. http://www.combatreform.org/MHAT_IV_Report_17NOV06.pdf

Olawade, D. B., David-Olawade, A. C., Wada, O. Z., Asaolu, A. J., Adereni, T., & Ling, J. (2024). Artificial intelligence in healthcare delivery: Prospects and pitfalls. *Journal of Medicine, Surgery, and Public Health*, *3*, 100108. DOI: 10.1016/j.glmedi.2024.100108

Olawade, D. B., Wada, O. Z., Odetayo, A., David-Olawade, A. C., Asaolu, F., & Eberhardt, J. (2024). Enhancing mental health with Artificial Intelligence: Current trends and future prospects. *Journal of Medicine, Surgery, and Public Health*, *3*, 100099. DOI: 10.1016/j.glmedi.2024.100099

Paliwal, S., Parveen, S., Singh, O., Alam, M. A., & Ahmed, J. (2023). The role of Ayushman Bharat Health Account (ABHA) in Telehealth: a new frontier of smart healthcare delivery in India. In *Lecture notes in networks and systems* (pp. 388–406). https://doi.org/DOI: 10.1007/978-3-031-47451-4_28

Panayiotou, S. (2024). Hegel and Levinas on the Notion of Esoteric Thought. In Pitkänen, O. (Ed.), *Philosophical Perspectives on Esotericism* (pp. 226–250). Routledge. DOI: 10.4324/9781003457046-12

Papagiannidis, E., Mikalef, P., & Conboy, K. (2025). Responsible artificial intelligence governance: A review and research framework. *The Journal of Strategic Information Systems*, *34*(2), 101885. DOI: 10.1016/j.jsis.2024.101885

Parums, D. V. (2023). Editorial: Infectious Disease Surveillance Using Artificial Intelligence (AI) and its Role in Epidemic and Pandemic Preparedness. *Medical Science Monitor*, *29*, e941209. DOI: 10.12659/MSM.941209 PMID: 37259578

Parvizpour, S., Beyrampour-Basmenj, H., Razmara, J., Farhadi, F., & Shamsir, M. S. (2023). Cancer treatment comes to age: From one-size-fits-all to next-generation sequencing (NGS) technologies. *BioImpacts*, *14*(4), 29957. DOI: 10.34172/bi.2023.29957 PMID: 39104623

Pashakhanlou, A. (2017). Fully integrated content analysis in International Relations. *International Relations*, *31*(4), 447–465. DOI: 10.1177/0047117817723060

Patra, J., Moulick, H. N., & Manna, A. K. (2013). Medical image processing in nuclear medicine and bone arthroplasty. *IOSR Journal of Computer Engineering*, *13*(2), 97–110. DOI: 10.9790/0661-13297110

Paziuk, A., Lande, D., Shnurko-Tabakova, E., & Kingston, P. (2025). Decoding manipulative narratives in cognitive warfare: A case study of the Russia-Ukraine conflict. *Frontiers in Artificial Intelligence*, *8*, 1566022. DOI: 10.3389/frai.2025.1566022 PMID: 41018734

Peoples, K. (2021). *How to write a phenomenological dissertation: A step-by-step guide* (Vol. 56). SAGE.

Peperzak, A. (1983). Phenomenology – Ontology – Metaphysics: Levinas' Perspective on Husserl and Heidegger. *Man and World, 16*(2), 113–127. DOI: 10.1007/BF01260324

Perelsztejn, N. (2024). Escape to Judaism: Levinas's First Steps toward Becoming a Jewish Thinker. *The Journal of Jewish Thought & Philosophy*, *32*(2), 267–291. DOI: 10.1163/1477285x-12341361

Petropoulos, G. (2021). Phenomenology and Ancient Greek Philosophy: An Introduction. *JBSP : The Journal of the British Society for Phenomenology. British Society for Phenomenology*, *52*(2), 95–97. DOI: 10.1080/00071773.2021.1899053

Phelps, W., & Grossman, D. (2021). *On killing remotely: The psychology of killing with drones*. Little, Brown and Company.

Pickel, A. (2022). *Handbook of Economic Nationalism*. Edward Elgar Publishing.

Pietrzak, P. (2019). "The Syrian Conflict (2011–ongoing): How a Perfectly Winnable Uprising has been transformed into a Civil War, Only to End up as a Ferocious Proxy War of Global Importance," *In Statu Nascendi – Journal of Political Philosophy and International Relations* Vol. 2, No. 1 2019, p. 34–95, ISBN: 9783838213095.

Pietrzak, P. (2023). Approaching regional conflicts through the prism of ontology in statu nascendi: The new compartmentalization of the IR theory. *In Statu Nascendi – Journal of Political Philosophy and International Relations, 6*(1).

Pietrzak, P. (2021). *On the Idea of Humanitarian Intervention: A New Compartmentalization of IR Theories* (1st ed.). Ibidem Verlag.

Pietrzak, P. (2022). *Introducing the idea of ontology in statu nascendi to the broader international relations theory. International Conference Proceeding Series: International Conference on Economics and Social Sciences*. ECLSS., https://www.eclss.org/publicationsfordoi/abst11act8boo8kIE%26SS2022_antalya.pdf

Pietrzak, P. (2024). The sixth great debate in international relations theory revolves around clarity: Ontology in statu nascendi as a cutting-edge tool for conflict management. In Pietrzak, P. (Ed.), *Dealing with regional conflicts of global importance* (pp. 1–27). IGI Global., DOI: 10.4018/978-1-6684-9467-7.ch001

Pinto-Coelho, L. (2023). How Artificial intelligence is shaping medical Imaging Technology: A survey of Innovations and applications. *Bioengineering (Basel, Switzerland), 10*(12), 1435. DOI: 10.3390/bioengineering10121435 PMID: 38136026

Png, M. T. (2022). At the tensions of South and North: Critical roles of global South stakeholders in AI governance. *ACM International Conference Proceeding Series*. DOI: 10.1145/3531146.3533200

Poalelungi, D. G., Musat, C. L., Fulga, A., Neagu, M., Neagu, A. I., Piraianu, A. I., & Fulga, I. (2023). Advancing Patient Care: How Artificial intelligence is transforming healthcare. *Journal of Personalized Medicine*, *13*(8), 1214. DOI: 10.3390/jpm13081214 PMID: 37623465

Poleshchuk, I. (2020). Heidegger and Levinas: Metaphysics, Ontology and the Horizon of the Other. *The Indo-Pacific Journal of Phenomenology*, *10*(2), 1–10. DOI: 10.2989/IPJP.2010.10.2.4.1085

Powers, M. (2025, March 3). Russia capitalizes on development of artificial intelligence in its military strategy. *Jamestown Foundation*. https://jamestown.org

Priyadharshini, K., Mohan, D., & Gokulakrishnan, D. (2023). Artificial intelligence-based multimodal molecular imaging fusion framework for cancer care. In *Proceedings of the 2023 International Conference on Innovative Computing, Intelligent Communication and Smart Electrical Systems (ICSES)* (pp. 1–6). DOI: 10.1109/ICSES60034.2023.10465530

Puri, G. (2024, September 20). How AI is ushering in a new era of value in US healthcare - healthcare business today. *Healthcare Business Today*. https://www.healthcarebusinesstoday.com/ai-new-era-value-us-healthcare/

Purves, D., Jenkins, R., & Strawser, B. J. (2015). Autonomous machines, moral judgment, and acting for the right reasons. *Ethical Theory and Moral Practice*, *18*(4), 851–872. DOI: 10.1007/s10677-015-9563-y

Radanliev, P. (2025). AI Ethics: Integrating transparency, fairness, and privacy in AI development. *Applied Artificial Intelligence*, *39*(1), 2463722. Advance online publication. DOI: 10.1080/08839514.2025.2463722

Rădulescu, B. (2026, March 26). Cognitive hegemony: the true objective of Russia's new active measures. Intelligence Info. https://www.intelligenceinfo.org/en/cognitive-hegemony-the-true-objective-of-russias-new-active-measures/

Rae, G. (2016). The Political Significance of the Face: Deleuze's Critique of Levinas. *Critical Horizons*, *17*(3–4), 279–303. DOI: 10.1080/14409917.2016.1190173

Raj, G. M., Dananjayan, S., & Agarwal, N. (2023). Inception of the Indian Digital Health Mission: Connecting. . .the. . .Dots. *Health Care Science*, *2*(5), 345–351. DOI: 10.1002/hcs2.67 PMID: 38938588

Raman, R., Kowalski, R., Achuthan, K., Iyer, A., & Nedungadi, P. (2025). Navigating artificial general intelligence development: Societal, technological, ethical, and brain-inspired pathways. *Scientific Reports*, *15*(1), 8443. DOI: 10.1038/s41598-025-92190-7 PMID: 40069265

Rashid, A. B., Kausik, A. K., Sunny, A. a. H., & Bappy, M. H. (2023). Artificial intelligence in the military: An overview of the capabilities, applications, and challenges. *International Journal of Intelligent Systems, 2023*(1). https://doi.org/DOI: 10.1155/2023/8676366

Rashid, A. B., & Kausik, M. K. (2024). AI revolutionizing industries worldwide: A comprehensive overview of its diverse applications. *Hybrid Advances*, *7*, 100277. DOI: 10.1016/j.hybadv.2024.100277

Raska, M. (2021). The sixth RMA wave: Disruption in military affairs? *The Journal of Strategic Studies*, *44*(4), 456–479. DOI: 10.1080/01402390.2020.1848818

Rawas, S. (2024). AI: The future of humanity. *Discover Artificial Intelligence*, *4*(1), 25. Advance online publication. DOI: 10.1007/s44163-024-00118-3

Ray, J. (2016). *China's industrial and military robotics development* (Report for the US–China Economic and Security Review Commission). Center for Intelligence Research and Analysis, Defense Group Inc. https://www.uscc.gov/sites/default/files/Research/DGI_China%27s%20Industrial%20and%20Military%20Robotics%20Development.pdf

Reiter, D., & Stam, A. (2002). *Democracies at war*. Princeton University Press. DOI: 10.1515/9781400824458

Riazi, M., Rezvani, R., & Ghanbar, H. (2023). Trustworthiness in L2 writing research: A review and analysis of qualitative articles in the Journal of Second Language Writing. *Research Methods in Applied Linguistics*, *2*(3), 100065. Advance online publication. DOI: 10.1016/j.rmal.2023.100065

Ribeiro, M. T., Singh, S., & Guestrin, C. (2016). "Why should I trust you?": Explaining the predictions of any classifier. *Proceedings of the 22nd ACM SIGKDD International Conference on Knowledge Discovery and Data Mining*. DOI: 10.1145/2939672.2939778

Ricoeur, P. (1980). Narrative time. *Critical Inquiry*, *7*(1), 169–190. DOI: 10.1086/448093

Riley, W. J. (2012). *Health disparities: Gaps in access, quality and affordability of medical care*. https://pmc.ncbi.nlm.nih.gov/articles/PMC3540621/

Roberts, H., Cowls, J., Morley, J., Taddeo, M., Wang, V., & Floridi, L. (2021). The Chinese approach to artificial intelligence: An analysis of policy, ethics, and regulation. *AI & Society*, *36*(1), 59–77. DOI: 10.1007/s00146-020-00992-2

Robertson, P. (2024, December 17). China's military rise: Comparative military spending in China and the US. CEPR. https://cepr.org/voxeu/columns/chinas-military-rise-comparative-military-spending-china-and-us

Robillard, M. (2018). No such thing as killer robots. *Journal of Applied Philosophy*, *35*(4), 705–717. DOI: 10.1111/japp.12274

Robinson, K., McKenna, B., & Rooney, D. (2021). The Relationship of Risk to Rules, Values, Virtues, and Moral Complexity: What We can Learn from the Moral Struggles of Military Leaders. *Journal of Business Ethics*, *179*(3), 749–766. DOI: 10.1007/s10551-021-04874-5 PMID: 34177014

Rogers, E. (2003). *Diffusion of innovations* (5th ed.). Free Press.

Rosário, A. T., & Boechat, A. C. (2025). How sustainable leadership can leverage sustainable development. *Sustainability (Basel)*, *17*(8), 3499. DOI: 10.3390/su17083499

Rosert, E., & Sauer, F. (2019). Prohibiting autonomous weapons: Put human dignity first. *Global Policy*, *10*(3), 370–375. DOI: 10.1111/1758-5899.12691

Roy, A. D., Das, D., & Mondal, H. (2023). The Tribal health system in India: Challenges in healthcare delivery in comparison to the global healthcare systems. *Cureus*, *15*(6), e39867. DOI: 10.7759/cureus.39867 PMID: 37404413

Royakkers, L., & van Est, R. (2015). A literature review on new robotics: Automation from love to war. *International Journal of Social Robotics*, *7*(5), 549–570. DOI: 10.1007/s12369-015-0295-x

RT. (2017, September 1). *'Whoever leads in AI will rule the world': Putin to Russian children on Knowledge Day*. https://www.rt.com/news/401731-ai-rule-world-putin/

Ruane, S., Sirinukunwattana, K., Kotanska, A., & Aberdeen, A. (2025). Developing AI-powered tools for histopathology: Opportunities, challenges and new friends along the way. *Diagnostic Histopathology*, *31*(5), 277–283. DOI: 10.1016/j.mpdhp.2025.03.003

Russon, J. (2021). To Account for the Appearances: Phenomenology and Existential Change in Aristotle and Plato. *JBSP : The Journal of the British Society for Phenomenology. British Society for Phenomenology*, *52*(2), 155–168. DOI: 10.1080/00071773.2020.1796515

Ryan, M. (2022). *War transformed: The future of twenty-first-century great power competition and conflict.* Naval Institute Press.

Rydmark, J., Kuylenstierna, J., & Tehler, H. (2021). Communicating uncertainty in risk descriptions: The consequences of presenting imprecise probabilities in time critical decision-making situations. *Journal of Risk Research, 24*(5), 629–644. DOI: 10.1080/13669877.2020.1801807

Saboury, B., Bradshaw, T., Boellaard, R., Buvat, I., Dutta, J., Hatt, M., Jha, A. K., Li, Q., Liu, C., McMeekin, H., Morris, M. A., Scott, P. J. H., Siegel, E., Sunderland, J. J., Pandit-Taskar, N., Wahl, R. L., Zuehlsdorff, S., & Rahmim, A. (2023). Artificial intelligence in nuclear medicine: Opportunities, challenges, and responsibilities toward a trustworthy ecosystem. *Journal of Nuclear Medicine, 64*(2), 188–196. DOI: 10.2967/jnumed.121.263703 PMID: 36522184

Saki, S., & Soori, M. (2025). Artificial intelligence, machine learning and deep learning in advanced transportation systems, a review. *Multimodal Transportation, 5*(1), 100242. DOI: 10.1016/j.multra.2025.100242

Sallis, J. (1998). Levinas and the Elemental. *Research in Phenomenology, 28*(1), 152–159. DOI: 10.1163/156916498X00092

Salmon, P. M., Walker, G. H., & Stanton, N. A. (2016). Pilot error versus sociotechnical systems failure: A distributed situation awareness analysis of Air France 447. *Theoretical Issues in Ergonomics Science, 17*(1), 64–79. DOI: 10.1080/1463922X.2015.1106618

Sandford, S. (1999). Levinas in the Realm of the Senses: Transcendence and Intelligibility. *Angelaki, 4*(3), 61–73. DOI: 10.1080/09697259908572056

Sarkar, I. N. (2022). Transforming health data to Actionable Information: Recent progress and future opportunities in health information exchange. *Yearbook of Medical Informatics, 31*(01), 203–214. DOI: 10.1055/s-0042-1742519 PMID: 36463879

Sarkar, S., Singh, Y. C., & Kaloiya, G. S. (2024). Psychotherapy and psychotropic drug treatment: Neurobiological and psychodynamic perspectives. *Indian Journal of Psychiatry, 66*, S126.

Sauer, F. (2016). Stopping 'killer robots': Why now is the time to ban autonomous weapons systems. *Arms Control Today.*

Sauer, F., & Schörnig, N. (2012). Killer drones: The 'silver bullet' of democratic warfare? *Security Dialogue, 43*(4), 363–380. DOI: 10.1177/0967010612450207

Scharre, P., & Horowitz, M. C. (2015). *Autonomy in weapon systems*. Center for a New American Security. https://www.cnas.org/publications/reports/an-introduction-to-autonomy-in-weapon-systems

Scharre, P. (2018). *Army of none: Autonomous weapons and the future of war*. W.W. Norton & Co.

Schneider, S. (2019). *Artificial you*. Princeton University Press.

Schulzke, M. (2011). Robots as weapons in just wars. *Philosophy & Technology*, *24*(3), 293–306. DOI: 10.1007/s13347-011-0028-5

Schulzke, M. (2013). Autonomous weapons and distributed responsibility. *Philosophy & Technology*, *26*(2), 203–219. DOI: 10.1007/s13347-012-0089-0

Schulzke, M. (2018). Drone proliferation and the challenge of regulating dual-use technologies. *International Studies Review*, *0*, 1–21.

Sedgwick, M. (2010). The concept of radicalization as a source of confusion. *Terrorism and Political Violence*, *22*(4), 479–494. DOI: 10.1080/09546553.2010.491009

Shajari, S., Kuruvinashetti, K., Komeili, A., & Sundararaj, U. (2023). The emergence of AI-based wearable sensors for digital health technology: A review. *Sensors (Basel)*, *23*(23), 9498. DOI: 10.3390/s23239498 PMID: 38067871

Sharkey, A. (2019). Autonomous weapons systems, killer robots and human dignity. *Ethics and Information Technology*, *21*(2), 75–87. DOI: 10.1007/s10676-018-9494-0

Sharma, R. S., Rohatgi, A., Jain, S., & Singh, D. (2023). The Ayushman Bharat Digital Mission (ABDM): Making of India's Digital Health Story. *CSI Transactions on ICT*, *11*(1), 3–9. DOI: 10.1007/s40012-023-00375-0

Shaw, I. (2017). Robot wars: US empire and geopolitics in the robotic age. *Security Dialogue*, *48*(5), 451–470. DOI: 10.1177/0967010617713157 PMID: 29081605

Shaw, M. (2005). The new Western way of war. *Polity*.

Shead, S. (2021, December 22). *UN talks to ban 'slaughterbots' collapsed*. CNBC. https://www.cnbc.com/2021/12/22/un-talks-to-ban-slaughterbots-collapsed-heres-why-that-matters.html

Shen, L., & Rangayyan, R. M. (1997). A segmentation-based lossless image coding method for high-resolution medical image compression. *IEEE Transactions on Medical Imaging*, *16*(3), 301–311. DOI: 10.1109/42.585764 PMID: 9184892

Sheringham, J., Kuhn, I., & Burt, J. (2021). The use of experimental vignette studies to identify drivers of variations in the delivery of health care: A scoping review. *BMC Medical Research Methodology*, *21*(1), 1–17. DOI: 10.1186/s12874-021-01247-4 PMID: 33888077

Shockey, R. M. (2010). Heidegger's Descartes and Heidegger's Cartesianism. *European Journal of Philosophy*, *20*(2), 285–311. DOI: 10.1111/j.1468-0378.2010.00408.x

Sigfrids, A., Leikas, J., Salo-Pöntinen, H., & Koskimies, E. (2023). Human-centricity in AI governance: A systemic approach. *Frontiers in Artificial Intelligence*, *6*, 976887. DOI: 10.3389/frai.2023.976887 PMID: 36872934

Skerker, M., Purves, D., & Jenkins, R. (2020). Autonomous weapons systems and the moral equality of combatants. *Ethics and Information Technology*, *22*(3), 197–209. DOI: 10.1007/s10676-020-09528-0

Sloan, E. (2015). Robotics at war. *Survival*, *57*(5), 107–120. DOI: 10.1080/00396338.2015.1090133

Smith, B. (2025, September 18). *Winning the AI race: Strengthening U.S. capabilities in computing and innovation*. Microsoft on the Issues. https://blogs.microsoft.com/on-the-issues/2025/05/08/winning-the-ai-race/

Smith, B., Miettinen, K., & Mandrick, W. (2009). The ontology of command and control (C2). *Proceedings of the 14th International Command and Control Research and Technology Symposium (ICCRTS)*.

Smith, F. (2016). Gadamer and Levinas on the Topic of Sociality. *Sophia Philosophical Review*, *9*(2), 95–109.

Smith, R. P. (1981). Boredom: A review. *Human Factors*, *23*(3), 329–340.1. DOI: 10.1177/001872088102300308 PMID: 7016721

Smith, T. (2002). The new law of war: Legitimizing hi-tech and infrastructural violence. *International Studies Quarterly*, *46*(3), 355–374. DOI: 10.1111/1468-2478.00237

Solomon, J., Dauber-Decker, K., Richardson, S., Levy, S., Khan, S., Coleman, B., Persaud, R., Chelico, J., King, D., Spyropoulos, A., & McGinn, T. (2023). Integrating clinical decision support into electronic health record systems using a novel platform (EvidencePoint): Developmental study. *JMIR Formative Research*, *7*, e44065. DOI: 10.2196/44065 PMID: 37856193

Sommerville, I. (2020). *Software Engineering* (10th ed.). Pearson.

Sparrow, R. (2007). Killer robots. *Journal of Applied Philosophy*, *24*(1), 62–77. DOI: 10.1111/j.1468-5930.2007.00346.x

Sparrow, R. (2009). Building a better WarBot: Ethical issues in the design of unmanned systems for military applications. *Science and Engineering Ethics*, *15*(2), 169–187. DOI: 10.1007/s11948-008-9107-0 PMID: 19048395

Sparrow, R. (2016). Robots and respect: Assessing the case against autonomous weapon systems. *Ethics & International Affairs*, *30*(1), 93–116. DOI: 10.1017/S0892679415000647

Spatola, N. (2024). The efficiency-accountability tradeoff in AI integration: Effects on human performance and over-reliance. *Computers in Human Behavior Artificial Humans*, *2*(2), 100099. DOI: 10.1016/j.chbah.2024.100099

Sperber, D., & Wilson, D. (1995). *Relevance: Communication and cognition* (2nd ed.). Wiley-Blackwell.

Stahl, B. C. (2021). *Artificial Intelligence for a Better Future*. Springer. DOI: 10.1007/978-3-030-69978-9

Stahl, R. M., & Popp-Madsen, B. A. (2022). Defending democracy: Militant and popular models of democratic self-defense. *Constellations (Oxford, England)*, *29*(3), 310–328. DOI: 10.1111/1467-8675.12639

Standish, P. (2008). Levinas and the Language of the Curriculum. In Egéa-Kuehne, D. (Ed.), *Levinas and Education* (pp. 56–66). Routledge.

Stanovich, K. E., & West, R. F. (2000). [3] Individual differences in reasoning: Implications for the rationality debate? *The Behavioral and Brain Sciences*, *23*(5), 645–726. DOI: 10.1017/S0140525X00003435 PMID: 11301544

Stanovich, K. E., West, R. F., & Toplak, M. E. (2016). *The rationality quotient: Toward a test of rational thinking*. MIT Press. DOI: 10.7551/mitpress/9780262034845.001.0001

Stanton, A. (Director). (2008). *Wall-E* [Film]. Walt Disney Pictures; Pixar Animation Studios.

Stapleton, T. J. (1983). *Husserl and Heidegger: The Question of a Phenomenological Beginning*. SUNY Press.

Stein, E. (2000). *Knowledge and Faith* (Publications, I. C. S., Trans.). (Original work published 1993)

Stockholm International Peace Research Institute. (2019). *Trends in world military expenditure 2018*. https://www.sipri.org/publications/2019/sipri-fact-sheets/trends-world-military-expenditure-2018

Stockholm International Peace Research Institute. (n.d.). *Stockholm International Peace Research Institute*. https://www.sipri.org/

Stop Killer Robots. (2022). *Stop killer robots*. https://www.stopkillerrobots.org/stop-killer-robots/

Storr, J. (2009). *The human face of war*. A&C Black.

Sullivan, Y., de Bourmont, M., & Dunaway, M. (2022). Appraisals of harms and injustice trigger an eerie feeling that decreases trust in artificial intelligence systems. *Annals of Operations Research*, *308*(1–2), 525–548. DOI: 10.1007/s10479-020-03702-9

Swanson, K., Wu, E., Zhang, A., Alizadeh, A. A., & Zou, J. (2023). From patterns to patients: Advances in clinical machine learning for cancer diagnosis, prognosis, and treatment. *Cell*, *186*(8), 1772–1791. DOI: 10.1016/j.cell.2023.01.035 PMID: 36905928

Switzer, R. (2000). Redeeming the Elemental: Levinas on Nature. *Interdisciplinary Environmental Review*, *4*(2), 23–35. DOI: 10.1504/IER.2002.053892

Szasz, T. S., & Hollender, M. H. (1956). The basic models of the doctor–patient relationship. *Archives of Internal Medicine*, *97*, 585–592. PMID: 13312700

Tai, M. C. (2020). The impact of artificial intelligence on human society and bioethics. *Tzu-Chi Medical Journal*, *32*(4), 339. DOI: 10.4103/tcmj.tcmj_71_20 PMID: 33163378

Takagi, K. (2025, January 27). Is the PLA overestimating the potential of artificial intelligence? National Defense University. https://www.ndu.edu/News/Article-View/Article/4116123/is-the-pla-overestimating-the-potential-of-artificial-intelligence

Talbert, M. (2016). *Moral responsibility: An introduction*. John Wiley & Sons.

Taminiaux, J. (1997). The Early Levinas's Reply to Heidegger's Fundamental Ontology. *Philosophy and Social Criticism*, *23*(6), 29–49. DOI: 10.1177/019145379702300603

Tapper, M. (1986). The Priority of Being or Consciousness for Phenomenology: Heidegger and Husserl. *Metaphilosophy*, *17*(2-3), 153–161. DOI: 10.1111/j.1467-9973.1986.tb00377.x

Taylor, R. D. (2020). Quantum artificial intelligence: A "precautionary" U.S. approach? *Telecommunications Policy*, *44*(6), 101909. DOI: 10.1016/j.telpol.2020.101909

Tegmark, M. (2017). *Life 3.0: Being human in the age of artificial intelligence*. Alfred A. Knopf.

Tewathia, N., Kamath, A., & Ilavarasan, P. V. (2020). Social inequalities, fundamental inequities, and recurring of the digital divide: Insights from India. *Technology in Society*, *61*, 101251. DOI: 10.1016/j.techsoc.2020.101251

Thacharodi, A., Singh, P., Meenatchi, R., Ahmed, Z. H. T., & Kumar, R. R. S. (2024). Revolutionizing healthcare and medicine: The impact of modern technologies for a healthier future—A comprehensive review. *Health Care Science*, *3*(5), 329–349. DOI: 10.1002/hcs2.115 PMID: 39479277

The 2025 AI Index Report | Stanford HAI. (2025). https://hai.stanford.edu/ai-index/2025-ai-index-report

Thomsen, M. D. T., & Knudsen, L. E. (2025). Editorial: Reviews in: regulatory science 2024-2025. *Frontiers in Medicine*, *12*, 1705977. DOI: 10.3389/fmed.2025.1705977 PMID: 41189882

Thornton, R. (2007). Asymmetric warfare: Threat and response in the twenty-first century. *Polity*.

Tooby, J., & Cosmides, L. (2025). The evolution of war and its cognitive foundations. *Evolution and Human Behavior*, *46*(3), 106687. DOI: 10.1016/j.evolhumbehav.2025.106687

Tryhorn, D., Dill, R., Hodson, D. D., Grimaila, M. R., & Myers, C. W. (2021). Modeling fog of war effects in AFSIM. *The Journal of Defense Modeling and Simulation*, *20*(2), 131–146. DOI: 10.1177/15485129211041963

Tsai, G. (2014). Rational persuasion as paternalism. *Philosophy & Public Affairs*, *42*(1), 78–112. DOI: 10.1111/papa.12026

Tsereteli, A. (2023, January 13). War in Europe: Use of Technologies in the Russia-Ukraine War. https://www.freiheit.org/south-caucasus/use-technologies-russia-ukraine-war

Tversky, A., & Kahneman, D. (1992). Advances in prospect theory: Cumulative representation of uncertainty. *Journal of Risk and Uncertainty*, *5*(4), 297–323. DOI: 10.1007/BF00122574

U.S. Congress. (2026, April 10). U.S. Security Assistance to Ukraine. https://www.congress.gov/crs-product/IF12040

U.S. Department of Defense, Office of the Under Secretary of Defense (Comptroller/Chief Financial Officer). (2026, July). *Program acquisition cost by weapon system: Fiscal year 2026 budget request.*

United Nations Institute for Disarmament Research. (2014). *Framing discussions on the weaponization of increasingly autonomous technologies*. http://www.unidir.org/en/publications/framing-discussions-on-the-weaponization-of-increasingly-autonomous-technologies

United Nations. (n.d.). *United Nations Charter*. https://www.un.org/en/about-us/un-charter/full-text

United States Conference of Catholic Bishops. (1993). *Centrality of conscience*. https://www.usccb.org/issues-and-action/human-life-and-dignity/war-and-peace/excerpts-from-the-harvest-of-justice-is-sown-in-peace-centrality-of-conscience-1993-11

US Department of Defense. (2007). *Unmanned systems roadmap 2007–2032*. https://www.hsdl.org/?abstract&did=481851

US Department of Defense. (2012). *Directive number 3000.09*. https://www.esd.whs.mil/portals/54/documents/dd/issuances/dodd/300009p.pdf

US Department of Defense. (2017a). *Department of Defense announces successful micro-drone demonstration*. https://dod.defense.gov/News/News-Releases/News-Release-View/Article/1044811/department-of-defense-announces-successful-micro-drone-demonstration/

US Department of Defense. (2017b). *Unmanned systems integrated roadmap FY2017-2042*. https://news.usni.org/2018/08/30/pentagon-unmanned-systems-integrated-roadmap-2017-2042

Vaidya, R., & Bobdey, S. (2021). Medical ethics during armed conflicts: Dilemmas of a physician soldier. *Medical Journal, Armed Forces India*, *77*(4), 377–381. DOI: 10.1016/j.mjafi.2021.08.013 PMID: 34594063

Vaishnnave, M. P., Devi, K. S., & Ganeshkumar, P. (2020). Automatic method for classification of groundnut diseases using deep convolutional neural network. *Soft Computing*, *24*(21), 16347–16360. DOI: 10.1007/s00500-020-04946-0

van Creveld, M. (1985). *Command in war*. Harvard University Press.

Varnosfaderani, S. M., & Forouzanfar, M. (2024). The role of AI in Hospitals and Clinics: Transforming Healthcare in the 21st century. *Bioengineering (Basel, Switzerland)*, *11*(4), 337. DOI: 10.3390/bioengineering11040337 PMID: 38671759

Vasey, C. R. (1980). *Existence and Existents*, by Emmanuel Levinas. *The Thomist*, *44*(3), 466–473. DOI: 10.1353/tho.1980.0022

Vasundhara. (2026, March 11). *Iron Dome Air Defence Missile System.* Army Technology. https://www.army-technology.com/projects/iron-dome/

Velez-Green, A. (2015, March 1). The foreign policy essay: The South Korean sentry—A "killer robot" to prevent war. *Lawfare.* https://www.lawfareblog.com/foreign-policy-essay-south-korean-sentry%E2%80%94-killer-robot-prevent-war

Vogel, R. J. (2011). Drone warfare and the law of armed conflict. *Denver Journal of International Law and Policy, 39*(1), 101–136.

von Clausewitz, C., Howard, M., & Paret, P. (1976). *On war.* Princeton University Press. DOI: 10.1515/9781400837403

Wakker, P. P. (2010). *Prospect theory: For risk and ambiguity.* Cambridge University Press. DOI: 10.1017/CBO9780511779329

Walzer, M. (2016). Just & Unjust Targeted Killing & Drone Warfare. *Daedalus, 145*(4), 12–24. https://www.jstor.org/stable/24916780. DOI: 10.1162/DAED_a_00408

Warren, A., & Hillas, A. (2020). Friend or frenemy? The role of trust in human-machine teaming and lethal autonomous weapons systems. *Small Wars & Insurgencies, 31*(4), 822–850. DOI: 10.1080/09592318.2020.1743485

Wasilow, S., & Thorpe, J. B. (2019). Artificial intelligence, robotics, ethics, and the military: A Canadian perspective. *AI Magazine, 40*(1), 37–48. DOI: 10.1609/aimag.v40i1.2848

Weitz, R. (2013). *South Korea's defence industry: Increasing domestic capabilities and global opportunities.* Korea Economic Institute of America, Academic Paper Series, 7, 4.

Wilkinson, T. M. (2012). Nudging and manipulation. *Political Studies, 61*(2), 341–355. DOI: 10.1111/j.1467-9248.2012.00974.x

Williams, E., & Standish, P. (2016). Sounds and Light: Levinas and the Elements of Thought. *Educational Philosophy and Theory, 48*(4), 360–373. DOI: 10.1080/00131857.2015.1047820

Williams, J. (2005). *The Transversal Thought of Gilles Deleuze.* Clinamen Press.

Wilson, D., & Sperber, D. (2002). Relevance theory. In *Handbook of pragmatics.*

Wither, J. K., Partnership for Peace Consortium of Defense Academies and Security Studies Institutes, & Creative Commons. (2023). Hybrid Warfare revisited: A Battle of 'Buzzwords.' Connections: *The Quarterly Journal*, 7–27. https://doi.org/.DOI: 10.11610/Connections.22.1.02

Withy, K. (2021). Thrownness. In Wrathall, M. (Ed.), *The Cambridge Heidegger Lexicon* (pp. 753–756). Oxford University Press. DOI: 10.1017/9780511843778.204

Wu, D., Tan, D. M., Baird, M., DeCampo, J., White, C., & Wu, H. R.Hong Ren Wu. (2006). Perceptually lossless medical image coding. *IEEE Transactions on Medical Imaging*, *25*(3), 335–344. DOI: 10.1109/TMI.2006.870483 PMID: 16524089

Xu, Y., & Wang, P. (2012). The frame problem, the relevance problem, and a package solution to both. *Synthese*, *187*(1), 43–72. DOI: 10.1007/s11229-012-0117-8

Yahil, A., Vija, A. H., & Hawman, E. G. (2006). Enhanced feature extraction in planar nuclear medicine using Pixon minimum-complexity image processing. In *Proceedings of the 2006 IEEE Nuclear Science Symposium Conference Record* (pp. 2037–2040). DOI: 10.1109/NSSMIC.2006.354314

Yang, J., Sohn, J. H., Behr, S. C., Gullberg, G. T., & Seo, Y. (2020). CT-less direct correction of attenuation and scatter in the image space using deep learning for whole-body FDG PET: Potential benefits and pitfalls. *Radiology. Artificial Intelligence*, *3*(2), e200137. DOI: 10.1148/ryai.2020200137 PMID: 33937860

Yan, X., & Qi, H. (2025). China–US competition, reverse globalization, and the regression of world politics. *China International Strategy Review*, *7*(1), 1–15. DOI: 10.1007/s42533-025-00187-8

Yildirim, J., & Sezgin, S. (2005). Democracy and military expenditure. *Transition Studies Review*, *12*(1), 93–100. DOI: 10.1007/s11300-005-0037-0

Youn, S., Ki, M. R., Abdelhamid, M. A., & Pack, S. P. (2024). Biomimetic materials for skin tissue regeneration and electronic skin. *Biomimetics*, *9*(5), 278. DOI: 10.3390/biomimetics9050278 PMID: 38786488

Zaidan, E., & Ibrahim, I. A. (2024). AI governance in a complex and rapidly changing regulatory landscape: A Global perspective. *Humanities & Social Sciences Communications*, *11*(1), 1121. Advance online publication. DOI: 10.1057/s41599-024-03560-x

Zarei, H., Baharmand, H., Bashiri, M., & Madanian, S. (2024). Technological advancements in humanitarian aid. *International Journal of Disaster Risk Reduction*, *109*, 104527. DOI: 10.1016/j.ijdrr.2024.104527

Zhang, C. A., Cho, S., & Vasarhelyi, M. (2022). Explainable artificial intelligence (XAI) in auditing. *International Journal of Accounting Information Systems*, *46*, 100572. DOI: 10.1016/j.accinf.2022.100572

Zhou, Q., Zuley, M., Guo, Y., Yang, L., Nair, B., Vargo, A., Ghannam, S., Arefan, D., & Wu, S. (2021). A machine and human reader study on AI diagnosis model safety under attacks of adversarial images. *Nature Communications*, *12*(1), 7281. DOI: 10.1038/s41467-021-27577-x PMID: 34907229

Zoubir, Y. H. (2023). *Routledge companion to China and the Middle East and North Africa*. DOI: 10.4324/9781003048404

About the Contributors

Piotr Pietrzak, Ph.D., is a co-founder of In Statu Nascendi Think Tank, a non-partisan, independent, and highly diversified network of scholars organized to study a contemporary conflict and political philosophy that provides advice and ideas on specific issues related to geoeconomic, political, and socio-economic problems. He immerses himself in geopolitics every day. As a political thinker, author, and trained ontologist, he takes pride in infusing a pragmatic perspective into IR theory, conflict management, and geopolitics. He is also the author of On the Idea of Humanitarian Intervention: A New Compartmentalization of IR Theories (2021) and editor of Dealing with Regional Conflicts of Global Importance (2024), Analyzing Global Responses to Contemporary Regional Conflicts (2024), Exploring the Implications of Local and Regional Conflicts (2025), and Strengthening International Relations Through Transformative Theory and Practice (2026).

Evangelos Koumparoudis, Ph.D., is the author of Medicine in the Post-consumerist Society: A Philosophical Overview (2023). He is associated with the Bulgarian Academy of Sciences (BAS) and Sofia University St. Kliment Ohridski. He held a visiting scholarship at the Sorbonne Université, Faculté des Lettres. A founding member of the Editorial Board of In Statu Nascendi – Journal of Political Philosophy and International Relations (2017-2023) and the cofounder of In Statu Nascendi Think Tank (2023-). He is fascinated by continental philosophy, ethics, AI, and new technologies. Dr. Koumparoudis contributed significantly to the publishing success of Dealing with Regional Conflicts of Global Importance (2024), Analyzing Global Responses to Contemporary Regional Conflicts (2024), Exploring the Implications of Local and Regional Conflicts (2025), International Relations Theory and Philosophical Political Insights into Conflict Management (2025) and Strengthening International Relations Through Transformative Theory and Practice (2026).

***

Akash Bag is a final-year Ph.D. student in Digital Forensics, currently pursuing his research with a strong focus on the intersection of technology and law. As an Assistant Professor at the School of Law and Justice, Adamas University in Kolkata, he fosters a dynamic learning environment where he shares his deep knowledge of international relations, national and international politics, international law, and international criminal law. Akash is passionate about exploring the complexities of the refugee crisis and the evolving global geopolitical landscape. Additionally, he dedicates himself to sharing stories of business and entrepreneurship, emphasizing the resilience and innovation that drive the global economy. Through his work, he aims to cultivate a community of curious minds, encouraging insightful discussions that drive forward-thinking solutions in both academic and real-world contexts.

Ananya Bharadwaj is a legal academician and researcher with over eight years of experience across academia, legal consultancy, and policy research. She has served as a Legal Consultant with the Directorate of Income Tax (Legal & Research), Delhi, and worked as a Research Associate at the Competition Commission of India (CCI). In academia, she has held positions as Assistant Professor at Amity University, Jharkhand, Indore Institute of Law, and the School of Law, KIIT University. She has also contributed as an Online Educator at UniqueLaw. She holds a Ph.D. in Law from KIIT University, with her research focused on Copyright infringement in cinematograph films and techno-legal countermeasures. A Gold Medallist in LL.M. (Corporate Laws) from NLU Odisha, she is also UGC NET qualified. She has authored and edited seven books, two legal magazines, and published 12 research papers and book chapters, including several in Scopus and UGC CARE-listed journals.

Raunaq Chawla is an academician and a researcher in psychology, currently serving as an Assistant Professor at O.P. Jindal Global University. With a Ph.D. from the University of Delhi and six years of teaching experience, Raunaq has made significant contributions to academia and research with his lectureship record proven in various colleges of Delhi and Mumbai. His expertise includes environmental psychology, organizational behavior, and consumer behavior. Raunaq has an impressive publication record, with several papers published in high-impact journals, including Scopus-indexed ones. His research spans diverse topics such as waste management behavior, the informal sector's role in solid waste management, and knowledge hiding in organizational settings. He has presented his work at various national and international conferences, showcasing his dedication to advancing psychological research. In addition to his research, Raunaq has conducted numerous workshops on topics like internet addiction, the scope of environmental psychology, and emotional intelligence. He also actively engages with students and the public through his YouTube channel, "Psychology Araamse," which aims to make psychology accessible and enjoyable. Raunaq's passion for education extends beyond his teaching responsibilities. He has held key roles in organizing workshops and seminars and has been a life skills trainer, counseling students across India. His dedication to psychology and education reflects his commitment to contributing to the betterment of society through his knowledge and expertise.

Anwesha Ghosh is an Assistant Professor at Amity Law School, Amity University Mumbai. Previously, she was a UGC-sponsored SRF at the Central University of Punjab, where she submitted her Ph.D. thesis on gender diversity in Indian corporate boards. A University Gold Medalist in LL.B. and LL.M., she has also received an Endowment Medal for Corporate and Commercial Laws. She has held research positions at GNLU and KIIT University, contributing to projects on HIV-affected children and the criminal justice system's response to rape. Her scholarly work is published in UGC Care and Scopus journals, and she authored An Introspection of Corporate Criminal Liability in India. She has presented at international conferences in Thailand and Turkiye, winning best paper awards for research on women's leadership and LGBTQ+ inclusion. With expertise in Corporate Law, mergers, and financial regulations, she actively mentors students, judges moot courts, and serves as a resource person in legal seminars.

Iskren Ivanov is an Associate Professor of International Security at the Department of Political Science of Sofia University St. Kliment Ohridski. Previously, Dr. Ivanov taught as Erasmus Professor at Macquarie University (2025), Visiting Scholar at the Hankuk University of Foreign Studies (2023), and Fulbright Lecturer at the University of Texas at Austin (2021 - 2022). Iskren holds a doctoral degree in Political Science from Sofia University and a postdoctoral degree in U.S. Foreign Policy from the University of Delaware.

Pranjal Khare is a distinguished corporate lawyer and is currently working as a law faculty at O.P. Jindal Global University, India. He specializes in Property Law, Corporate laws, Intellectual Property Rights, and International law. Mr. Pranjal is a highly regarded scholar and has published on numerous issues related to international affairs, economy, politics, data & privacy, artificial intelligence and IP disputes. He is also a frequent speaker at legal conferences and has served as a consultant to numerous law firms and corporations.

Hanh Le is a passionate researcher and accomplished Full Stack Software Developer with deep expertise in software development, data analytics, data science, and cloud computing. She holds a Ph.D. in Information Technology and a Master's in Computer Science, bringing both academic rigor and a solution-driven mindset to her work. With extensive experience in front-end and back-end development, she specializes in scalable, database-driven applications that support evidence-based decision-making. Dr. Le has led transformative tech initiatives across healthcare, law enforcement, and retail, translating complex needs into high-performing systems. She values interdisciplinary collaboration, believing that impactful innovation often arises at the intersection of fields. Her blend of technical skill, research excellence, and collaborative spirit continues to drive progress in academia and industry alike.

Anushree Mukte is Academic professional with 12+ years' teaching experience in Business Law and other law related modules for management, chartered accountancy and company secretaryship students and research expertise in Intellectual Property Law (IPR), specializing in pharmaceutical patents, public health policy, and innovation management. Completed doctoral studies with critically analysing public policy frameworks for compulsory licensing of pharmaceutical patents within India's healthcare sector. She has authored multiple SCOPUS indexed & UGC CARE Listed publications addressing intersections of IPR and pharmaceutical policy.

Stavros S. Panayiotou, Ph.D., is associated with the University of Athens, Greece. He holds a PhD degree in Philosophy from Sofia University "St. Kliment Ohridiski". He graduated from University of Cyprus obtaining a BA degree in History and Archaeology in the School of Philosophy, he has undertaken an MA in Byzantine Culture, History and Religion at Cardiff University and an MA in Philosophy from the Open University of London. Panayiotou is a philologist who teaches Modern Standard Arabic Language and Philosophy in the educational center of the Ministry of Education and Culture in Cyprus. His research interests cover themes on Metaphysics, ethics and Ontology of personhood and relationalizm in Levinas' work Totality and Infinity and Otherwise than Being.

Naresh Prajapati is an Assistant Professor of Law at Bennett University, Greater Noida. He researches the intersection of international environmental law, public international law, human rights, and comparative constitutional law, with a sustained focus on climate change. His doctoral thesis, "Judicial Approach for Human Rights Protection Amid Climate Change: A Comparative Analysis Across India, Bangladesh, Sri Lanka, and Maldives," examines how South Asian judiciaries translate climate harms into enforceable rights and remedies, informing adaptation planning, loss-and-damage pathways, and environmental constitutionalism across the region. Methodologically, he blends doctrinal analysis with socio-legal inquiry to connect legal texts and institutions to lived realities and policy outcomes. Mr. Prajapati has presented internationally in Kazakhstan, Türkiye, Bangladesh, Sri Lanka, the Maldives, and India on the ICJ's climate advisory proceedings, transboundary water governance between India and Bangladesh, ocean decarbonisation under the IMO 2023 greenhouse gas strategy, climate security, and the emerging debate on ecocide and international criminal law. Recent publications include a Q1 study on citation dynamics in climate law in Global Knowledge, Memory, and Communication, as well as work on women's leadership in climate resilience in the Journal of International Women's Studies. He serves as Lead Editor for a JIWS special issue, Senior Editor for the Africa University Law Review, and reviewer for multiple Scopus-indexed journals and conferences. At Bennett University, he teaches International Law, International Environmental Law, Human Rights, and Legal Research Methodology, delivers invited lectures, and chairs sessions on climate governance and research design. He is UGC-NET JRF qualified in Law and in Human Rights and Duties, and undertakes advanced training on the Paris Agreement, the Sustainable Development Goals, and open research practices.

Regan R, B.E., M.Tech., Ph.D., MISTE, is an accomplished academician and researcher serving as Assistant Professor in the Department of Computer Science and Engineering at the University College of Engineering, Villupuram (a Constituent College of Anna University, Chennai). He obtained his Doctor of Philosophy in Information and Communication Engineering from Anna University in 2022, M.Tech from Bharath University in 2010, and B.E. from Mailam Engineering College in 2007. With over 15 years of teaching and research experience, Dr. Regan's areas of expertise include Information Security, Computer Networks, and Wireless Mesh Networks. He has authored numerous international journal papers (including SCIE and Scopus-indexed), presented papers in IEEE and IET conferences, and published several technical books with reputed publishers such as Springer, Lambert Academic Publishing, Walter de Gruyter, and IGI Global. His research contributions also include multiple patents published and granted in India and abroad, along with a funded project under the Tamil Nadu State Council for Science and Technology. A life member of ISTE and other professional bodies such as IAENG and IRED, Dr. Regan has actively organized conferences, FDPs, and social outreach programs, while serving in key administrative roles including Placement Coordinator, Estate Officer, and Deputy Warden. Dedicated to both academic excellence and institutional development, he continues to contribute significantly to the advancement of computer science education and research.

Anuj Sinha is a legal academic and researcher specializing in Competition Law, Constitutional Law, and Jurisprudence. He was awarded a Ph.D. in Law by Hemwati Nandan Bahuguna Garhwal (A Central) University, Uttarakhand, in December 2024. His doctoral dissertation, titled "Combating Cartelization and Bid-Rigging: A Critical Study with Reference to The Competition Act, 2002," provides a doctrinal and policy-oriented analysis of anti-competitive agreements, with specific focus on collusive bidding practices and cartel conduct. The study critically evaluates the effectiveness of the Competition Act, 2002, the role of the Competition Commission of India (CCI), and judicial pronouncements addressing market fairness, regulatory efficacy, and consumer welfare. Prior to this, Dr. Sinha completed his LL.M. from the same university, with a dissertation titled "Euthanasia and Right to Die Under Article 21 of the Indian Constitution: An Analysis," where he explored the constitutional scope of personal liberty and the evolving legal discourse on end-of-life rights in India. He holds a B.A. LL.B. (Hons.) from IMS, Dehradun, affiliated to HNB Garhwal University, where he consistently performed at the top of his class and actively participated in moot courts, debates, and legal awareness initiatives. Dr. Sinha is a UGC-NET qualified scholar in Law (2019, 2021, 2022) and was awarded the Junior Research Fellowship (JRF), upgraded to Senior Research Fellow (SRF) in 2024.

Tulip Suman, a distinguished law professor, brings over a decade of teaching and research expertise in intellectual property law, public international law, human rights, geopolitics, foreign affairs, and diplomacy.

Nithya V, is a Teaching Fellow in the Department of Computer Science and Engineering at the University College of Engineering, Villupuram Campus, Anna University. She is completed a Ph.D. in Information and Communication Engineering at Anna University, Chennai. Her research interests include web application security, cryptography, and network security. She has published multiple research papers in international journals and conferences. Nithya has authored academic books and contributed to various faculty development programs. She holds an M.Tech in Information Security from Pondicherry Engineering College and a B.E. in Computer Science from Mailam Engineering College.

Index

A

B

C

D

E

H

I

J

K

L

M

N

O

P

R

S

T

U

W

www.ingramcontent.com/pod-product-compliance
Lightning Source LLC
LaVergne TN
LVHW080909300926
849162LV00003B/618